Emerging Secure Networks, Blockchains and Smart Contract Technologies

Jong-Moon Chung

Emerging Secure Networks, Blockchains and Smart Contract Technologies

Springer

Jong-Moon Chung
Department of Electrical and Electronic
Engineering
Yonsei University
Seoul, Korea (Republic of)

ISBN 978-3-031-65865-5 ISBN 978-3-031-65866-2 (eBook)
https://doi.org/10.1007/978-3-031-65866-2

This Springer imprint is published by the registered company Springer Nature Switzerland AG
The registered company address is: Gewerbestrasse 11, 6330 Cham, Switzerland

To my father and mother for their love and devotion, to my wife Seung-Won and daughters Yonhee and Yonju for their love and care, and to God for all these blessings and love.

Acknowledgments

I deeply thank the Springer Nature executive editor Anthony Doyle for inviting me and supporting the preparation of this book, and I sincerely thank the production editors Arumugam Deivasigamani and Sharmila Anbu, and production supervisor Madanagopal Deenadayalan for all the hard work on this book. I thank my graduate research assistant Jungmin Seo for helping me with the final draft review and file format conversion. I also thank my graduate research assistants Min Soo Joo for drawing the two mouses in Fig. 11.1, Jaewook Jung for reconfirming the network projects, and Minsuk Ko for rechecking the security architectures. I also thank my graduate research assistants Minsu Choi, Yunyeong Goh, Jaewook Jung, Sangsoo Park, and Gangwoo Lee for reviewing selected chapters of the final draft of my book. I am also very grateful to Yonsei University, Oklahoma State University, and Pennsylvania State University and to all my current graduate research assistants Wonsuk Yoo, Wonsik Yang, Minsu Choi, Yunyeong Goh, Jaewook Jung, Sangdo Kim, Min Soo Joo, Dongjun Jung, Jeongyoon Shin, Byounghoon Son, Jungmin Seo, Juyeong Hwang, Sunghyun Kim, Myoungbo Kim, Dr. Jae-Dong Kim, Minsuk Ko, Youngwook Kim, Giyoung Hwang, Daeun Ko, Jungsuk Baik, Haeri Kim, Gangwoo Lee, Minseung Park, Shinhyeok Oh, Daekwang Kim, Gunkee Lee, Hyunho Jang, Sangsoo Park, Hyuckjun Kim, Hwankyu Song, Wonjun Lee, Janggeun Yoo, Yeeun Park, Yerin Sung, Youngmin Lee, Inbum Lee, Eunsu Lee, Junwon Kim, and Byungwook Oh for their research effort to advance science and technology with me. I also thank Dr. Taeyoung Ha, Dr. Hyoungjun Cho, Dr. Seungwoo Seo, Dr. Younghwan Shin, Dr. Sooeun Song, Dr. Changsung Lee, and Dr. Junsung Kim for their help earlier as my former graduate assistants for my Coursera courses. Especially, I deeply thank my father Chin O. Chung and mother Young C. Chung, wife Seung-Won H. Chung, and daughters Yonhee Chung and Yonju Chung, and most of all I cordially thank God for all these blessings and love.

Introduction

Currently, almost all financial services and merchandise purchasing can be conducted through the Internet. In addition, the emergence of Bitcoin and cryptocurrencies have changed the way payments and exchanges can be conducted and has paved the way for smart contracts to become a reality and serve as an effective and valuable platform of automated crowdsourcing applications and functionalities. In addition, the popularity of smartphones, laptops, and tablet computers has made all information services accessible wirelessly and portably. As a result, almost all of our personal and professional information, money, securities, and assets are controllable and transferable through online networked services. Due to this increasing value and dependence on network services, the number and severity of cyber-attacks are growing. In addition, the rapid advancement of Artificial Intelligence (AI) technologies is making cyber-attacks significantly more sophisticated, precise, automated, and intelligent. The only way to continue to evolve is to make our systems and networks more secure and intelligent. The objective of this book is to provide the reader with the necessary knowledge to create new technologies that can enhance the network performance while making everything more secure and safe. For this purpose, this book covers the essential network technologies, important cyber-attack cases, and explains the essential technologies used in hash functions, encryption, blockchains, Bitcoin, smart contracts, and Ethereum. In addition, deep reinforcement learning, federated learning, and digital twin technologies are described in this book along with new methods of how network security and blockchains can be enhanced. In addition, this book describes how ChatGPT, GPT, GPT2, GPT3, DALL-E, DALL-E 2, Imagen, Stable Diffusion, and BERT are created using generative AI technologies like Variational Auto-Encoder (VAE), Autoregressive Model (ARM), Flow Model (FM), Diffusion Model (DM), and Generative Adversarial Network (GAN). This book ends with an overview of emerging technologies with some predictions on how these technologies may evolve and recommendations for future endeavors.

Who This Book Is For

This book is for anyone interested in the Internet, TCP/IP, security, deep reinforcement learning, generative Artificial Intelligence (AI), blockchains, Bitcoin, smart contracts, Ethereum, as well as wired and wireless communications. You do not need to be a student in this field or an engineer or computer scientist to understand the material in this book. Only your interest is needed. This book is for business and marketing experts who realize the market potential and may have plans to use networks, security, AI, blockchain, or smart contract technologies in their organization and need to work with engineers, programmers, and scientists. This book is also for government, academia, and public administrators who plan to use networks, security, AI, and blockchain services to enhance education, job preparation, re-education, and field training. This book is also for entertainment artists, musicians, scriptwriters, and designers who see the potential in new trends of networking, digital twin modeling, generative AI, and blockchain technology, which can be used to significantly grow all aspects of their businesses. This book also introduces how ChatGPT, GPT, GPT2, GPT3, DALL-E, DALL-E 2, Imagen, Stable Diffusion, and BERT are created using generative AI technologies. Especially, this book is for students who are seeking knowledge in state-of-the-art technologies that have longevity and are extremely exciting, fun, admirable, and full of creativity. Indeed, the networking, AI, digital twin, security, encryption, and blockchain technology field has great value and is worthy of investment. Not all chapters of this book may be needed for now, so selective reading of chapters is welcome. The contents of the chapters are interconnected and can provide a synergistic effect in understanding the technical details. However, the chapters are written in a way such that selected chapters can be read without needing to read all chapters. The chapters initially skipped may become useful later when more technical details are needed. The following chapter description may help you in selecting the chapters that you may want to read first. Thank you for choosing this book and I hope you enjoy it and that it helps you in your studies and work.

Chapter Overview

The subject of this book is emerging technologies used in security, TCP/IP, Internet, hash functions, cryptography, blockchains, Bitcoin, smart contracts, and Ethereum. In addition, advanced Artificial Intelligence (AI) technologies like deep reinforcement learning, generative AI, federated learning, and digital twin technologies are introduced and relations to network security and blockchains are described. This book provides an easy-to-understand overview of emerging cyber-attacks in Chap. 1, which is based on the IBM 'X-Force Threat Intelligence Index 2022' report, where the recent trends in advanced cyber-attacks, global distributions of cyber-attacks, and industry types that receive the most cyberattacks are introduced. Next, the most advanced AI and computer modeling technologies are introduced in Chap. 2, which include Deep Reinforcement Learning (DRL), Deep Q-Network (DQN), Deep Deterministic Policy Gradient (DDPG), Federated Learning (Collaborative Learning), and Digital Twin technologies, where considerations to network security and blockchains based on these advanced technologies are also discussed in Chap. 11. Chapter 2 also introduces various generative AI technologies, including, Variational Auto-Encoder (VAE), Autoregressive Model (ARM), Flow Model (FM), Diffusion Model (DM), and Generative Adversarial Network (GAN). Chapter 2 explains how VAE is used in the famous DALL-E image creation system; ARM is used in ChatGPT, GPT, and DALL-E; and DM is used in DALL-E 2, Imagen, and Stable Diffusion generative AI systems. Chapters 3–5 describe the details of TCP/IP networks and provides projects that can be easily conducted on a Personal Computer (PC) or laptop, which can significantly help the reader understand how networks actually operate and identify the security features that are currently being used on a computer and smartphone. More details on the projects are provided in the following. Chapter 3 includes a project to setup a network connecting a total of 175 PCs that are located in four rooms that need to have access to other PCs and the Internet through a gateway/router. In this project, the IP addresses of the 175 PCs need to be assigned, subnets have to be configured, and the gateway/router IP routing table needs to be setup. In addition, Chap. 5 introduces a project to check Internet connection configurations, conduct Internet ping, and trace route measurements. In addition, the Wireshark Internet analysis project of Chap. 5 shows the methods of how to check a transmitted or received IP packet's source and destination IP address, observe the TCP three-way handshake parameters, check the TCP window size and transport security protocols being used, generate file transfer I/O graphs, analyze the data rate of the network, and conduct IP address based physical location (e.g., mailing address) identification. Next, Chap. 6 provides details on hash technology, Chap. 7 describes cryptography technology, and Chap. 8 introduces various blockchain technologies. Knowing these technologies together is very important, because, only when the reader understands how the fundamental technologies actually work, then new solutions to improve the performance and enhance the security can be effectively designed and developed. Blockchains, Bitcoin, cryptocurrencies, smart contracts, and Ethereum are all distributed systems that are operated over the global Internet. So, knowledge of networking, security, and AI automated control technology is essential, which is covered in this book. Based on this information, the details of Bitcoin are introduced in Chap. 9. An additional important feature of this book is that it includes recent updated information on Ethereum's Proof-of-Stake (PoS) technology, which began to be used in September of 2022. This is important because the world's first and largest smart contract company Ethereum changed from using the Proof-of-Work (PoW) based Ethash to the new PoS. In addition, essential parts of Ethereum PoS were recently upgraded (Edition 0.3 Capella) on July 1st of 2023, which are described in this book. This is very significant considering the level of improvement in energy efficiency, performance, and safety, which is highly desired. Because so many operations and services are based on blockchain-operated smart contracts, it is important to understand this new PoS technology in detail, as the same or similar transitions to PoS are expected to occur for almost all cryptocurrencies and smart contract platforms. The main benefit of reading the book is that the reader can easily see the relation of the Internet, IoT, blockchains, AI, and cyber-attacks. This is why these technologies are taught together in this book. In addition, considering the complexity of each topic, these contents are explained in very simple terms and in an easy-to-learn sequence considering the build-up relation of the chapters and are presented to the reader in a way that synergistic effects can be obtained. However, each chapter can be easily read and understood separately.

Contents

About the Author

Jong-Moon Chung received B.S. and M.S. degrees in electronic engineering from Yonsei University (Seoul, South Korea), and at the Pennsylvania State University, he received his Ph.D. in electrical engineering and served as an instructor and assistant professor from 1997 to 1999. From 2000 to 2005, he served at the Oklahoma State University as a tenured associate professor in the School of Electrical and Computer Engineering and director of the Oklahoma laboratory for communications, networking, and bioengineering. Since 2005, he has been with Yonsei University, where he serves as a tenured professor in electrical and electronic engineering, associate dean of the College of Engineering, and (joint appointment) professor of the Department of Emergency Medicine in the College of Medicine and Department of Periodontics in the College of Dentistry. Dr. Chung received the '(Haedong) Award of Excellence' from the National Academy of Engineering of Korea (NAEK) in 2022, four (in 2023, 2021, 2019, and 2012) top national awards from the South Korea government, and numerous awards for outstanding research and teaching from Yonsei University. From the Oklahoma State University, he received the 'Distinguished Faculty Award' in 2003 and 'Technology Innovator Award' in 2004, and the Halliburton 'Outstanding Young Faculty Award' (from the Halliburton Corporation) in 2005. In 2005, he also received the 'Regents Distinguished Research Award,' which is the highest research award of the Oklahoma State Board of Regents and commonly considered the highest lifetime research achievement award in the

State of Oklahoma. He received this award at the age of 37, where he was the youngest recipient of this award. In 2000, he received the 'First Place Outstanding Paper Award' at the IEEE EIT 2000 conference held in Chicago, IL, USA. His courses in Coursera (www.coursera.org) on deep learning, big data, TCP/IP, Internet, security, IoT technologies, 4G, 5G, Wi-Fi, Bluetooth, AR, MR, multimedia streaming, MPEG-DASH, cloud and edge computing, smartphones, and smartwatches, have been accessed by over 2 million users and listed among the top 33 industry recommended courses in Coursera. He is currently the Vice President of the IEEE Consumer Technology Society (CTSoc), IEEE Product Safety Engineering Society (PSES), Korean Society for Internet Information (KSII), Korean Society for Engineering Education (KSEE), and serves as Senior Editor of the IEEE Transactions on Consumer Electronics, Section Editor of the Wiley ETRI Journal, Chair Editor-in-Chief of the KSII Transactions on Internet and Information Systems, and served as Editor of the IEEE Transactions on Vehicular Technology from 2011 to 2021. Dr. Chung also serves as an IEEE Distinguished Lecturer and Secretary General of the Electrical and Electronic Engineering Division of the National Academy of Engineering of Korea. His book titled 'Emerging Metaverse XR and Video Multimedia Technologies' was published by Apress of Springer Nature, (c) 2023. He is also a member of the Eta Kappa Nu (HKN) honor society and a Fellow of the IEEE.

Chapter 1
Emerging Cyber-Attacks

This chapter is focused on emerging cyber-attacks, in which the recent global trend is reviewed and details of the most prominent issues are covered followed by the details of emerging ransomware, botnet attacks, and cryptojacking technologies. In the first section, the most recent trends in advanced cyber-attacks are introduced, where the leading cyber-attack types and initial access infection techniques are the main focus. In addition, the global distribution and industry types that receive the most cyber-attacks are described. The second section focuses on ransomware, which currently is the most influential type of cyber-attack in the world. Details of ransomware history and classifications are introduced followed by ransomware attack procedures and examples based on Emotet and Trickbot are provided. In the third section, botnet attacks using Internet of Things (IoT) devices are introduced. Botnet centralized and distributed operational structures are introduced followed by a description of the two most representative IoT botnets Mirai and Mozi is provided. The fourth section focuses on cryptojacking, which is a malware based cyber-attack on blockchains, cryptocurrencies (e.g., Bitcoin (BTC)), and smart contracts (e.g., Ethereum (ETH)). In addition, cryptojacking defense strategies are introduced. The understanding of this chapter can be enhanced by the other chapters of this book that explain the details of the Internet, TCP/IP operations, IoT networks, cryptography, hashing, blockchains, Bitcoin, and Ethereum.

1.1 Recent Trends in Advanced Cyber-Attacks

Advancements in data science, Artificial Intelligence (AI), and big data technology has enabled many new services and devices to become highly intelligent and accurate in making versatile and complex control decisions. These technologies have made industry, businesses, governments, and consumer electronic devices much more effective and efficient, where nowadays many aspects of life depend on these versatile

J.-M. Chung, *Emerging Secure Networks, Blockchains and Smart Contract Technologies*, https://doi.org/10.1007/978-3-031-65866-2_1

and adaptive systems. These advanced systems rely on control through Information Technology (IT) networks and databases as well as the electromechanical hardware and software based on Operational Technology (OT) to run manufacturing factories as well as power and public supply facilities.

The reliance on IT networks and databases is truly significant, as AI and big data systems need to process the data and exchange information to properly support the manufacturing and public service facilities as well as various consumer electronic services. Therefore, it is evident that network security, big data processing, and database protection is critical to maintain. This is more important than ever because the number of cyber-attacks has increased a lot and the level of malicious abuse to its victims (e.g., businesses, governments, and users) is significant.

The evolution of AI technologies (including machine learning, deep learning, and reinforcement learning) also made cyber-attack systems highly intelligent, autonomous, faster, and adaptive (more details are provided in Chaps. 2 and 11). Therefore, this book is focused on introducing various emerging cyber-attack methods and new intelligent defense techniques that can strengthen network and blockchain database security.

Based on IBM's 'X-Force Threat Intelligence Index 2022' report [1], the leading cyber-attack types in 2021 are presented in Fig. 1.1 and described in the following. The number one cyber-attack type is ransomware at 21%, rank 2 is server access at 14%, rank 3 and 4 are Business Email Compromise (BEC) and data theft, which are both at 8%, rank 5 is credential harvesting at 7%, and ranks 6, 7, and 8 are Remote Access Tool (RAT), misconfiguration, and malicious insider, which are each at 5% [2].

In both 2020 and 2021, the most frequent types of cyber-attack were ransomware. Among all types of cyber-attacks, ransomware attacks composed of 23% in 2020 and 21% in 2021. The top ranked ransomware group REvil (which is also known

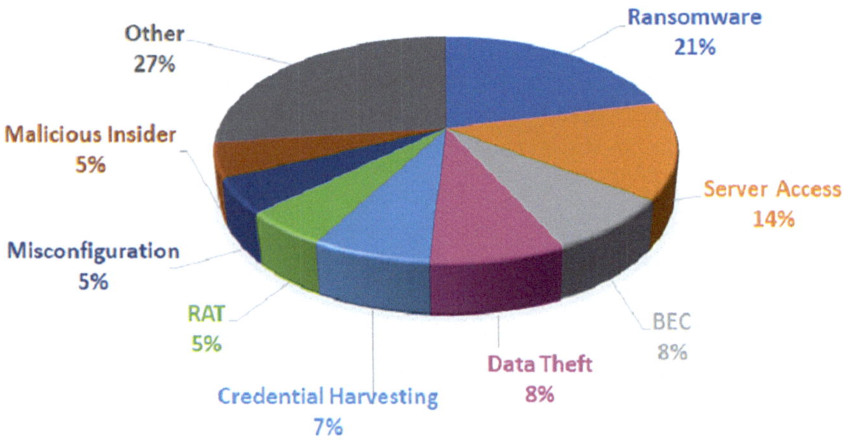

Fig. 1.1 Leading cyber-attack types in 2021

as Sodinokibi) composed of 37% of all ransomware attacks, followed by the Ryuk group which composed of 13% of all ransomware attacks. The third top ranked ransomware group was LockBit 2.0 at 7%, which was followed by Atom-Silo, AvosLocker, BitLocker, BlackMatter, Conti, CryptoLocker, Crystal, DarkSide, Eking, Medusa, Nefilim, Ragnar Locker, and the Xorist group, which each composed of approximately 3% of all ransomware attacks in 2021.

REvil is known to have operated until October of 2021 for a duration of 31 months, which is significantly longer than the 17 month average duration of a ransomware to be used before they transform into another form (i.e., rebranded) or stopped being used (i.e., disbanded).

Ransomware attacks as well as almost all cyber-attacks start with procedures to obtain initial access, which use various infection vectors and exploitation techniques. The most popular initial access techniques used in 2021 are presented in Fig. 1.2 and listed in the following. The top ranked method used to gain initial access for a cyber-attack is phishing, which occupies a massive 41% of attack types. For a targeted phishing attack, the click rate (i.e., success rate) is approximately 17%, but when accompanied by a phone call (which is called 'voice phishing' or 'vishing' for short) the click rate radically increases by approximately 3 times, reaching a 53.2% click rate [1]. Vulnerability exploitation schemes are second on the list occupying 34%, followed by stolen credentials at 9%, brute force techniques at 6%, remote desktop and removable media techniques are each at 4%, and password spraying is at 1%.

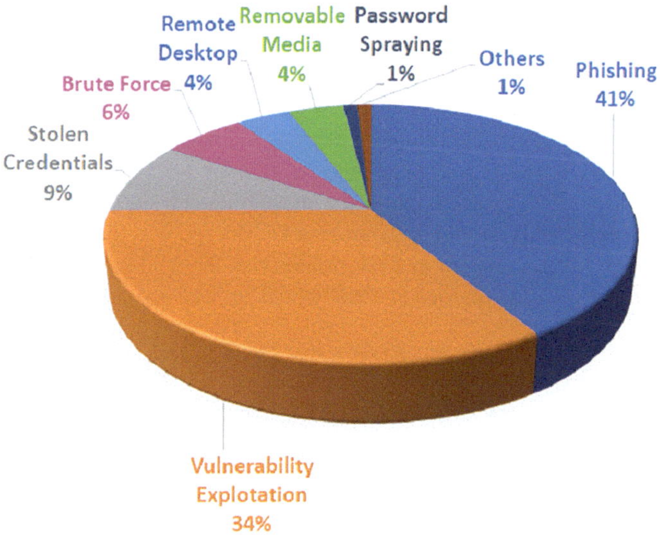

Fig. 1.2 Cyber-attack initial access infection techniques used in 2021

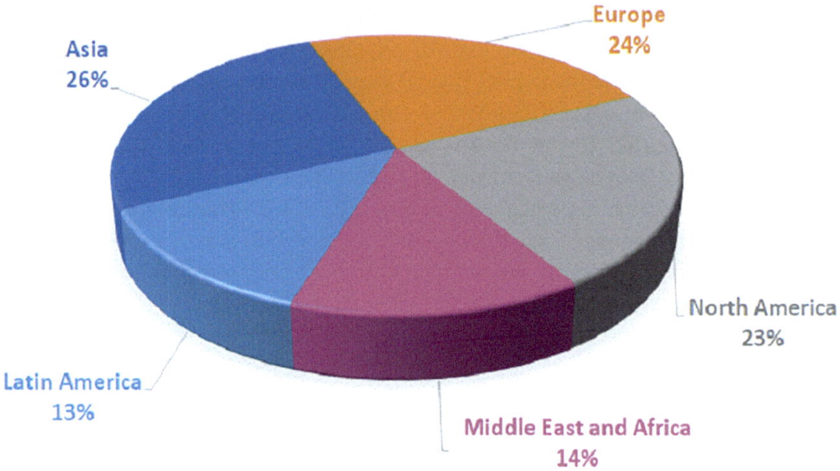

Fig. 1.3 Global regional distributions of cyber-attacks

1.1.1 Global Distributions of Cyber-Attacks

The global regional distributions of cyber-attacks are also important [1]. In Asia, Japan, Australia, and India were attacked the most. One of the reasons Japan was a frequent target is due to the 2020 Summer Olympic Games, which were postponed and opened on July 23rd and closed on August 8th of 2021 in Tokyo, due to the COVID-19 pandemic. In addition, Asia's top attacked industries leads off with the finance and insurance domain, which received the most cyber-attacks that composed of 30%. The second highest are the manufacturing industries at 29%, and third highest are professional and business services at 13%.

As a result, in 2021, Asia became the most frequently attacked region in the World, reaching 26% of all global cyber-attacks. Europe was the second highest at 24%, third was North America at 23%, fourth was the Middle East and Africa at 14%, and fifth was Latin America at 13%, which are illustrated in Fig. 1.3.

1.1.2 Industry Types that Receive the Most Cyber-Attacks

Figure 1.4 presents the industry types with the most cyber-attacks in 2021 [1]. Due to an increase in ransomware attacks and server access attacks, the manufacturing industries became the top industry type among all industries, receiving 23.2% of all attacks. Especially, Asia's manufacturing industries received the most attacks at 32% among all cyber-attacks of the world's manufacturing industries. Infections on the manufacturing industries were mostly initiated through vulnerability exploitations at 47%, which are followed by phishing attacks at 40%.

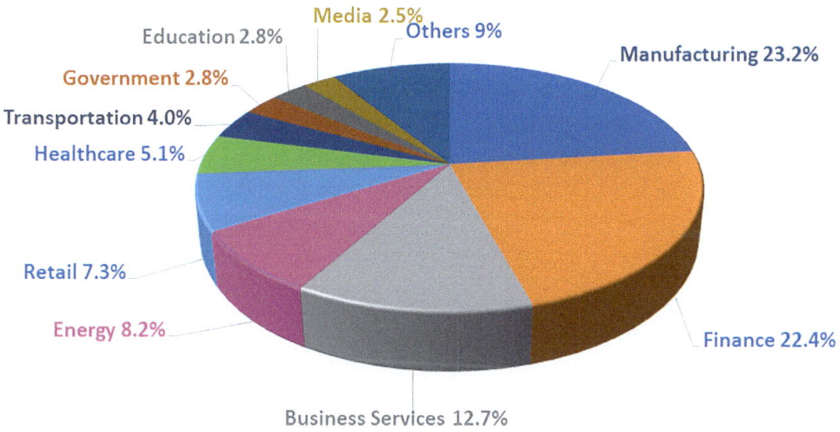

Fig. 1.4 Industry types with the most cyber-attacks in 2021

The second highest attacked industry type are finance and insurance organizations at 22.4%, in which among these cyber-attacks, banks received the most attacks reaching 70%, followed by insurance companies at 16%. Most initial infections were initiated using phishing attacks, reaching a gross 46% of attacks.

At 12.7%, professional and business services rank the third highest cyber-attacked industry type during 2021, in which ransomware attacks were the most, composing of 32% of all cyber-attacks. Infections were mostly initiated using vulnerability exploitation techniques, which account for 50% of all incidents. Professional and business services are very diverse. Professional services considered in these statistics consist of information technology service providers, law firms, architects, accountants, and consulting firms. Business services consist of office administration firms, human resource firms, security service companies, travel agencies, and landscape firms.

Fourth on list are the energy organizations at 8.2% in which ransomware attacks were 25% of all attack types. Notably, the DarkSide ransomware attack that the United States of America (USA) Colonial Pipeline experienced in May of 2021 is a representative example [3].

The Colonial Pipeline has a length of approximately 46,400 km (i.e., 29,000 miles), which is used to deliver petroleum to 45% of the USA East costal area. The ransomware attack resulted in a fuel outage for 87% of the Washington D.C. area gas stations. The Colonial Pipeline ransom was 75 Bitcoins (BTC), which was approximately worth $4.4 million US dollars (USD) at that time. After this event, ransomware attacks on the energy sections were less frequent for a few months, possibly due to counter-attack retaliation fear.

Fifth on the raking list is retail and wholesale at 7.3%, sixth is healthcare at 5.1%, seventh is transportation at 4%, eighth is the government organizations at 2.8%, ninth is education organizations also at 2.8%, and tenth is media companies at 2.5%.

1.2 Ransomware

As mentioned above, based on the X-Force Threat Intelligence Index 2022 report from IBM Security, among the top cyber-attack type in 2020 and 2021, ransomware ranks at the top. Ransomware occupied 23% in 2020 and 21% in 2021 among all types of cyber-attacks [1]. Therefore, it is necessary to study more on ransomware attacks.

1.2.1 Ransomware History

The influence of ransomware has become widely acknowledged. A chronical list of selected ransomware events is provided below [4].

- The first ransomware emerged in 1989, when Personal Computers (PCs) first came out and the Internet was in its early stages. In 1989, AIDS Trojan, which is also known as PC Cyborg, emerged as the first ransomware. The AIDS Trojan malware had gotten on to the 20,000 floppy disks that were mailed to the AIDS conference participants, which is how the infection began. The AIDS Trojan malware would use symmetric encryption on the 'c:' drive file names and then requested for a ransom payment. More details on symmetric and asymmetric encryption algorithms are explained in Chap. 7 of this book. Based on the AIDS Trojan influence, several years later the cryptovirology concept was introduced to alert the public about ransomware related extortion cybercrimes.
- In 2005, the Internet exceeded one billion user connections, which is when a more modernized ransomware named GPCode emerged. GPCode was spread using email phishing and used custom made symmetric encryption to lock the victim's computer data.
- In addition, during the 2005 and 2006 period, asymmetric encryption began to be used by ransomwares like CryZip, Archiveus, and Krotten. The asymmetric encryption technology made data recovery almost impossible without paying the ransom.
- In 2008, new types of ransomwares emerged, which included Randsom.C and Seftad. Randsom.C was a locker ransomware that used a phishing approach pretending to be an official contact from the Windows Security Center. Seftad would hinder the booting process by controlling the Master Boot Record (MBR) of the malware contaminated computer.
- Seftad and other ransomwares would use Paysafecard, which is a prepayment system, as one of the ways to receive the ransom payment. Up to 2009, the method of how the ransom payment would be paid was a limitation in ransomware attacks. This is because most payment methods made it possible to track its user. This totally changed as Bitcoin and other cryptocurrencies emerged since 2009. The (pseudo) anonymity of account holders in Bitcoin and other cryptocurrencies enabled ransomware attackers the perfect means to collect the ransom payments

and remain hidden and unknown. As a result, the emergence of cryptocurrencies was one of the reasons ransomware attacks rapidly grew. More related details on Bitcoin and cryptocurrencies (Chap. 9) as well as blockchains (Chap. 8) are provided in this book.

- In 2012, the locker ransomware Reveton, which is also known as Police ransomware, emerged.
- In 2013, the CryptoLocker ransomware emerged, which used 2048 bit RSA to encrypt files and received ransom payments via Bitcoin. More details on RSA encryption (including examples) are provided in Chap. 7 of this book.
- In 2014, the Curve-Tor-Bitcoin (CTB) Locker ransomware emerged, which used Elliptic Curve Cryptography (ECC) for encryption of files and used The Onion Router (TOR) for secrete web browsing to receive the ransom payment, which were commonly requested in Bitcoins.
- In 2014, the Cryptowall cryptographic ransomware emerged and infected over 600,000 systems, which also used TOR for secrete website access and commonly received ransom payments in Bitcoins. TOR (or Onion) is a dark web website that can be accessed only on the dark web (which should not be visited), but cannot be found through public search engines like Google.
- In 2014, the Android Defender ransomware emerged, which was the first ransomware for mobile phones that use the Android operating system (OS).
- In 2015, the Linux.Encoder emerged as the first GNU and Linux platform and OS ransomware, which used hybrid encryption technology. GNU is a group of free software (which are licensed under the General Public License of the GNU Project) in which the library of GNU tools are used in many OSs including Linux [5].
- Up to 2015, ransomware attacks were mostly conducted by expert hackers, but the emergence of Ransomware-as-a-Service (RaaS) kits made ransomware applications very easy to use. For example, the platform-agnostic feature of RaaS enabled ransomwares to be modified to fit various OSs and platforms. In addition, RaaS kits were also publicly available in underground/dark websites. RaaS is one of the reasons ransomware attacks could rapidly grow.
- In 2016, the KeRanger emerged as the first macOS ransomware, which used hybrid encryption technology.
- In 2017, WannaCry emerged to influence over 250,000 computing systems in 150 countries worldwide. WannaCry exploited the vulnerability of the Server Message Block (SMB) Server Remote Code Execution process of the Microsoft Windows OS. SMB is a protocol used to share files developed by IBM. WannaCry applied multi-stage key encryption technology, in which files were encrypted with different Advanced Encryption Standard (AES) keys, and then 2048 bit RSA was used to encrypt the AES keys again. More details on encryption are provided in Chap. 7 of this book.
- In 2018, PureLocker emerged, which used hybrid encryption. To be platform-agnostic, PureLocker was designed with PureBasic programming language.
- In 2019, Ryuk emerged as a very powerful enterprise cyber-attack ransomware. Ryuk commonly used TrickBot malware to infect computing systems. A TrickBot

is a botnet malware that processes banking Trojan cyber-attacks in multiple phases, resulting in compromised networks, credential stealing, account manipulation, data exploitation, etc.

- In 2020, the Corona ransomware emerged and attacked hospitals by ransoming patient health records. As this was during the COVID-19 global pandemic, the Corona ransomware had a horrific influence.

1.2.2 Ransomware Classification

Ransomware types can be classified in several ways [4], which are summarized in Fig. 1.5.

One way is to use the ransoming object, which may be files of the platform system. In Cryptographic ransomware, the attacker will encrypt files and use this as a hostage until a ransom payment is made. In Locker ransomware, the attacker will lock the access system, screen, browser, etc. (but may not encrypt the files) until a ransom payment is made.

Infection methods could be another way to classify ransomware. First, phishing is the most common malware infection method, in which e-mails, phone calls, Short Message(ing) Service (SMS), Instant Messenger (IM), Social Networking Service

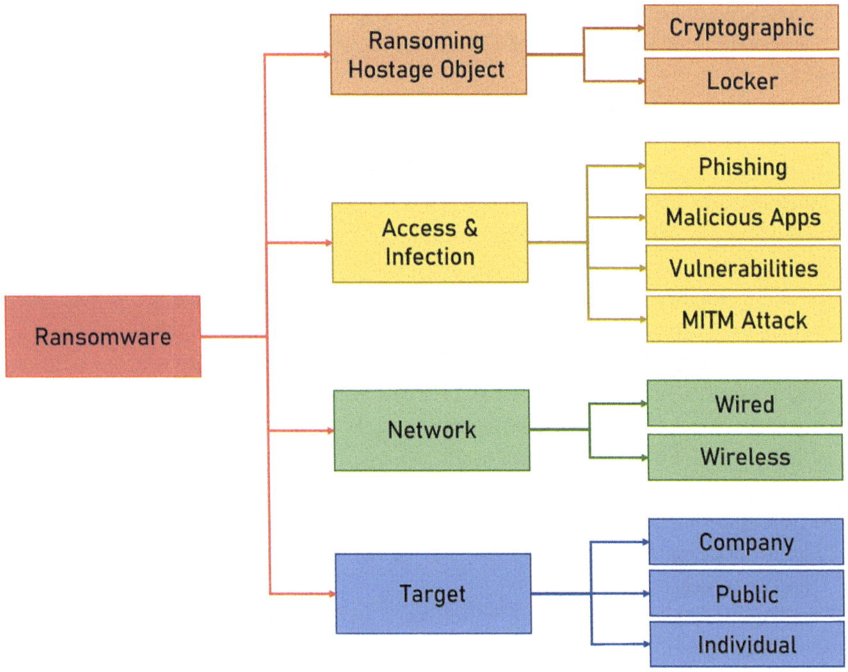

Fig. 1.5 Ransomware classification

(SNS), etc. are used. Second, malicious apps such as advertisements or infected websites could be used as camouflaged sources that trigger malware downloads into a system. Third, vulnerabilities in the OS and network protocol can be exploited until a weak point is detected and used to penetrate the systems to plant ransomware malware. Fourth, Man-In-The-Middle (MITM) techniques could be used to obtain system login information such that the attacker can access the system and take control over the system's platform and files. More details on MITM attacks are provided in Chap. 7 of this book.

Another way to classify ransomware is based on network type of wired and wireless. Wired networks are connected through pre-installed servers and routers located at fixed locations that connect users to the Internet and intranet servicing companies, public facilities, home appliances, the military, etc. Wireless networks include smartphones, portable computers (e.g., laptops, tablets, pads, slates, and smart devices), autonomous vehicles, drones, as well as various Internet of Thing (IoT) systems. For example, a future ransomware attack could be made to hijack a wireless network that is servicing an autonomous driving vehicle with passengers onboard as hostages. Such a situation would be horrific, which is why ransomware attacks of such kinds must be considered in advance and prevented by all means. Since the Android Defender ransomware emerged in 2014, various types of ransomwares for mobile systems have emerged and are expected to grow. Fundamentally, the OS and system operational pattern, requirements, power supply, and environment of wireless and wired systems are different, which make their selection of hostage type and ransoming payment methods different.

The ransomware target could be another way to classify ransomware types. First, many companies have been ransomware victims, which includes manufacturing, financial, sales, retail, logistics, etc. Second, public facilities and infrastructures have been targets. Such examples include the Corona ransomware that attacked hospitals during the COVID-19 pandemic in 2020 and the Colonial Pipeline ransomware attack on the southeastern area of the United States that occurred in 2021. Third, individuals are always targets for ransomware attacks, as defense mechanisms and retaliation methods are very limited, making individuals easy victims.

1.2.3 Ransomware Attacks Procedures

Ransomware attacks can be classified into 5 stages, based on IBM's 'X-Force Threat Intelligence Index 2022' report [1], where stage 1 is 'Initial Access,' stage 2 is 'Post Exploitation,' stage 3 is 'Understand and Expand,' stage 4 is 'Data Collection and Exfiltration,' and stage 5 is 'Ransomware Deployment.' The five stages of a ransomware attack are presented in Fig. 1.6 and explained in the following.

Ransomware attack stage 1 (as well as almost all cyber-attacks) starts with procedures to obtain initial access, which use various infection vectors and exploitation techniques. As presented in Fig. 1.2, the top ranked methods used to gain initial

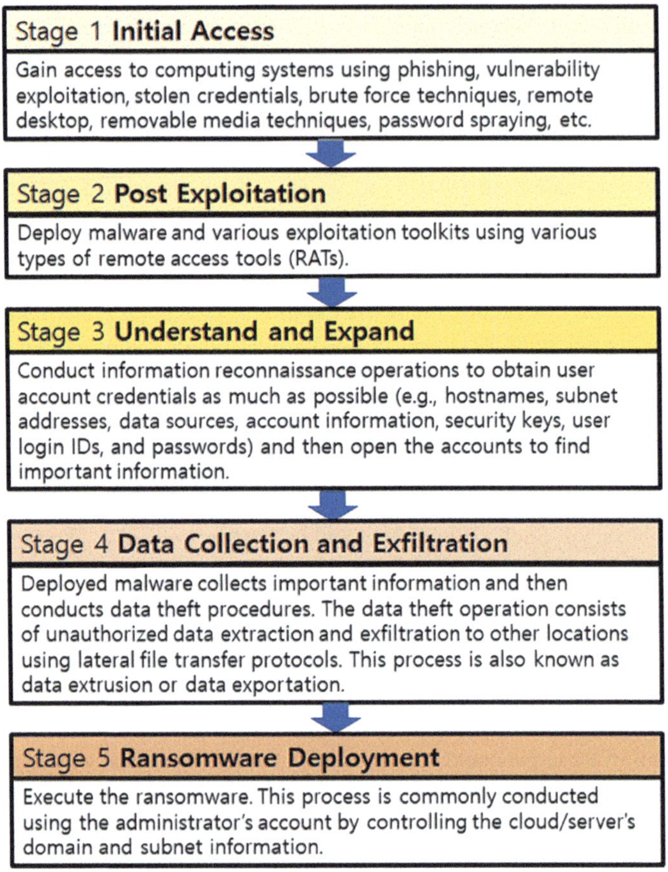

Fig. 1.6 Five stages of a ransomware attack

access for a cyber-attack is phishing, vulnerability exploitation, stolen credentials, brute force techniques, remote desktop, removable media, and password spraying.

Ransomware attack stage 2 techniques depend on the stage 1 initial access type established. For example, if a remote desktop breach was established, the stage 2 post exploitation process may include deployment of post exploitation toolkits like a Remote Access Tool (RAT) and deployment of malware, which will be used in the stage 4 exfiltration process.

Stage 3 is focused on understanding the breached system and expanding on collecting important information to use in stage 4 and stage 5 of the ransomware attack. For this purpose, reconnaissance operations are conducted to collect important credentials, which include hostnames, important subnets to use, data sources, account information, security keys, user login IDs, and password information.

In stage 4, data is collected, and the exfiltration process is conducted. Exfiltration is a process of unauthorized data extraction, which is commonly conducted using malware, which most likely was deployed in stage 2. Almost all ransomware attack incidents since 2019 conduct both ransomware activity as well as data theft, resulting in ‘double extortion’ damage to their victims [1]. In this stage, functions like Microsoft’s Server Message Block (SMB) may be used for lateral file transfers.

The final stage of a ransomware attack is stage 5, which is when the ransomware is deployed. The ransomware is executed using the domain administrator’s credentials, which most likely would have been obtained in stage 3 through the reconnaissance process of searching for essential credentials as well as domain and subnet information.

Understanding these ransomware attack stages can significantly help in development of countermeasure technologies. If the ransomware kill chain is well understood, it can be detected more easily, and therefore, can be blocked more effectively.

1.2.4 Ransomware Attack Example

An example of a ransomware attack using Emotet and Trickbot is presented in the following to demonstrate the network procedures and characteristics [2, 3].

Ryuk, Emotet, and Trickbot are all very famous cyber-attack malware in the security world. Emotet first emerged in 2015 as a banking trojan horse malware, which was reconfigured in loader malware in 2016, and then was reformed into a Loader as a Service (LaaS). The LaaS model of Emotet made it easy to use, which made it widespread. Emotet was upgraded many times and is still used in combinations of various cyber-attacks and ransomware operations. During 2018 to 2019, Emotet was used in data exfiltration cyber-attacks in cities like Frankfurt (Germany), Allentown (Pennsylvania, USA), and Berlin (Germany). Many efforts were made to eliminate Emotet, which were effective for short periods, but more powerful upgraded versions of Emotet have consistently emerged, in which Emotet is still in use in many cyber-attacks. In late 2020, an upgraded version of Emotet was used to attack major government and private enterprise networks, which shown that it had evolved into the first enterprise level big loader malware. Emotet is continuously evolving and upgraded versions still have significant influence on cyber networks.

A loader is a Trojan horse malware used in the initial procedures to gain access of a computer/server such that it can be compromised. This is why loaders are also called Remote Access Trojans (RATs). A Trojan horse is also called a Trojan for sort, which is a malware that is made to seem legitimate and safe to the computer/server’s user such that it will be allowed into the user’s computer or server. After it has entered the computer, it will compromise the system such that the malware can control the computer. The term Trojan horse originates from the story that a large wooden Trojan horse filled with Greek solders was used to fool the leaders of Troy, who brought the Trojan Horse in to the gates of the city of Troy, which led to the city’s destruction and defeat in war.

Trickbot is a Trojan horse malware that can operate on various OSs including Microsoft Windows. Trickbot first emerged in 2016 and spread rapidly through various routes using email phishing, Google Docs, executable programs, batch files, etc. [6].

Trickbot was first used to exfilterare bank data and account credentials, where upgrades evolved it into a powerful modular Trojan horse malware that can be easily used with other malware.

The combination of Trickbot and Emotet being used together became apparent in 2018. Trickbot and Emotet have been frequently used by the Ryuk ransomware in a triad fashion. In addition, Trickbot and Emotet are also used by the Conti ransomware.

In Fig. 1.7, an example ransomware attack using Emotet and Trickbot is presented. The procedures start with uploading a Emotet Trojan horse malware spam (malspam) program to a website link. The web link is included in phishing emails and various other distributions, in which a computer is contaminated when the user clicks on the malspam web link. This is when the Emotet loader malspam gains initial access to eventually compromise the computer and the Emotet loader is used to conduct data exfiltration to steal information and move it to a remote location. Then the Trickbot is executed to enable the ransomware program (e.g., Ryuk or Conti) to be installed and the ransom attack to begin. The Trickbot also uses its access to the compromised computer to gain access to the network's active directory and compromise the network's (subnet) domain controller. This enables the Trickbot to assist the ransomware program to contaminate the network and enable a network level ransom attack to be initiated. When a network level ransom attack is possible, major government and private enterprise networks can be compromised and exfiltrated.

1.3 Botnet Attacks Using IoT Devices

The number of Internet of Things (IoT) devices and size of IoT networks are growing rapidly. Currently, every modernized building, office, home, facility, industry, and factory uses massive amounts of IoT devices interconnected by high-speed networks. The processing capability and memory size of many IoT devices are relatively small compared to Personal Computers (PCs) or servers, which is why it is very difficult to install a sophisticated security system on an IoT device. Therefore, cyber-attacks to compromise an IoT device has always been a major threat. To make matters even worse, botnets can be used to compromise an IoT device, which will help to compromise additional neighboring IoT devices, and help launch a Denial of Service (DOS) or massive Distributed DoS (DDoS) attack on target servers, systems, and networks [7].

For example, in 2016, using thousands of IoT devices with limited processing capability and small memory, the Mirai botnet was able to execute a DDoS attack on the Krebs on Security company, starting with an initial traffic size of 620 Gbps (Giga bit/s $= 10^9$ bits/s). Since its emergence in 2016 August, the Mirai botnet was used to conduct powerful DDoS attacks on many large organizations including OVH,

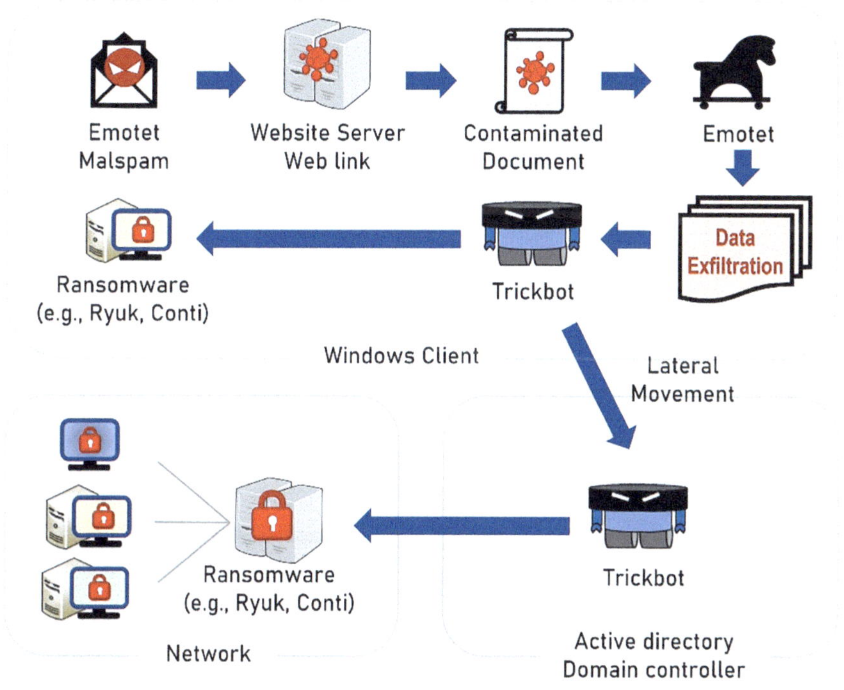

Fig. 1.7 Ransomware attack example using Emotet and Trickbot

Dyn, and Liberian Lonestar [8]. Incidents of Mirai botnet attacks rapidly grew from approximately 143,000 cases in 2018 to approximately 225,000 cases in 2019.

Botnets commonly use a centralized network structure or a decentralized Peer-to-Peer (P2P) network structure.

Figure 1.8 presents a botnet centralized Command and Control (C&C) network model, where the botnet Master controls a network of botnets through the C&C Server. The IoT devices that are compromised by malware and used for botnet attacks are called 'bots' [7]. The Botmaster can control the C&C Server to conduct various types of distributed DDoS attacks on one or multiple target servers or networks, which makes the centralized botnet structure very effective and efficient. One way to defend against a centralized botnet can be to find and disable the Botmaster or C&C Server.

Figure 1.9 shows a botnet decentralized C&C network example, where the Botmaster controls the bots directly communicating with them using a P2P network communication protocol (and does not use any C&C Server). The famous Mozi botnet uses this type of a P2P network, where more details are provided in the following section. It may be more difficult to defend against a decentralized botnet because there is no C&C Server that can be traced and all communication between the bots are P2P communication based.

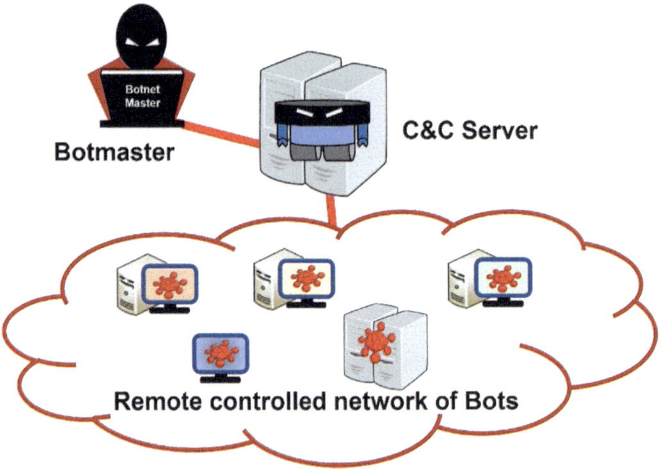

Fig. 1.8 Botnet centralized C&C example

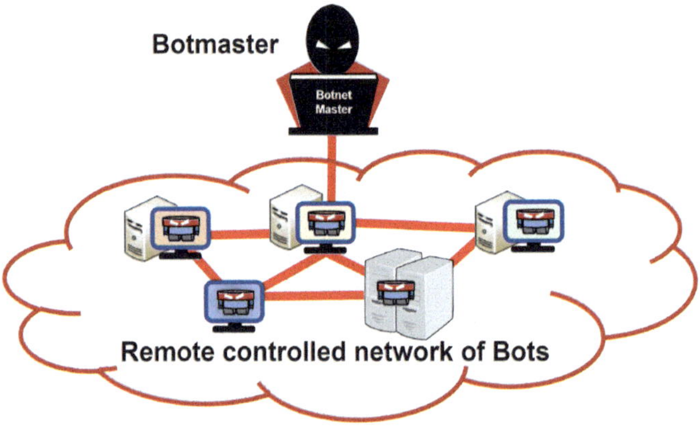

Fig. 1.9 Botnet decentralized C&C example

In addition, for the Mozi botnet case, the Botmaster uses asymmetric key based encrypted messages when communicating with the decentralized bots based on P2P communication, which makes the content of the messages unknown to network firewalls and cyber-attack anomaly detection systems. More details on asymmetric key cryptography technology is presented in Chap. 7.

Two representative IoT botnets are the Mirai (centralized C&C server-based structure) and Mozi (distributed P2P structure), which are described in further detail in the following sections.

1.3.1 Mirai Botnet

A Mirai botnet network consists of Botmaster, C&C Server, Report Server, Loader, and many bots in a centralized botnet structure. The bots are the IoT devices compromised by the Mirai malware.

Figure 1.10 presents a Mirai botnet DDoS attack operation example. Bots will continuously try to find new IoT devices that are vulnerable such that initial access can be made into the device's system. In this process, various username and password pairs will be tested [2, 8].

A predecessor of the Mirai botnet is the BASHLITE DDoS malware, which used a list of 14 generic username and password pairs to access IoT devices. The Mirai botnet was released with a more extensive list of 62 generic username and password pairs to gain easier access to a wider group of IoT devices. Mirai is equipped with a much more advanced faster vulnerability scanner in addition to using various credentials used by routers of popular consumer electronic products. This is why it is important to change the username and password of a new IoT device when conducting installation to a network, as default usernames and passwords make the IoT device vulnerable to botnet attacks.

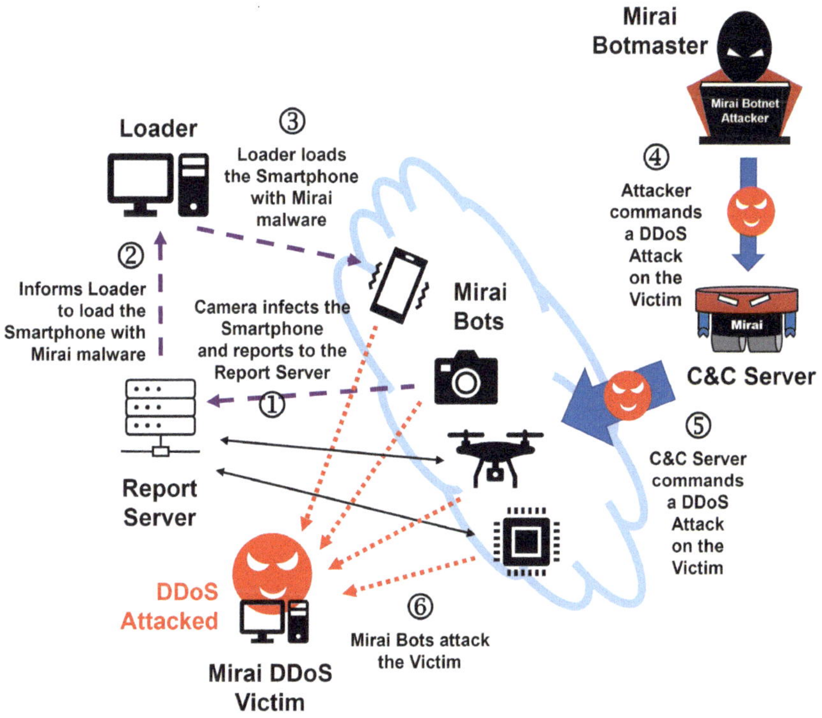

Fig. 1.10 Mirai botnets DDoS attack operation example

To avoid detection and counter retaliation, Mirai has a blacklist of IP addresses to avoid attacking, which include the US Department of Defense, US Postal Service, Hewlett-Packard, General Electric, and the Internet Assigned Numbers Authority (IANA).

In Fig. 1.10, a Mirai botnet DDoS attack operation example is presented. If a Bot (e.g., camera) succeeds in initial access (step ① in Fig. 1.10) to another IoT device (e.g., smartphone), then the Bot will report the smartphone's IP address to the Report Server (step ② in Fig. 1.10). Then the Report Server will contact the Loader (step ③ in Fig. 1.10) such that the Mirai malware can be fully loaded to the IoT device (e.g., smartphone). The Loader will make the IoT device download the Mirai malware using the Trivial File Transfer Protocol (TFTP) or GNU Wget protocol. The Mirai malware will make the IoT device shut down the Telnet port (port numbers 23 and 2323, which supports TCP connections) and Secure Shell (SSH) port (port number 22) and will turn off processes that other malwares (e.g., Qbot or.anime) may use. The Loader continuously monitors the compromised IoT devices and will reinstall the malware if the device gets rebooted.

Various DDoS attacks (e.g., ACK, DNS, Generic Routing Encapsulation (GRE)-IP, HTTP, Simple Text Oriented Messaging Protocol (STOMP), SYN, Valve Source Engine (VSE), UDP, and UDPPlain) can be conducted by the Botmaster at any time by commanding an attack through the C&C Server, which will activate the Bots to flood a targeted victim's system or network (steps ④, ⑤, and ⑥ in Fig. 1.10).

1.3.2 Mozi Botnet

The Mozi botnet emerged in December of 2019, and many variants have been discovered ever since [9]. A difference with the Mozi botnet to the Mirai botnet is that the Mozi botnet uses a distributed P2P structure and the Mirai botnet uses a centralized structure based on a C&C Server.

Figure 1.11 illustrates a Mozi botnet DDoS attack operation example, where the Botmaster sends configuration files to the Distributed Sloppy Hash Table (DHT), and the DHT communicates with the Mozi Bots using asymmetric key encrypted P2P messages that others cannot read. The Public DHT Node supports the startup of the Mozi Bots. The Mozi Bots repeatedly conduct vulnerability exploitation, weak password attacks, and brute force cracking to gain initial access to neighboring IoT devices.

The initialization stage of the Mozi malware will disguise the Mozi Bot by closing the watchdog functionality and renaming the process to SSH or dropbear, based on the host system type. The Mozi malware will block the SSH and Telnet ports (just like the Mirai malware did) and open some random UDP and TCP ports for its DHT to use and communicate through. In addition, the ports will be scanned and closed to prevent other intrusions and securely maintain the compromised status of the Mozi botnet functionality. Due to these procedures, it is very difficult to recover an IoT

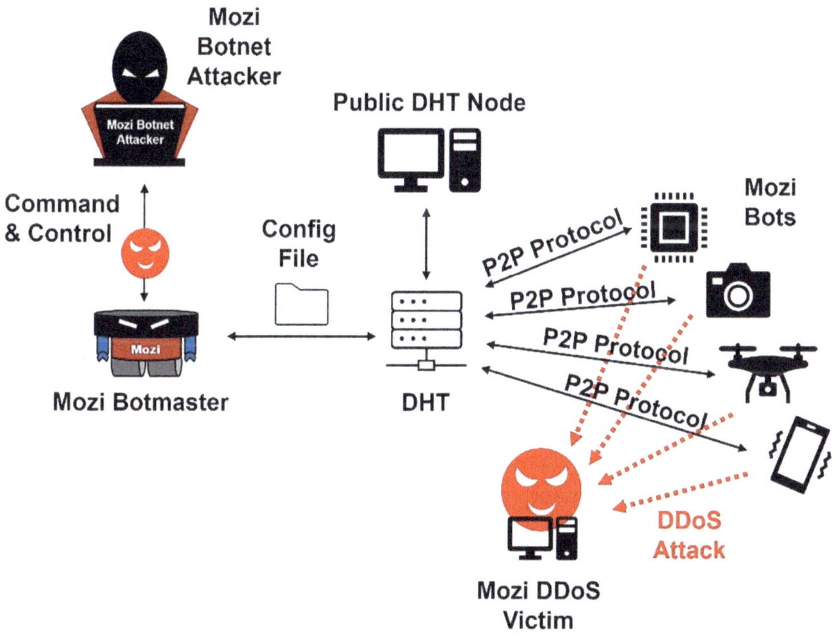

Fig. 1.11 Mozi botnet DDoS attack operation example

device that is compromised by Mozi malware. Other botnets use similar procedures to maintain their compromised status.

The Mozi botnet attacker can trigger a DDoS attack on a target system or network by controlling the Botmaster, which will use the DHT to individually control the Mozi Bots through encrypted P2P messages.

1.4 Cryptojacking on Blockchain Cryptocurrencies

Cryptojacking is another form of a malware based cyber-attack, but with much different characteristics [10]. Due to the immense value of cryptocurrencies (e.g., Bitcoin's BTC (Chap. 9) and Ethereum's ETH (Chap. 10)), numerous miners around the world attempt to compute hash values to receive cryptocurrency rewards. Because acquiring the computational equipment and network infrastructure is so expensive, a new form of mining that takes place on a victim's computer without the owner's knowledge or authorization is called cryptojacking. The word 'cryptojacking' could be seen as a nickname for 'cryptocurrency reward hijacking' because the victim's computer conducts the mining work in which the computer's owner should receive the mining reward, but this reward is hijacked by the cryptojacking control entity.

Cryptojacking is very hard to detect, because it is made to always operate in stealth mode, so it will try to remain undetected while conducting its cryptocurrency mining operation for as long as possible. In addition, cryptojacking malware has been embedded into many trusted websites, and, as a result, the cryptojacking malware spread extensively.

For example, it was reported in 2017 that an Italian bank was infected with Bitcoin cryptojacking malware. In 2018, it was reported that the Google advertisement programs on YouTube contained cryptojacking malware. In 2018, it was also reported that the UK Government had provided a plugin program that had been contaminated with cryptojacking malware. Another incident in 2018 related the popular game platform Steam being influenced with cryptojacking malware. In 2020, it was reported that the US Department of Defense (DoD) servers had been contaminated with cryptojacking malware from the service provider Coinhive (and mined 35.4 Monero coin), which were discovered during the Bug-Bounty Challenge. In addition, in 2020 it was also reported that the Zoom online meeting application software had been contaminated with cryptojacking malware [10].

Service Providers are the entities that develop cryptojacking programs and distribute them. In 2017, Coinhive emerged as the first cryptojacking website in-browser malware service provider, which became very popular. Now there are many cryptojacking Service Providers, which include Authedmine, Browsermine, Coinhave, Coinimp, Coinnebula, Cryptoloot, DeepMiner, JSECoin, Monerise, Nerohut, Webmine, WebminerPool, and Webminepool [10].

There are two major types of cryptojacking, in which Fig. 1.12 presents a host-based cryptojacking operation example and Fig. 1.13 describes a website-based cryptojacking operation example.

Figure 1.12 presents the procedures used in host-based cryptojacking operations. The cryptojacking script owner (which will be called the 'Cryptojacker' for short) attempts to gain initial access (step 1) to a victim's computer (e.g., using phishing), and once access is obtained, the cryptojacking malware will be uploaded (step 2). Without any knowledge or authorization of the computer's owner, the victim's computer will connect to the mining pool (step 3) to receive a mining assignment (step 4). The victim's computer will conduct the mining process (step 5) and report its results to the mining pool (step 6). Any rewards issued by the mining pool will be delivered to the Cryptojacker (step 7).

Figure 1.13 illustrates the operation of a website-based cryptojacking process. The Cryptojacker will register with a Service Provider to gain the credentials to access its website (step 1). Once the access credentials are obtained (step 2), the Cryptojacker uploads the cryptojacking malware to the website (step 3). When the victim's computer connects to the cryptojacking contaminated website (step 4), it receives both the website information along with the hidden cryptojacking malware (step 5). Without any knowledge or authorization of the computer's owner, the victim's computer will connect to the mining pool (steps 6 and 7) to receive a mining assignment (steps 8 and 9). Steps 6 through 9 may need to pass through the Service Provider, based on configuration of the network operation. The victim's computer will conduct the mining process (step 10) and report its results to the mining pool (steps 11 and

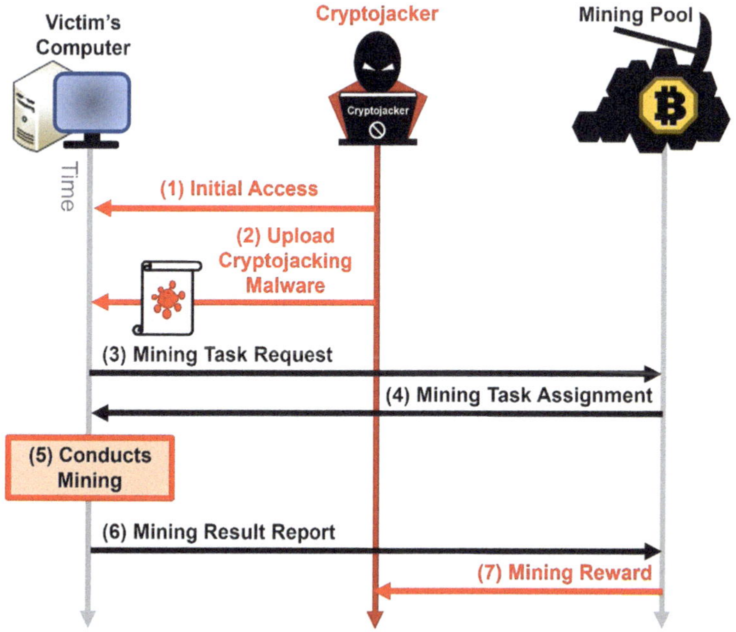

Fig. 1.12 Host-based cryptojacking operation example

12). Steps 11 and 12 may need to be pass through the Service Provider, based on the configuration of the network operation. Any rewards issued by the mining pool will be delivered to the Service Provider (step 13), which will be sent to the Cryptojacker (step 14).

1.4.1 Cyber-Attack and Cryptojacking Adaptive Defense Strategies

Technical strategies need to be developed to defend against these types of cyber-attacks. Considering that new AI techniques are added to cyber-attack algorithms to make them more intelligent and adaptive to the defense scheme, the defense schemes have to be able to quickly evolve as well. Therefore, there is no single solution to prevent cyber-attacks. Instead, defense systems need to be able to learn attack Tactics, Techniques, and Procedures (TTPs) and autonomously deploy defense mechanisms. The following tips are suggested to enhance the cyber security level.

First, a multi-layered security defense system must be employed. For this purpose, various layer-based diagnostics tools need to be used to search potential vulnerability points, which need to be self-tested. Abnormal behavior must be monitored where

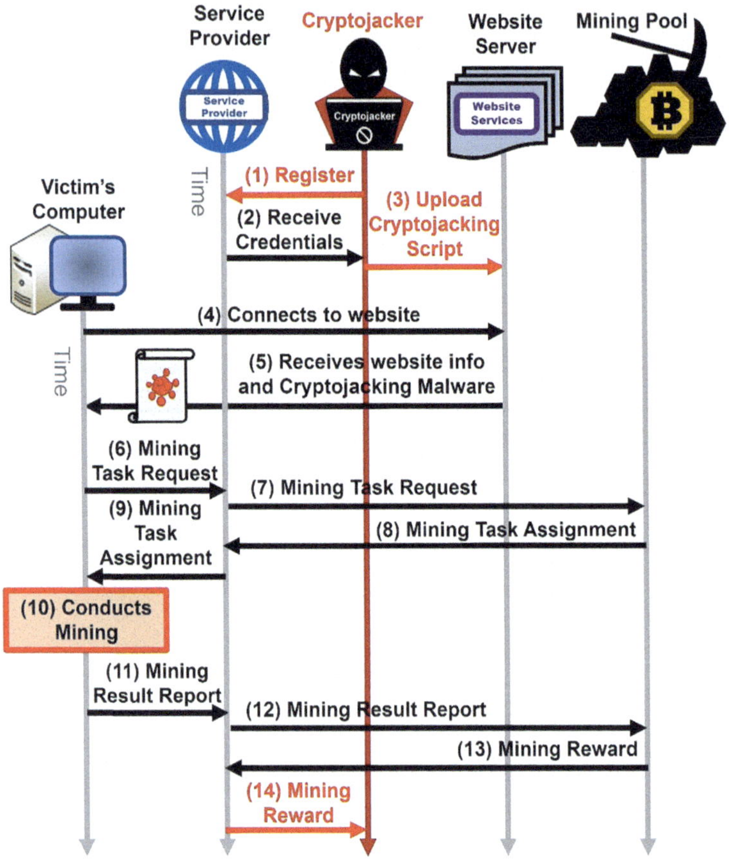

Fig. 1.13 Website-based cryptojacking operation example

the Demilitarized Zones (DMZs) and multi-layered firewalls need to be used. More details on firewalls and DMZs are introduced in the following sections of this chapter.

Second, to prevent ransomware attacks and database malware exfiltration, consistent monitoring of lateral data transfers is needed. As mentioned above, commonly the stage 4 of a ransomware attack includes a data exfiltration process. Therefore, if an unauthorized or undesired data transfer or exfiltration process is detected, then it can be considered that a ransomware attack may be in progress, and the characteristics of ransomware attack stages 1 through 3 also need to be carefully examined in the network routers, gateways, computers, and servers.

Third, the network and database defense technologies need to be self-learning and autonomously adaptive to potential cyber-attack patterns. This is because cyber-attack techniques are evolving using new AI technologies. For this purpose, AI defense technologies that use Generative Adversarial Network (GAN), Deep Reinforcement Learning (DRL), and Deep Deterministic Policy Gradient (DDPG) can be very effective. More details on GAN, DRL, and DDPG are introduced in Chaps. 2 and 11 of this book.

1.5 Recent Evolution of Cyber-Attacks

Taking a look back at 2015, the evolution of cyber-attacks becomes more evident and many of these cyber-attack techniques are still frequently used [11]. Based on the 2016 McAfee Labs Threat Report, the top 7 network attack types in 2015 Q4 are summarized below and presented in Fig. 1.14 [12].

The most frequent cyber-attack type was **Browser attacks** at 36%. In browser attacks, attackers disguise malware as an application or an update (e.g., phishing), where browser users are tricked into downloading the hidden malware. The browser intruded malware attacks the Operating System (OS) or applications of the user's computer. Browser attacks can be defended by updates to the browser, OS, and applications.

Zero-day attack is an attack on the vulnerability of a website that commonly results in disabling of web services. The term 'zero-day' refers to the day that a critical system, network, software, or platform vulnerability was patched, such that the web operation was restored.

Attackers are always looking for vulnerabilities in browsers and website plugins, in which vulnerabilities have been found in three quarters of all websites. As browsers and website attacks are very frequently attempted, a high-level of security has to be maintained at all times.

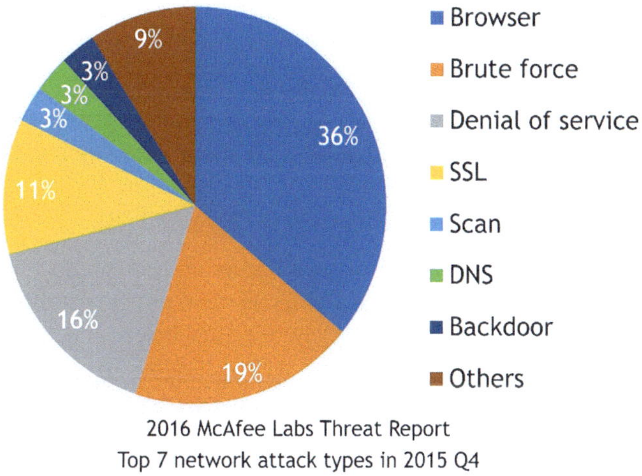

Fig. 1.14 Top 7 network attack types in 2015 Q4

The second most frequent cyber-attack type was **Brute force attacks** at 19%, where repeated attempts to decode a password or pin number are attempted. Easy common passwords and Dictionary attacks are commonly attempted first. Brute force attacks can be defended by frequent and well-selected password changes and by avoiding logging in too frequently.

The third most frequent cyber-attack type was **Denial of Service (DoS) attacks** at 16%. DoS attacks a computer, server, or network to block communications and services, where attackers overload the target with more requests than the target system can handle. Common targets are online banking, email, and commercial websites. A larger scale coordinated DoS attack is call a Distributed DoS (DDoS), which uses multiple (distributed) computers (infected by malware and botnets or in a zombie computer status) to conduct the DoS attack. DoS and DDoS can be defended by anti-virus software, firewalls, and email filters. As botnets use multiple devices to conduct Internet attacks (e.g., DoS, DDoS, steal data, send spam, intrusions) the effect of DoS or DDoS are still very significant. In addition, zombie computers are frequently used as botnets. A zombie computer is a hacker compromised computer connected to the Internet, where in most cases the user of the computer is unaware of this issue.

Many DoS and DDoS attacks use **buffer overflow** (a.k.a., buffer overrun) techniques. Buffer overflow occurs when an anomaly program (malware) overruns the buffer's boundary and overwrites into adjacent memory locations. Common defense schemes are listed below.

- Randomizing the layout of memory
- Deliberately leaving space between buffers
- Monitor actions that write into adjacent memory spaces.

Two representative types of Buffer overflow methods are stack overflow and heap overflow, which are described below.

Stack buffer overflow uses the following manipulation methods.

- Manipulates a local variable related to the vulnerable buffer on the stack.
- Manipulates the return address in a stack frame.
- Manipulates the function pointer or exception handler to create operation malfunction.
- Manipulates the stack frame's local pointer or local variable.

Heap overflow is a buffer overflow that occurs in the heap data area. Heap is used in dynamic memory allocation (of program data) of applications in run-time. Protection methods to prevent heap overflow are listed below.

- Separate the code and data (to prevent execution of the payload)
- Randomize heap location, so it is not located at a fixed offset position
- Periodically check the condition of the heap.

All these methods are commonly applied in OSs and Apps.

The fourth most frequent cyber-attack type was **Secure Sockets Layer (SSL) attacks** at 11%. SSL is used to setup a secured (encryption protected) link between 'website and browser' and 'email server and email user application.' In the early connection stage, SSL attackers intercept user information (e.g., access passwords, cookies, authentication tokens, etc.) before it is encrypted. Attackers try to gain sensitive data, such as, credit card information, social security numbers, etc.

The fifth most frequent cyber-attack type was **Scans** at 3%, where an attacker scans for an open computer port that can be used to gain access to the computer. Attackers repeatedly send messages to computer ports to find security vulnerabilities, where scans are commonly used before launching an attack.

The sixth most frequent cyber-attack type was **Domain Name Server (DNS) attacks** at 3%. DNS is used to change domain names into IP addresses. In a DNS spoofing attack (a.k.a. DNS cache poisoning) the cyber-attacker attempts to change the domain name's IP address to a wrong address. DNS spoofing is used for DNS hijacking, which redirects a user to a bogus website or a computer controlled by the attacker/hacker. DNS attacks can be defended by using a random source port and updating server security patches.

The seventh most frequent cyber-attack type was **Backdoor attacks** at 3%. A backdoor is a computer remote access application that is used by developers or administrators to bypass the security system to enable system fixes, upgrades, and add patches or filters. Hackers try to gain access through malicious backdoors in hardware or software components, which can be defended by updating security patches.

Various other types of cyber-attacks add up to 9%. It is important to note that companies attacked once are almost always attacked multiple times again, and bigger companies receive more attacks. Considering that currently ransomware attacks are increasing rapidly, a high-level of protection needs to be provided at all times. It is also a concern that companies are not reporting all security breaches and attacks, as such events can lead to billions of personal records being lost or stolen, and result in serious damage in the company's reputation and trust. In addition, a scarier thing is that the company may not know that they have been breached or how to recover.

1.5.1 Phishing

Phishing is one of the most common and frequently used malware contamination methods. Phishing is the act of a hacker fishing (phishing) around (commonly through email or message emphasizing urgency, importance, and/or opportunity) to obtain resources and information (e.g., usernames, passwords, credit card info, and money) by disguising as a trustworthy entity. Phishing emails commonly contain attachment files and website links that are infected with malware. Phishing is one of the most common ways to infect a computer with malware. There are various types of phishing, which are described below.

- **Spear phishing** is phishing directly to an individual or company. Using personal information of the individual significantly increases the success probability. This is the most common and successful type of a security attack setup method.

- **Clone phishing** uses a legitimate previously delivered email by resending it to the receiver containing an attachment file or a website link that is infected with malware. This fools the receiver as if it is an update or reply email.
- **Whaling** is a phishing attack made to a high-ranked executive (the Whale) of the company to attack important business files using executive-level access.
- **Link manipulation** is an act of phishing a website link that is infected with malware or connects to a hacker website. Link manipulation makes a website link look like a popular legitimate website link, but the hacker's web address is disguised under text or a tab that is not shown.
- **Filter evasion** is the act of phishing emails that use images (instead of text) to avoid anti-phishing filters used in security systems.
- **Website forgery** is the act of using a phishing website that replaces the original address bar or website with the hacker's address bar or website. Users are fooled into providing important login, password, account information to the hacker.
- **Covert redirect** is a phishing technique of corrupting a website to have a malicious login popup dialogue box that covertly redirects a login to the hacker's website.
- **Social engineering** is a phishing technique that provokes a user (using situating technical and social factors, e.g., fake news) to click on a malicious link or hacker website.
- **Phone phishing** is conducted through a telephone call or short message service (SMS) text message or a social network service (SNS) message to trick people into giving up personal information.

1.5.2 Man-in-The-Middle (MITM) Attacks

Another serious type of an attack is a **Man-in-the-Middle (MITM) attack**, where an attacker secretly relays and manipulates packets between communicating users and servers. MITM attacks use active eavesdropping and manipulation of information. Defense against MITM attacks include using enhanced authentication based on a trusted 3rd party Certificate Authority (CA) that can verify certificates, latency examination-based tamper examination, and HTTP public key pinning (a.k.a., certificate pinning), where the HTTP server first announces (pins up) a list of public key hashes that can be used for message/data encryption. More details on the MITM attack procedures are provided in Chap. 7.

1.6 Cyber-Attack Defense Systems

The most effective cyber-attack defense systems include updated software and OSs in addition to advanced Intrusion Detection System (IDS), Firewalls, Demilitarized Zone (DMZ), Transport Layer Security (TLS), Wired Equivalent Privacy (WEP), and Secure Shell (SSH) technologies, which are described in the following.

1.6.1 Intrusion Detection System (IDS)

An IDS is a system that detects malicious activity and policy violations in network devices, where the detection results are reported to the network administrator or Security Information and Event Management (SIEM) system. As the detection sensitivity is increased, the number of false alarms also increases, in which false alarms are removed by the alarm filters.

IDS system types can be classified into Host IDS (HIDS) and Network IDS (NIDS). An HIDS monitors all Internet packets sent and received, where the status snapshots are taken and compared with its former status. HIDS operates on PCs, laptops, smartphones, etc. On the other hand, a NIDS consistently monitors to detect abnormal behavior on network systems like gateways, routers, and switches.

IDS detection schemes include Signature-based IDS and Anomaly-based IDS. A Signature-based IDS scheme searches for specific intrusion patterns (e.g., malware sequences). On the other hand, an Anomaly-based IDS scheme detects new unknown attacks (e.g., new malware behavior).

Intrusion Prevention System (IPS) or Intrusion Detection and Prevention System (IDPS) are a combination of an IDS and Intrusion Counter Response System. Modern IPS and IDPS systems use a lot of machine learning and deep learning monitoring techniques to detect new attack patterns. Therefore, Chap. 2 of this book covers the most advanced artificial intelligence (AI) technologies, which are 'Generative AI' and 'Deep Reinforcement Learning' to assist future IDS development.

1.6.2 Firewall

Firewall is a network security monitoring and control system, that monitors all transmitted and received packets. Firewalls includes IDS mechanisms. Firewalls can be classified into network-based and host-based systems. A network-based firewall is used in network gateways, routers, and switches to protect the network. Host-based firewalls are used in computer OSs for endpoint (i.e., PC, smartphone) protection. As firewall technology has evolved significantly, generations of firewalls are divided as follows.

First generation (1G) firewalls were based on packet filtering of network addresses and ports. **Second generation (2G) firewalls** added on stateful filters for IP packet and transport (i.e., TCP and UDP) protocol inspection in addition to 1G firewall technologies. 2G firewalls track all state changes in IPv4, IPv6, TCP, UDP, etc. **Third generation (3G) firewalls** add various application layer (e.g., HTTP, DNS, FTP, etc.) filtering to the 2G and 1G firewall technologies. Advanced technologies included in 3G firewalls include Deep Packet Inspection (DPI), IPS, IDPS, device MAC address and reputation monitoring, as well as Web Application Firewall (WAF) technology.

1.6.3 Demilitarized Zone (DMZ)

A more advanced form of a firewall protection mechanism is to use two firewalls to form a Demilitarized Zone (DMZ). The term DMZ is originally used to represent an area in which there is agreement to not conduct any military operations (e.g., like the DMZ between North Korea and South Korea). In network security, a DMZ is a physical and/or logical network subarea that is between the external network (e.g., Internet) and internal network (e.g., local area network (LAN)), which is a buffer area that is used to carefully monitor all network connections before enabling entrance to the internal network, thereby significantly enhancing the security level of the internal network [13].

DMZs are commonly formed using two firewalls, one at the exiting edge of the external network and one at the entrance of the internal network as presented in Fig. 1.15. The network area between these two firewalls becomes the DMZ. DMZs add an additional layer to protect the internal network (e.g., LAN). The firewall at the exiting edge of the external network is called a 'perimeter firewall' (a.k.a., front-end firewall) and the firewall at the entrance of the internal network is called the 'internal firewall' (a.k.a., back-end firewall). There are benefits if the vendors of the perimeter firewall and internal firewall are different, as the two firewalls would not have the same vulnerabilities. However, this would increase the purchasing cost and maintenance fees.

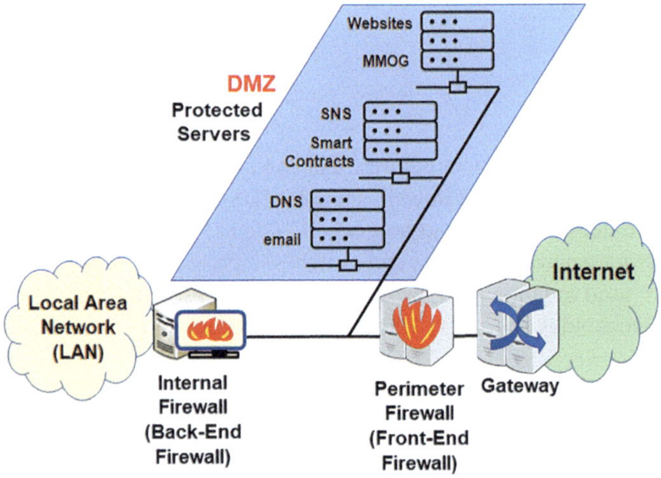

Fig. 1.15 DMZ using two firewalls

1.6.4 Transport Layer Security (TLS)

TLS uses network cryptographic protocols to enable secure communications. TLS was defined in 1999 (RFC 2246 defines TLS 1.0) and updated in 2006 (RFC 4346 defines TLS 1.1) and 2008 (RFC 5246 defines TLS 1.2) and 2018 (RFC 8446 defines TLS 1.3), where use of TLS 1.3 is demonstrated in the experiments presented in Chap. 5 Figs. 5.20, 5.21, 5.23, and 5.26.

TLS replaced Secure Sockets Layer (SSL) defined and updated in 1994, 1995, and 1996. TLS 2048 bit encryption has been used since 2013, where extensions will be made in the future. TLS provides privacy and data integrity between networked applications, where symmetric cryptography is used with encryption keys generated uniquely for each connection at session setup using the TLS handshake protocol. Public-key cryptography is used to authenticate the identity of the communicating systems and a message authentication code is used to ensure data integrity (prevent alterations). More details on cryptography technology are presented in Chap. 7.

1.6.5 Wi-Fi Protected Access 2 (WPA2)

WEP security algorithm was developed for the Institute of Electrical and Electronics Engineers (IEEE) 802.11 Wireless Local Area Networks (WLANs, a.k.a., Wi-Fi) to provide data confidentiality equivalent to the security level used in wired networks (ratified in 1997) using encryption and authentication techniques. In 2003, the Wi-Fi Alliance announced that WEP will be replaced (superseded) with Wi-Fi Protected Access (WPA). WPA is a Wi-Fi Alliance developed (in 2003) security protocol and security certification program that uses the Temporal Key Integrity Protocol (TKIP) to dynamically generate new encryption keys for each packet. WPA is an improvement from WEP, but TKIP can now be broken, so new enhancements were required. WPA2 is the current WPA version commonly used, in which WPA2 certification is mandatory (since March 13, 2006) for all Wi-Fi devices. WPA uses Counter Mode Cipher Block Chaining Message Authentication Code Protocol (CCMP) which is an Advanced Encryption Standard (AES) technology. WPA2 includes all mandatory elements of the IEEE 802.11i standard and requires Wi-Fi Alliance testing and certification.

1.6.6 Secure Shell (SSH)

SSH is a cryptography protocol to enable secure services over unsecured networks. SSH supported applications include remote command-line login, remote command execution, as well as various other secured network services. The upgraded version OpenSSH is commonly supported by all servers.

1.7 Summary

Every network connected electronic device and system needs to be equipped with a security system, as the types of emerging cyber-attacks are becoming more diverse, versatile, and powerful, as illustrated in this chapter. Even the small computing resources of Internet of Things (IoT) devices are being used for botnet attacks that have resulted in large network distributed denial of service (DDoS) disruptions. One of the reasons that the level of cyber-attack danger is increasing is because all of our new business and home appliances are IoT devices that have limited network security protection, due to the limited computing and memory resources these devices are equipped with. In addition, considering that almost all of our personal information is in our smartphones, computers, laptops, and tablets/pads, we are at high risk of data exploitation and becoming a victim of ransomware attacks. In addition, a massive amount of personal and business information is stored in cloud computing centers, local edge computing clouds, and big data systems, which are vulnerable to various cyber-attacks as they are open-accessible systems to the public. Another factor that must be included is the rapid advancements of Artificial Intelligence (AI) technology. Especially, generative AI, Generative Adversary Networks (GAN), and Deep Reinforcement Learning (DRL) systems can be programmed to automatically find Point-of-Failures (PoFs), loop holes, and security niches in network security systems and device Operating Systems (OSs). Considering this diverse range of cyber-attack methods, the reader is recommended to read the chapters on cryptography (Chap. 7), hash technology (Chap. 6), deep reinforcement learning, generative AI, federated learning, and digital twin technology (Chap. 2), blockchains (Chap. 8), Bitcoin (Chap. 9), smart contracts and Ethereum (Chap. 10), and future emerging technologies (Chap. 11) of this book, as many valuable technologies used in networking and security systems are introduced in detail.

References

1. IBM Security, X-force threat intelligence index 2022 (2022). https://www.ibm.com/downloads/cas/ADLMYLAZ
2. M. Ko (advisor J.-M. Chung), Trust-based Sybil Attack Detection and Avoidance Routing Technology to Improve IoT Network Security, MS thesis, Graduate School, Yonsei University (2023)
3. S. Cisco, B. Hazel (Editor-in-Chief), Defending against critical threats: analyzing key trends (2022). https://www.cisco.com/c/en/us/products/security/defending-against-critical-threats.html
4. H. Oz, A. Aris, A. Levi, A.S. Uluagac, A survey on ransomware: evolution, taxonomy, and defense solutions. ACM Comput. Surv. 54(11), article 238 (2022) https://doi.org/10.1145/3514229
5. Wikipedia, GNU. https://en.wikipedia.org/wiki/GNU
6. Wikipedia, Trickbot. https://en.wikipedia.org/wiki/Trickbot

7. S.N. Thanh Vu, M. Stege, P.I. El-Habr, J. Bang, N. Dragoni, A survey on botnets: incentives, evolution, detection and current trends. Future Internet, **13**(8), 198 (2021). https://www.mdpi.com/1999-5903/13/8/198
8. A. Borys, A. Kamruzzaman, H.N. Thakur, J.C. Brickley, M.L. Ali, K. Thakur, An evaluation of IoT DDoS cryptojacking Malware and Mirai Botnet, in *Proceedings of the IEEE World AI IoT Congress (AIIoT)*, June 6–9 (2022). https://ieeexplore.ieee.org/document/9817163
9. T.-F. Tu, J.-W. Qin, H. Zhang, A comprehensive study of Mozi botnet. Int. J. Intell. Syst. (2022)
10. E. Tekiner, A. Acar, A.S. Uluagac, E. Kirda, A.A. Selcuk, SoK: cryptojacking Malware, in *Proceedings of the IEEE European Symposium on Security and Privacy (EuroS&P)*, Sept. 6–10 (2021). https://ieeexplore.ieee.org/document/9581251
11. Coursera, J.-M. Chung, Introduction to TCP/IP https://www.coursera.org/learn/tcpip
12. McAfee Labs, 2016 McAfee Labs Threat Report, the top 7 network attack types. https://www.infopoint-security.de/medien/rp-quarterly-threats-sep-2016.pdf
13. Wikipedia, DMZ (computing). https://en.wikipedia.org/wiki/DMZ_(computing)
14. Calyptix, Top 7 Network Attack Types in 2016, June 13, 2016. https://www.google.co.kr/amp/www.calyptix.com/top-threats/top-7-network-attack-types-2016/amp/
15. Semantic, 2016 Internet Security Threat Report, vol. 21 (2016). https://www.symantec.com/security-center/threat-report
16. Cloudbric, A year in review: the top ten cyber attacks of 2016 in the gaming industry (2017). https://www.cloudbric.com/blog/2016/12/a-year-in-review-the-top-ten-cyber-attacks-of-2016-in-the-gaming-industry/
17. Calyptix, Ransomware prevention: 5 ways to avoid a crisis, February 23 (2015). http://www.calyptix.com/malware/ransomware-prevention-5-ways-to-protect-your-business/
18. Cloudbric, What is zero-day attack? (2017). https://www.cloudbric.com/blog/2016/05/what-is-zero-day-attack/
19. Internet Engineering Task Force. www.ietf.org
20. W. Stallings, *Data and Computer Communications*, 10th edn. (Pearson, 2014)
21. L.L. Peterson, B.S. Davie, *Computer Networks: A Systems Approach*, 5th edn. (Morgan Kaufmann, 2011)

Chapter 2
Deep Reinforcement Learning, Generative AI, Federated Learning, and Digital Twin Technology

This chapter introduces the most popular state-of-the-art artificial intelligence (AI) technologies, which include Deep Reinforcement Learning, Generative AI, and Federated Learning. In addition, a description of Digital Twin technology is provided, which is an important design, modeling, and testing platform that uses many AI applications. Section 2.1 covers the basics of deep reinforcement learning technology by providing an overview of supervised learning, unsupervised learning, and reinforcement learning. In addition, details on how more advanced deep reinforcement learning systems like Deep Q-Network (DQN) and Deep Deterministic Policy Gradient (DDPG) are designed and operate are introduced. Section 2.2 focuses on generative AI technologies, which introduces the details of Variational Autoencoder (VAE), Autoregressive Model (ARM), Flow Model, Diffusion Model, and Generative Adversarial Network (GAN) technology. Section 2.3 introduces federated learning technology, where collaborative learning techniques are introduced and federated learning types are explained. Section 2.4 provides details on digital twin technology, covering both digital twin system types and applied technologies. Examples of how deep reinforcement learning, generative AI, federated learning, and digital twin technology can influence networks and blockchain systems are provided in Chap. 11 of this book.

2.1 Deep Reinforcement Learning

Machine learning is a type of AI system that can adapt its decision algorithm to new environments and circumstances based on the information it has learned in the past. Deep learning is a type of machine learning technology that uses multiple processing layers that consist of a large number of computing parameters (such as a deep neural network (DNN)) to enable very accurate complex operations that are also adaptable to situation changes.

J.-M. Chung, *Emerging Secure Networks, Blockchains and Smart Contract Technologies*, https://doi.org/10.1007/978-3-031-65866-2_2

Based on the method of information learning and training, AI systems can be divided into three categories, which include, supervised learning, unsupervised learning, and reinforcement learning. A reinforcement learning system that uses deep learning technology is called a **Deep Reinforcement Learning** system.

The deep learning technologies used by deep reinforcement learning systems commonly include deep neural networks. For 2D or 3D image and video data type processing, the Convolutional Neural Network (CNN) type of deep neural network is most popular. On the other hand, voice and acoustic signal recognition, sensor and text data analysis, and computer program generation require a deep neural network that is capable of serial data sequential analysis and processing, in which a Recurrent Neural Network (RNN) is the most popular form of deep neural network technology to apply.

For more details on CNN and RNN type deep neural network technologies, please refer to Chap. 4 (Deep Learning for Metaverse XR and Multimedia Systems) of the author's book titled "**Emerging Metaverse XR and Video Multimedia Technologies**" published by Apress (Springer Nature, (c)2023) [1].

In order to describe the technical details of deep reinforcement learning systems, first, information on the related base technologies need to be introduced. Therefore, in this section, a description of supervised learning, unsupervised learning, and reinforcement learning are first introduced, and then the details of deep reinforcement learning technologies are provided.

2.1.1 Supervised Learning and Unsupervised Learning

Supervised learning systems will learn the characteristics of the dataset through a supervised approach that uses labeled data as input. **Labeled data** is data that has the desired output known when it is used as the input to an AI system. The most common AI system used in supervised learning is an artificial neural network (ANN), which has an input layer of artificial neurons, inner hidden layers of artificial neurons, and an output layer of artificial neurons as presented in Fig. 2.1. An ANN is also called a neural network (NN) for short. Each artificial neuron is represented by a circle and its connecting lines used to receive and send signals.

In Figs. 2.1 and 2.2, the circle is modeled after the neuron cell body (i.e., soma) and the connecting lines perform the role of the dendrites and axon terminals of a neuron cell, where more details are presented in Figs. 4.2 and 4.4 of [1].

Because all arrowed lines point in the same direction (from the input to the output direction) in Fig. 2.1, this is a forward propagation process. In comparison, Fig. 2.2 shows the back propagation part (from the output error back into the ANN).

The lines that interconnect the artificial neurons have weight values that can be trained (i.e., adjusted), which are used for the ANN to learn information and produce specific outputs for certain inputs. The input layer is where the input data to the ANN is inserted and the output layer is where the output of the ANN is delivered [1].

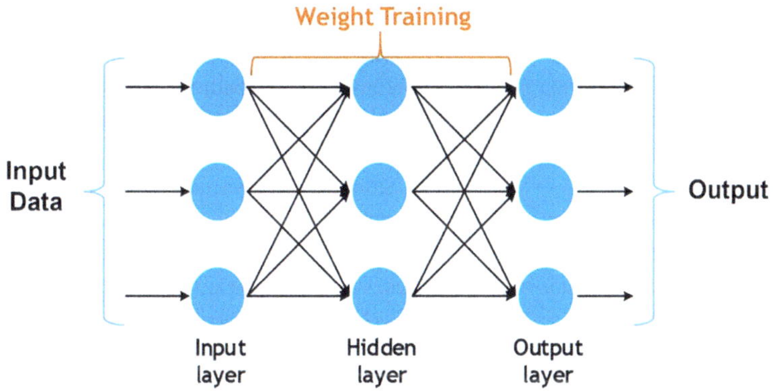

Fig. 2.1 Example of an artificial neural network (ANN)

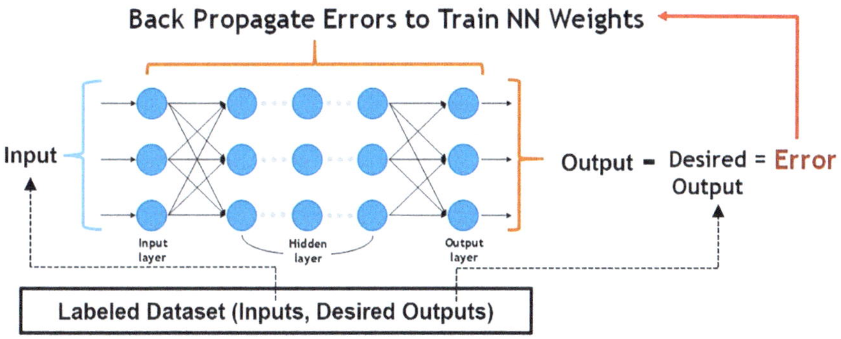

Fig. 2.2 Example of back propagation applied to an ANN or DNN

If the ANN system does not provide the desired output corresponding to the label data input, then the ANN system will make internal adjustments based on the size of the error of the output. The error of the output is the difference between the desired output and the actual output of the ANN system, which is based on the labeled data used to train the ANN system.

This type of an ANN system is also called a **Multi-Layer Perceptron (MLP)**. A perceptron is a supervised learning program or system that can perform basic classification operations. In an MLP, many perceptrons are used together to conduct complex classification and selection operations. In addition, an ANN that has many layers (i.e., an input layer, multiple hidden layers, and an output layer) where each layer consists of numerous artificial neurons is called a **Deep Neural Network**.

Supervised learning of ANNs or DNNs commonly use back propagation technology to update the weights of the artificial neurons in the neural network (NN) as presented in Fig. 2.2 [1].

Supervised learning is effective in solving classification problems and conducting regression-based analysis of data. Classification problems occur in cases where a mixed dataset needs to be divided based on specific parameters. Regression is a statistical tool used in data analysis to determine the relationship of how one parameter influences (or has a relationship) with the other parameters in the dataset. Commonly, the selected parameter is a dependent variable and the other parameters are the independent variables in the regression analysis process.

Unsupervised learning systems are trained without labeled data, which make them very useful and widely applicable. Unsupervised learning is used to find hidden patterns in a dataset, which can be used to conduct classification or clustering of data, or reduce the dimension of the dataset, or apply to future incoming data. An unsupervised learning algorithm will analyze the dataset and find unique patterns in the data that can help extract data with common characteristics.

In data analysis and problem-solving AI systems, supervised learning and unsupervised learning are used in many different ways. In addition, an advanced AI system like ChatGPT uses both supervised learning and unsupervised learning in its training process, which will be described in the following sections.

2.1.2 Reinforcement Learning

There are two basic reinforcement mechanisms used in training and education. One is 'Positive Reinforcement' and the other is 'Negative Reinforcement.' Positive reinforcement is a mechanism that the trainer/teacher will provide a desirable incentive/reward to the trainee/student when a proper/desired behavior is conducted. On the other hand, negative reinforcement is a mechanism that the trainer/teacher will remove an undesirable disadvantage/punishment from the trainee/student when a proper/desired behavior is conducted.

A desirable incentive/reward used in positive reinforcement would be like providing a compliment, something delicious to eat, desirable prize, or some type of favorable treatment. On the other hand, negative reinforcement examples include canceling a debt when a worthy act is conducted, or a bad grade is removed when a student submits a good makeup report.

When training a pet or conducting education, penalties/punishment as well as a combination of both positive and negative reinforcement are commonly used, which need to be very carefully and sensitively applied such that the training or education does not result in any mental or physical harm to the trainee. In addition, positive and negative reinforcement can be misused such that it becomes an abuse or torment to the trainee/student, so significant caution is always needed by the trainer/teacher and the influence of the positive and negative reinforcement on the trainee/student needs to be carefully monitored as every living being will receive the reinforcement treatment in a different way.

2.1.3 Reinforcement Learning System Architecture

Reinforcement learning systems obtain information by taking actions in the environment to learn the best way to achieve its goal, which include completing a task while maximizing the reward it receives. Supervised learning systems required labeled data. However, reinforcement learning systems do not need labeled data as they use the rewards based on their actions (at a given state) to learn about the environment. In reinforcement learning, the entity that will learn (be trained) what actions to take based on its experience is called the **agent**. Based on the accumulated trial-and-error experience, a well-trained agent can conduct the best actions (i.e., make optimal state transitions) to achieve the desired goal of the operation. Reinforcement learning is different from supervised and unsupervised learning based on this aspect. However, a reinforcement learning system can use supervised and unsupervised learning to be more efficient.

As the reinforcement learning system obtains information through interaction with the environment, it builds a policy that the agent will use to find the best way to map states to actions, in order to maximize the sum of the **rewards**. In reinforcement learning, the reward sum is also called the value. Reinforcement learning systems will setup a **value function** to monitor the accumulated rewards based on the agent's actions. Due to this self-learning capability that does not need labeled data, reinforcement learning has become a very popular and powerful AI tool. Figure 2.3 presents a reinforcement learning based computer system concept diagram.

The original design of reinforcement learning systems is based on the **Bellman equation**, which is a necessary condition used in mathematical optimization algorithms like dynamic programming. The Bellman equation divides the dynamic optimization process into simpler and smaller problems that can be solved in a sequence. For example, in a reinforcement learning system, the overall objective would be to build a policy that could maximize the reward sum while achieving the operation goal. The agent will try to accomplish this by solving a sequence of simpler and

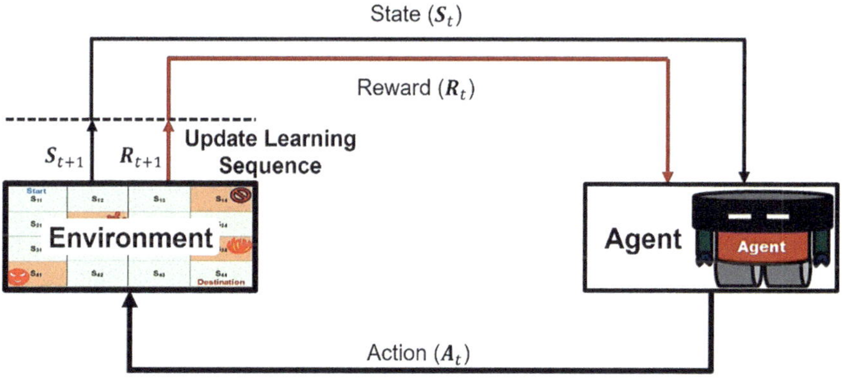

Fig. 2.3 Reinforcement learning based computer system concept diagram

smaller problems. The agent will try to find the best way to map the agent's current state to an action that can provide the largest reward, while moving itself in the proper direction [2–5].

The reinforcement learning based computer system in Fig. 2.3 includes the following 6 basic parts: agent, action, environment, internal state, observation, and reward.

Environment: The environment space is where the actual system design is included, as well as all of the physical limits and constraints. An actor will attempt a series of actions within the environment to reach the final goal while trying to learn what series of actions would provide the largest value (i.e., sum of rewards), and will record the trial-and-error attempts (and corresponding rewards) in its internal state memory so future actions can be made more efficiently and the rewards the agent receives will be maximized.

Agent and Action: The agent is the mobile entity that takes actions within the environment to learn the most rewarding way to accomplish the operation goal.

Observation and Internal State: The agent's position, actions, and rewards have to be consistently observed and recorded in the agent's internal state memory. The agent will observe and keep track of what it has experienced so far in its internal state memory so it can learn from its trial-and-error to more effectively maximize the rewards it receives in future actions within the environment.

Reward: The environment needs to be setup to provide rewards for specific actions taken at certain locations within the environment. The rewards will be given to the agent based on the sequence of actions conducted and locations visited within the environment.

Reinforcement learning is effective to use in project planning, puzzle solving, trajectory control, path finding in a maze, as well as various dynamic programming tasks. Reinforcement learning will use information of the environment dynamics by collecting data of rewards obtained through the activities of the agent. The reinforcement learning system will collect the following information based on repeated trials of actions taken. In the environment at time t, if the agent at state S_t conducts the action A_t, then it will obtain a reward of R_t.

Using this information, an agent control policy π is setup during the reinforcement learning process, which is based on the conditional probability $P(S_{t+1}, R_{t+1}|S_t, R_t)$ that represents the probability of the agent being at state S_t and having received reward R_t, and making a transition to state S_{t+1} and receiving reward R_{t+1} as a result of action A_t taken at time t at state S_t.

The reinforcement learning system is focused on creating a policy of how to control the actions of the agent such that the value V (which is the sum of the rewards, $V = \sum_{t=1}^{\infty} R_t = R_1 + R_2 + \ldots$) is maximized while achieving the project goal.

If the reinforcement learning system does not have sufficient conditional probability $P(S_{t+1}, R_{t+1}|S_t, R_t)$ data accumulated (which commonly happens at the beginning of a process), the system will have to learn the characteristics of the new environment and setup its policy. The policy will govern the control algorithm used by the agent to decide the next action to take.

A reinforcement learning algorithm can be designed based on several principles, which are described in the following.

2.1.3.1 Model-Based Versus Model-Free Reinforcement Learning

One way to classify a reinforcement learning algorithm is model-based versus model-free, which are described in the following [2–5].

In a **model-based** reinforcement learning system, first an attempt to learn the model of the trajectories is conducted, and then the system will try to setup an optimal action policy. A model-based policy will have higher priorities assigned to more preferred actions based on its accumulated interaction with the environment. Problematic cases have occurred in the past where a model-based reinforcement learning system resulted in a greedy agent behavior that performed actions to maximize the reward regardless of what the action may result in, which in some cases accidents have occurred.

In a **model-free** reinforcement learning system, the policy will execute actions multiple times for each state and will adjust itself to obtain a maximized reward as the agent attempts to achieve the operation goal while following the constraints imposed. Model-free policy reinforcement learning has also been referred to as **direct reinforcement** learning or **explicit trial-and-error reinforcement learning** in some papers. Model-free reinforcement learning algorithms include **Q-Learning** and **Policy Gradient**, which both will be explained in the following sections.

A model-based reinforcement learning agent will make a prediction of the reward (based on its current state/location) before conducting an action, and will execute the action that provides it with the highest reward. On the other hand, in model-free reinforcement learning, commonly the agent will need to conduct the action to learn what type of a reward it will get.

A model-based reinforcement learning system will commonly perform with high efficiency if the environment and operation rules are known in advance. So, for a game or factory robot, where the environment is known and the rules and rewards are fixed, a model-based reinforcement learning system would be preferred due to the maximized value it can provide in a very efficient way.

Model-based reinforcement learning systems have modular designs that are commonly based on a library of independent functions. So, compared to model-free reinforcement learning systems, model-based reinforcement learning systems are easier to design, debug, and add new functions. In addition, there are many papers and programs that can be used to design new functions, which makes the system design, development, and upgrading tasks easier and more efficient.

However, model-based reinforcement learning systems do not use task-based gradient feedback information from the newer action-based explorations of the environment. Therefore, it is more challenging to make a model-based reinforcement learning system learn new dynamics and plans. So, if the environment changes, the results from a model-based reinforcement learning system can be biased such that the performance range is constrained and operations are limited.

A model-free reinforcement learning system will commonly perform better when the agent's action space and environment is larger such that there is much more diversity in the results and rewards. Model-free reinforcement learning systems are commonly used in autonomous driving vehicles and autonomous flying drones as the agent's action space and environment are extremely large, diverse, and dynamic. As can be speculated, for autonomous driving vehicles and autonomous flying drones, finding a sufficiently accurate reinforcement learning model for such a diverse range of situations and environment conditions would be unrealistic and impractical, which is why model-free reinforcement learning systems are commonly used. For these reasons, in this book we will focus more on model-free reinforcement learning systems.

Model-free policy reinforcement learning systems include Deep Q Network (DQN), Double DQN (DDQN), Deep Deterministic Policy Gradient (DDPG), Asynchronous Advantage Actor-Critic Algorithm (A3C), Trust Region Policy Optimization (TRPO), Proximal Policy Optimization (PPO), Twin Delayed Deep Deterministic Policy Gradient (TD3), Soft Actor-Critic (SAC), and Distributional Soft Actor-Critic (DSAC) [5, 6].

2.1.3.2 On-Policy Versus Off-Policy Reinforcement Learning

Another way that reinforcement learning systems can be classified is based on the on-policy and off-policy algorithm used by the agent.

In **on-policy** reinforcement learning, the agent's own experience is used to form a statistical control algorithm, which is used in future action control. On-policy reinforcement learning algorithms include A3C, PPO, and TRPO.

In **off-policy** reinforcement learning, the agent's experience and the behavior and experience from observing other behavior (as well as exploratory or random behavior) is used to form a statistical control algorithm to use in future action control. Due to the larger size of data used in off-policy reinforcement learning, a higher performing system will proportionally require more memory and processing capability.

In addition, on-policy reinforcement learning is a special-case subset of off-policy reinforcement learning. Off-policy reinforcement learning algorithms include DQN, DDQN, TD3, DDPG, SAC, and DSAC.

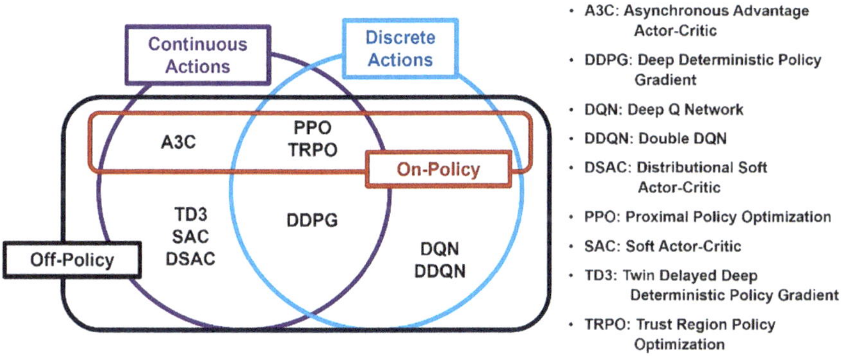

Fig. 2.4 Model-free reinforcement learning types

2.1.3.3 Continuous Versus Discrete Action Reinforcement Learning

Another way that reinforcement learning systems can be classified is based on the actions being continuous or discrete. Discrete action reinforcement learning algorithms can conduct actions among the predefine discrete states. On the other hand, continuous action reinforcement learning algorithms can conduct an action that can be in any direction and have any size. DQN and DDQN are representative discrete action reinforcement learning algorithms. Continuous action reinforcement learning algorithms include A3C, TD3, SAC, and DSAC. In addition, DDPG, PPO, and TRPO can support both continuous and discrete action reinforcement learning.

Figure 2.4 summarizes the model-free reinforcement learning types [6].

In the following, the Q-learning algorithm is introduced, which is one of the essential reinforcement learning algorithms.

2.1.4 Q-Learning

Q-learning is an off-policy reinforcement learning algorithm, where the agent uses a finite **Markov decision process** to learn the optimal action policy to take based on the state in the environment. Based on the agent's current state, the Q value represents the expected reward of the action that can be taken. An algorithm that applies 'greedy action' will make the agent perform the action that can result in the largest Q value [2, 3, 7].

To help describe the Q-learning process, a hazard field reinforcement learning environment example is presented in Fig. 2.5. Please note that this section presents a simplified example of the Q-learning process and hazard field crossing path reinforcement learning operation to make it easier to understand the fundamental background technologies applied.

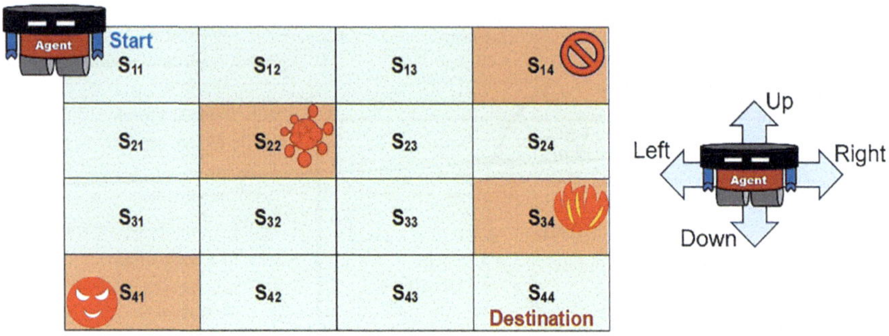

Fig. 2.5 Hazard field crossing path reinforcement learning environment example and possible agent movements

Figure 2.5 also includes the possible agent movements on the right side, which include 4 types of possible movements, which are Up, Down, Right, or Left. All movements would make the agent move to a neighboring block (state) corresponding to the direction of the action that is selected.

The state of the agent can be at one of the 16 locations in the hazard field, which are $\mathbf{S_{11}}$, $\mathbf{S_{12}}$, $\mathbf{S_{13}}$, $\mathbf{S_{14}}$, $\mathbf{S_{21}}$, $\mathbf{S_{22}}$, $\mathbf{S_{23}}$, $\mathbf{S_{24}}$, $\mathbf{S_{31}}$, $\mathbf{S_{32}}$, $\mathbf{S_{33}}$, $\mathbf{S_{34}}$, $\mathbf{S_{41}}$, $\mathbf{S_{42}}$, $\mathbf{S_{43}}$, and $\mathbf{S_{44}}$. Among the states, $\mathbf{S_{11}}$ is the starting state of the agent and $\mathbf{S_{44}}$ is the final state that the agent will try to arrive at. Among the states, $\mathbf{S_{14}}$, $\mathbf{S_{22}}$, $\mathbf{S_{34}}$, and $\mathbf{S_{41}}$, have hazards, so if the agent enters these states, then the agent will lose its mobility and will not be able to make any further actions, so the agent will not be able to reach the final destination $\mathbf{S_{44}}$.

For each state there is a reward that can be obtained for arriving at that location. For example, the reward for going to state $\mathbf{S_{12}}$ is $\mathbf{R_{12}}$, the reward for going to state $\mathbf{S_{13}}$ is $\mathbf{R_{13}}$, so in generic terms, the reward for going to state $\mathbf{S_{XY}}$ is $\mathbf{R_{XY}}$, where $\mathbf{X}$ and $\mathbf{Y}$ can only be one of the numbers of 1, 2, 3, and 4 in Fig. 2.5. So, the possible rewards are denoted as $\mathbf{R_{11}}$, $\mathbf{R_{12}}$, $\mathbf{R_{13}}$, $\mathbf{R_{14}}$, $\mathbf{R_{21}}$, $\mathbf{R_{22}}$, $\mathbf{R_{23}}$, $\mathbf{R_{24}}$, $\mathbf{R_{31}}$, $\mathbf{R_{32}}$, $\mathbf{R_{33}}$, $\mathbf{R_{34}}$, $\mathbf{R_{41}}$, $\mathbf{R_{42}}$, $\mathbf{R_{43}}$, and $\mathbf{R_{44}}$.

Because the states $\mathbf{S_{14}}$, $\mathbf{S_{22}}$, $\mathbf{S_{34}}$, and $\mathbf{S_{41}}$ have hazards, their corresponding rewards $\mathbf{R_{14}}$, $\mathbf{R_{22}}$, $\mathbf{R_{34}}$, and $\mathbf{R_{41}}$, will be negative values (i.e., penalty values) and the agent will not be able to make any further actions due to the hazards influence. Negative reward $\mathbf{R_{14}}$ is given for arriving at state $\mathbf{S_{14}}$, negative reward $\mathbf{R_{22}}$ is given for arriving at state $\mathbf{S_{22}}$, negative reward $\mathbf{R_{34}}$ is given for arriving at state $\mathbf{S_{34}}$, and negative reward $\mathbf{R_{41}}$ is given for arriving at state $\mathbf{S_{41}}$.

From the start state $\mathbf{S_{11}}$, the agent can be made to take random actions so a variety of states can be visited. The random actions will enable the agent to visit many states, which will help the agent to learn the rewards due to the various actions that will enable the agent to reach the final destination $\mathbf{S_{44}}$. Among the many action sequences that the agent experiences, the one with the highest value will be selected to be included in the optimal policy (π^*) for the agent.

For example, Fig. 2.6 presents an agent at state $\mathbf{S_{32}}$ trying to make a decision on where to go based on the Q value options available.

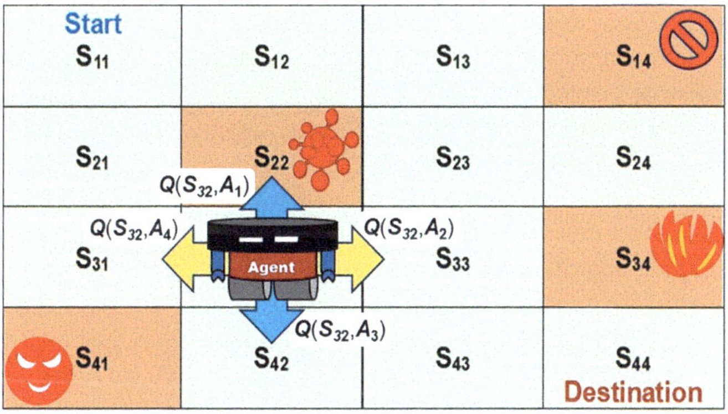

Fig. 2.6 Example of the Agent at state $\mathbf{S_{32}}$ in the hazard field environment trying to make a decision on where to go based on the Q value options (i.e., actions $\mathbf{A_1}$, $\mathbf{A_2}$, $\mathbf{A_3}$, and $\mathbf{A_4}$) available

Next, formulas are applied to the Q-learning process so a computer program example can be presented. The objective of Q-learning is to build a policy π to control the sequence of actions (A_t) that will take the agent from the current state (S_t) to the next state (S_{t+1}) in which the goal of the learning is achieved and the overall value V (i.e., sum of rewards $V = \sum_{t=0}^{\Psi} R_t$) is maximized (where Ψ is the maximum number of iterations, which can be a finite number, or set to ∞).

Figure 2.7 presents a basic Q-learning based computer program example (when the learning rate is one), which can be considered a high-level program pseudocode. Many reinforcement learning algorithms use a Q function. The Q function has two input values, which are the state S and the action A.

1. Set the agent at the starting point.
2. For each state S and action A, initialize the Q function table entry $Q(S, A)$ to zero.
3. Observe the current state S.
4. Repeat until the destination is reached or the number of permitted actions expires.
   i. Select an action A for the agent and have the agent execute the action (where the selected action is represented as A').
   ii. Observe the new state S' the agent has arrived at.
   iii. Receive the reward R for the new state S' that the agent arrived at.
   iv. Update the table entry for $\hat{Q}(S, A)$ using the Q function in (2.1).

$$\hat{Q}(S, A) \leftarrow R + \gamma \max_{A'} \hat{Q}(S', A') \quad (2.1)$$

   v. Update the state $S \leftarrow S'$.

Fig. 2.7 Reinforcement learning example implementation of a basic Q-learning algorithm

State S can be seen as the location of the agent in the environment, and action A is based on the direction and amount of movement the agent has made in one step when moving away from current state S.

The Q function's output is the reward R that the agent would obtain for the action A conducted at state S in the environment. So, for the input values S and A into the Q function, the output is the reward value R, which is represented as $Q(S, A) = R$ in the reinforcement learning based computer algorithm.

The Q-learning algorithm introduced in Fig. 2.7 is one of the fundamental reinforcement learning algorithms. The Q-learning algorithm starts with step 1, where the agent is set at the starting point. In step 2, all Q table states and all actions will be set to 0 in the initialization process (i.e., $Q(S, A) = 0, \forall S, \forall A$). After observing the current state S in step 3, the repeated loop process of step 4 is executed, which is repeated until the agent arrives at the target destination, or the number of permitted actions (Ψ) is reached (to avoid infinite program loops).

In sub-step i, an action A for the agent is selected and executed. The new state S' that the agent has arrived at is recorded in sub-step ii. In sub-step iii, the agent receives a reward R for arriving at the new state S'. In sub-step iv, the term $R+\gamma \max_{A'} \hat{Q}(S', A')$ is a summation of the reward R and the term $\gamma \max_{A'} \hat{Q}(S', A')$. In $\gamma \max_{A'} \hat{Q}(S', A')$, the γ term is a discount factor that is a constant value in the range of $0 \leq \gamma < 1$, and the term $\max_{A'} \hat{Q}(S', A')$ is the largest Q function value obtainable. In sub-step v, which is the last line of the Q-learning algorithm, the term $S \leftarrow S'$ makes the new state S' to become the current state S in the next round of the agent's action selection and execution in the Q-learning algorithm. The **update learning sequence** line in Fig. 2.3 shows where the Q-learning algorithm transitions to a new round.

Figure 2.8 illustrates the results of the Q-learning algorithm applied to the hazard field reinforcement learning example environment shown in Fig. 2.5. The results show a sequence of successful actions and state transitions.

There are many ways the agent can reach to the destination state $\mathbf{S}_{44}$ from the starting state $\mathbf{S}_{11}$ through a sequence of state transitions. Among these cases, Fig. 2.8 presents two shortest state transition sequences that the agent can reach the final destination based on the hazard field environment. The agent has to avoid the hazard states $\mathbf{S}_{14}$, $\mathbf{S}_{22}$, $\mathbf{S}_{24}$, and $\mathbf{S}_{34}$ to successfully reach the final destination.

In ①, a sequence of actions resulted in the following state changes $\mathbf{S}_{11} \rightarrow \mathbf{S}_{21} \rightarrow \mathbf{S}_{31} \rightarrow \mathbf{S}_{32} \rightarrow \mathbf{S}_{42} \rightarrow \mathbf{S}_{43} \rightarrow \mathbf{S}_{44}$ for the agent to reach the destination $\mathbf{S}_{44}$, which results in a total reward value of $\mathbf{R}① = \mathbf{R}_{11} + \mathbf{R}_{21} + \mathbf{R}_{31} + \mathbf{R}_{32} + \mathbf{R}_{42} + \mathbf{R}_{43} + \mathbf{R}_{44}$.

In ②, a sequence of actions resulted in the following state changes $\mathbf{S}_{11} \rightarrow \mathbf{S}_{12} \rightarrow \mathbf{S}_{13} \rightarrow \mathbf{S}_{23} \rightarrow \mathbf{S}_{33} \rightarrow \mathbf{S}_{43} \rightarrow \mathbf{S}_{44}$ for the agent to reach the destination $\mathbf{S}_{44}$, which results in a total reward value of $\mathbf{R}② = \mathbf{R}_{11} + \mathbf{R}_{12} + \mathbf{R}_{13} + \mathbf{R}_{23} + \mathbf{R}_{33} + \mathbf{R}_{43} + \mathbf{R}_{44}$.

After multiple trial-end-error attempts, it is assumed that the agent learned that the sequence of state changes of ① and ② result in successful arrivals to the destination state $\mathbf{S}_{44}$ with the shortest state transition sequences and highest values (i.e., highest

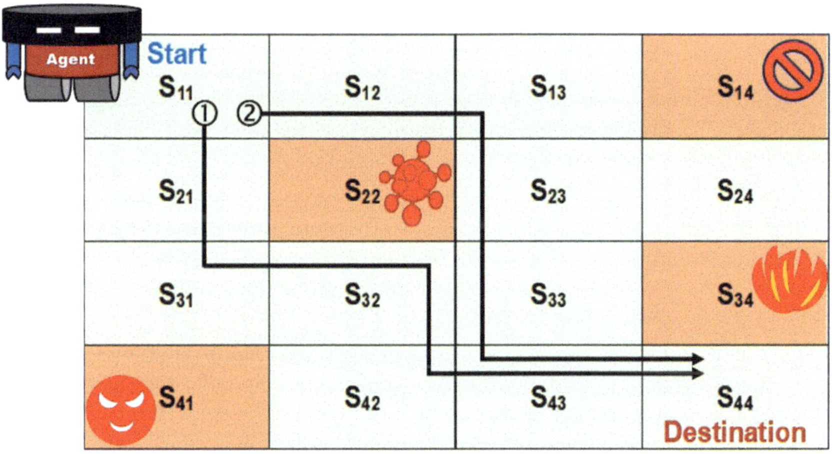

- State transition sequence ① is $S_{11} \rightarrow S_{21} \rightarrow S_{31} \rightarrow S_{32} \rightarrow S_{42} \rightarrow S_{43} \rightarrow S_{44}$.
- State transition sequence ② is $S_{11} \rightarrow S_{12} \rightarrow S_{13} \rightarrow S_{23} \rightarrow S_{33} \rightarrow S_{43} \rightarrow S_{44}$.

Fig. 2.8 Hazard field reinforcement learning example of a sequence of successful actions and state transitions

sum of rewards). So, in this example, among these two routes, the agent will compare the combined rewards of **R**① and **R**②, where the one with the highest value will be chosen as the optimal path to use by the Q-learning system in the future.

2.1.4.1 Limitations of the Hazard Field Type Reinforcement Learning

In the hazard field example, the environment was small and known, so the agent could easily visit all possible states in order to find the policy of actions that would maximize the final value (i.e., sum of the rewards).

However, if the environment is much larger and many unknow states exist, then it would be very difficult and inefficient to use trial-and-error and test all states to figure out the policy of actions that would maximize the final value. So, enhancements to the Q-learning mechanism are needed, which is why deep reinforcement learning algorithms were created.

2.1.5 Deep Reinforcement Learning

Deep reinforcement learning combines reinforcement learning with deep learning technologies. Traditional reinforcement learning systems show limitations in performance when the number of states in the environment grows, which exponentially increases the number of possible action sequences, making it very difficult to establish a high-performance action control policy. To overcome this issue, deep reinforcement learning combines reinforcement learning with a deep learning policy neural network so the agent can more accurately estimate the optimal actions to take [2–4].

Figure 2.9 presents a deep reinforcement learning system model in which the agent is controlled by a deep neural network (DNN) that learns the policy π based on past training data. The agent will make a decision based on its current state S_t at time t and the policy π of the DNN.

One of the most popular deep reinforcement learning algorithms is the Deep Q Network (DQN) mechanism, which is described in the following section.

2.1.5.1 Deep Q Network (DQN)

The performance of reinforcement learning systems is influenced by the feature representation's quality. DQN uses experience replay techniques to more effectively solve high-correlation problems. DQN replaces the Q-learning **Q value table** with a **deep neural network (DNN) based Q value** to be able to more effectively extract high-level representations from the data, as presented in Fig. 2.10. The Q value of Q-learning is estimated through a DNN. The Q value estimation error is used to adjust the weights in the DNN using back propagation procedures. A comparison of tabular Q-learning and DQN deep Q-learning is provided in Fig. 2.10 [2–4].

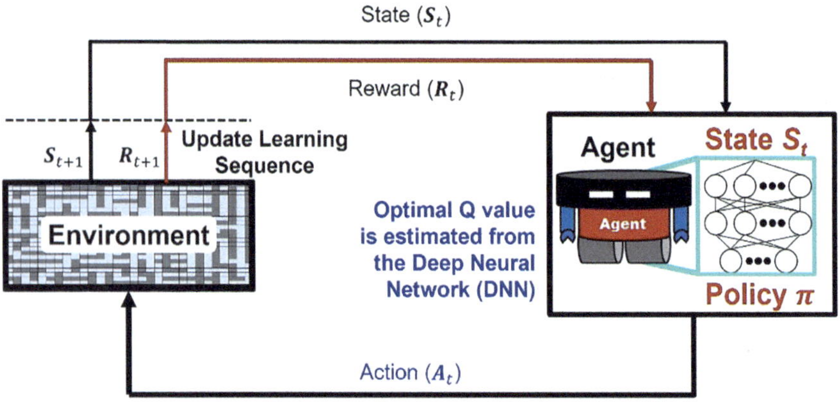

Fig. 2.9 Deep reinforcement learning system model

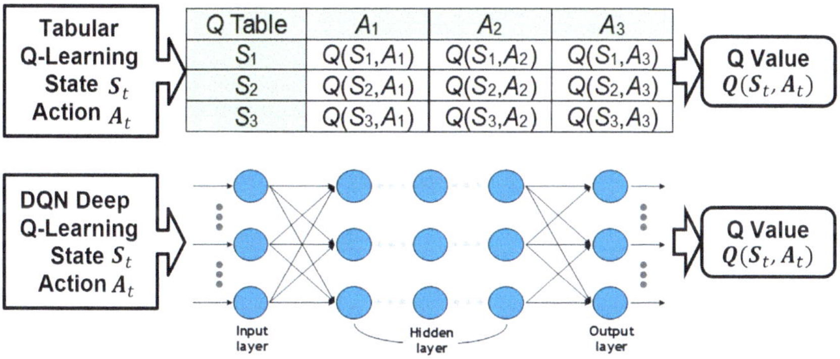

Fig. 2.10 Comparison of tabular Q-learning and DQN deep Q-learning

Unlike the simple hazard field example in Fig. 2.5, if the environment is much larger and more complex, like in Fig. 2.11, it may be difficult or impractical for the agent to visit all states in the environment. Therefore, the DQN agent will try to find the best policy π^* to control the agent's sequence of actions (A_t) with only limited knowledge of the states of the environment using the DNN.

The DQN will be focused on obtaining the optimal policy π^* that can control the agent's sequence of actions (A_t) based on the state (S_t) in the environment that will maximize the overall value while taking the agent to its desired destination. The policy π^* that controls the agent's sequence of actions (A_t) considers the accumulated knowledge of the environment and the current state (S_t) of the agent.

The DQN reinforcement learning system will train the prediction DNN $\theta_t^{Prediction}$ (to minimize the loss function) so the optimized policy function π^* in (2.2) can be obtained.

$$\begin{aligned}\Pi^* &= \min_{\theta} \sum_{t=0}^{T} [\hat{Y} - Y]^2 \\ &= \min_{\theta} \sum_{t=0}^{T} \left[\hat{Q}\left(S_t, A_t|\theta_t^{Prediction}\right) - \left(r_t + \gamma \max_{A'} \hat{Q}\left(S_{t+1}, A_t'|\theta_t^{Target}\right)\right)\right]^2 \end{aligned} \quad (2.2)$$

The prediction value $\widehat{Y} = \widehat{Q}\left(S_t, A_t|\theta_t^{Prediction}\right)$ is the prediction made at time t based on the prediction DNN $\theta_t^{Prediction}$. The target value $Y = \left(r_t + \gamma \max_{A'} \hat{Q}\left(S_{t+1}, A_t'|\theta_t^{Target}\right)\right)$ depends on the target state at time t based on the environment DNN θ_t^{Target}.

To build the Experience Replay capability, the DNN conducts trial-and-error actions of the agent and records the States S_t, Actions A_t, and resulting Rewards R_t for each time instant t in its database. From this database, randomly selected samples of <State, Action, Reward> data tuples will be used to train the prediction

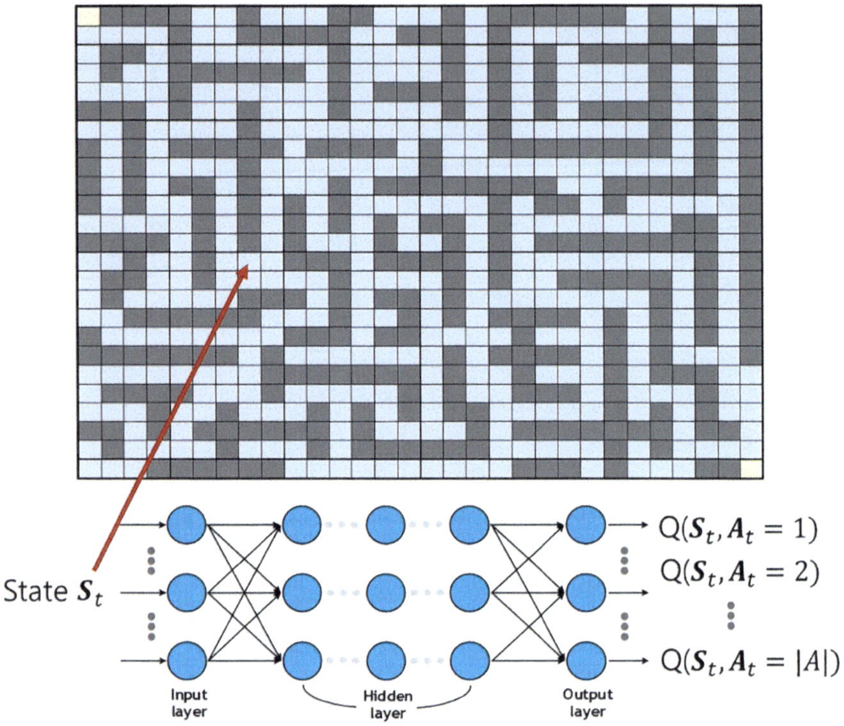

Fig. 2.11 Deep Q network (DQN) system structure

DNN $\theta_t^{Prediction}$ such that the prediction value $\widehat{Y}$ can become as close as possible (or the same) as the target value Y. For this purpose, the weight values that control the artificial neurons in the prediction DNN $\theta_t^{Prediction}$ will be adjusted for each of the randomly selected <State, Action, Reward> data tuples based on the difference between the prediction value $\widehat{Y}$ and the target value Y. This process is repeated until the permitted number of iterations expires or until the average difference between the prediction value $\widehat{Y}$ and the target value Y fall below a preset threshold level, thereby achieving a desired level of accuracy.

2.1.5.2 Double Deep Q-Networks (DDQN)

DQN will update its weights using the largest Q value based on the possible actions that can be taken at the next state. Commonly the Q values of the next state's action are small, so an irregularly large Q value can result in an overestimation issue in DQN systems. Therefore, to avoid this overestimation issue, the DDQN system uses two DNNs, where the Q network uses a DNN to train and make optimal action selections and the Target Q network uses a separate DNN to make evaluations. The DQN Loss

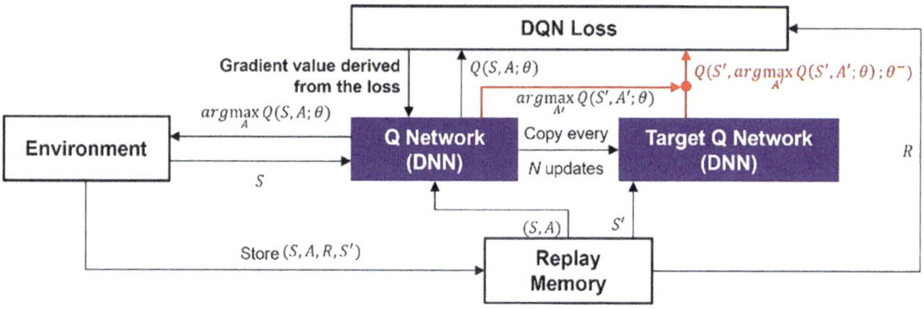

Fig. 2.12 Double deep Q network (DDQN) deep reinforcement learning system

is computed using the difference of these two values, which are marked by the red lines and red Q function in Fig. 2.12 [2–4].

2.1.5.3 Deep Reinforcement Learning Pros and Cons

Pros of deep reinforcement learning technology are described in the following. Deep reinforcement learning is effective in solving complex problems and does not require to be trained in advanced as it can learn from exploration of actions in the environment. Therefore, unlike supervised learning and unsupervised learning systems, deep reinforcement learning systems do not need a large labeled dataset, nor is a dataset of system parameters required. Deep reinforcement learning will attempt to solve a complex problem by dividing the overall problem into smaller problems which can be sequentially solved using a Markov decision process. In the process of solving a problem, the system will attempt to estimate the optimal value function and optimal policy to control the agent such that the objective of the operation is accomplished while maximizing the obtained value (i.e., the sum of rewards).

Cons of deep reinforcement learning technology are described in the following. For successful results, the deep reinforcement learning system needs to have the actions and rewards properly set in the environment, which may be difficult to implement. In addition, if the overall problem cannot be easily divided into smaller problems that can be sequentially solved using a Markov decision process, then it might not be suitable to apply deep reinforcement learning. As a result, deep reinforcement learning systems may have convergence instability problems.

To overcome some of the limitations that deep reinforcement learning systems have, DDPG was created, which is an extension technology to deep reinforcement learning systems, as described in the following.

2.1.5.4 Deep Deterministic Policy Gradient (DDPG)

DQN performs well when the actions of the Q function are discrete. This means that DQN performs well when the environment has subdivided domains that form individually divided states, then the DQN discrete actions can be used to make transitions between the discrete states. However, if the environment is not formed of individually divided discrete states, then DQN will not be effective [2–4, 8].

If the environment is a continuous domain, then the agent will need to conduct actions that are continuous (analog values) and can be in any direction within the environment. This is the case for many operational environments.

Among the continuous domain reinforcement learning system models, DDPG is one of the most popular and effective, so it will be described in further detail in this section.

As mentioned above, model-free reinforcement learning algorithms use action exploration-based policy searching methods to setup their action policy, and model-based reinforcement learning algorithms control their actions in a way that will maximize their value. The value is the sum of the rewards obtained from the sequence of actions conducted by the agent.

As described above, DDPG is a model-free Actor-Critic reinforcement learning system that uses continuous value actions. DDPG uses two deep learning networks, one for the Actor and one for the Critic. The Critic network is trained using DQN or DDQN technology. The Actor's deep learning policy network is called the π network, as it will check the current state of the agent and conduct an action that will attempt to maximize the Q function, as presented in Fig. 2.13.

An actor-critic reinforcement learning system is a combination of both the policy search and the value maximization schemes. Actor-critic reinforcement learning algorithms include DDPG, TD3, A3C, SAC, IMPALA, etc. The policy gradient algorithm can be changed into an actor-critic reinforcement learning algorithm by replacing the real value with the estimated Q value function.

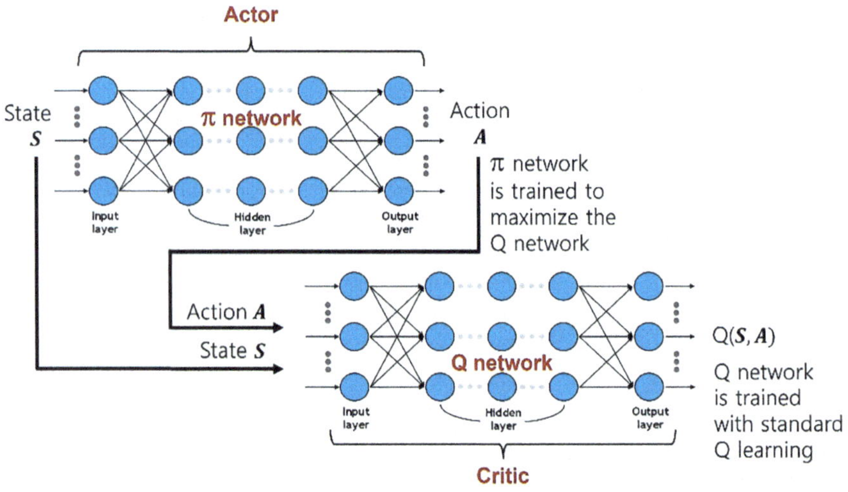

Fig. 2.13 DDPG actor-critic system structure

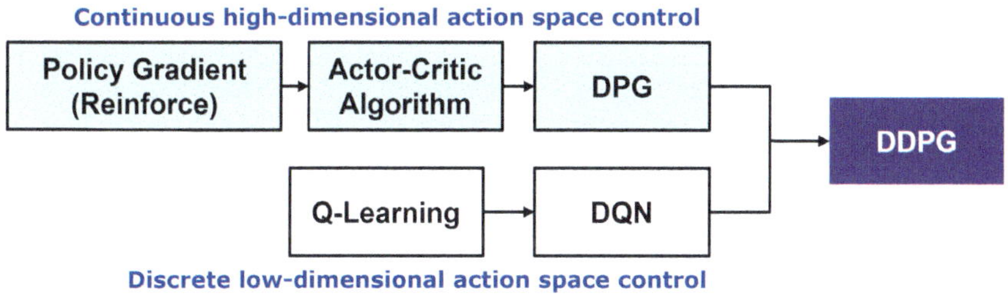

Fig. 2.14 DDPG deep reinforcement learning system architecture

Figure 2.14 presents a DDPG deep reinforcement learning system architecture. The DDPG deep reinforcement learning system architecture combines two systems. The DDPG system uses an actor-critic algorithm policy gradient based DQN system for continuous and high-dimensional action space control and a separate Q-learning DQN system is used for discrete low-dimensional action space control. The actor-critic algorithm uses an actor-network to select the actions and a critic-network to evaluate the value of the action, which are used by the Deterministic Policy Gradient (DPG) control [2–4, 8].

2.1.5.5 Comparison of DQN, DDQN, and DDPG

A comparison of some of the characteristics of different deep reinforcement learning models is summarized in Fig. 2.15.

Future technologies will be developed to improve DQN, DDQN, and DDPG, which is one of the purposes of this book. In the following section, a new field of advanced AI technology is introduced, which is called 'generative AI.'

2.2 Generative AI

Generative AI models can be classified in many ways, considering their complex system nature. One basic method of dividing generative AI systems is Deep Generative Models (DGMs) and Probabilistic Graphical Models (PGMs) [1]. DGM systems will map the input data into a latent vector using a deep neural network (DNN). Before DGMs became popular, PGMs were frequently used to map the input data into multiple interconnected latent variables using multiple mapping functions. The functions used in PGMs have much simpler forms compared to the DNNs used in DGMs [2].

A latent variable is an internal variable in an AI system that is used to compute, transform, or map input data to help prepare the proper system output. A simple system may not need a latent variable, as it may be able to directly map the input

Deep Q-Network (DQN)	Double Deep Q-Network (DDQN)	Deep Deterministic Policy Gradient (DDPG)
DQN is the first reinforcement learning system to use a DNN. The DNN is used to estimate the optimal Q value of the Q-learning system.	DDQN solves the overestimation problem of DQN and enhances the overall performance. DDQN uses two DNNs. One DNN is used to train the policy network such that optimal actions are chosen, and the other DNN is used to make accurate target network evaluations, which will help prevent overestimation problems.	DDPG combines DQN/DDQN and Policy Gradient technology to properly provide continuous and high-dimensional action space control. The Policy Gradient technology uses an Actor-Critic algorithm for optimal action selection and to adjust the updates.

Fig. 2.15 Comparison between deep reinforcement learning models

data to an output value. But because generative AI systems are very complex, latent variables are frequently used.

The remarkable performance of generative AI systems currently available are based on DGM technology. Therefore, this section focuses on the newest DGM technologies, which include Variational Auto-Encoder (VAE), Autoregressive Model (ARM), Flow Model (FM), Diffusion Model (DM), and Generative Adversarial Network (GAN). Among these generative AI systems, VAE, FM, DM, and GAN use latent variables.

Variational Auto-Encoder (VAE) technology is used in the famous DALL-E image creation generative AI system as well as other systems. VAE systems use latent variable embedding to maximize the variational lower bound.

Embedding is used by computers to make various types of data (e.g., words, symbols, images, video, and signal) easily processible by computer programs. Generative AI systems use embedded data for data analysis, similarity matching, search engines, etc. In the embedding process, various types of encoding and mappings will be conducted. The embedding process may use various types of metadata. Metadata is data extracted from a dataset. This means that metadata is a processed, mapped, clustered/grouped, classified, prioritized, and/or transformed data of another dataset. Metadata helps to provide information and characterizes other datasets, making the analysis and processing of the dataset more effective and efficient.

Autoregressive Model (ARM) technology is used in ChatGPT, GPT, and DALL-E, and BERT uses transformer generative AI systems as well as other systems. ARM systems attempt to learn the conditional probabilistic distribution information of each variable based on past inputs and generated outputs.

Flow Model (FM) technology is a foundational technology for generative AI systems. FMs use invertible transforms of probabilistic distributions and latent variable embedding processes.

Diffusion Model (DM) technology is used in DALL-E 2, Imagen, and Stable Diffusion generative AI systems. DMs use multiple stages to gradually add noise (i.e., noisify) to input data/images and then reverses this process through multiple stages to gradually remove the added noise (i.e., denoisify) to generate new content.

Generative Adversarial Network (GAN) uses adversarial training to enhance the quality, resolution, scale, and/or precision of contents created from various generative AI models as well as probabilistic/stochastic models used in computer-based system analysis or designs, as well as digital twin or Turing machine modeling.

In this section, the variable x will be used to represent the input and z will be used to represent the latent variable. The probability distribution of the input variable x is $p(x)$ and the probability distribution of the latent variable z is $p(z)$.

Generative AI systems like VAE, FM, DM, and GAN will generate a latent variable value z based on each given input x to the system, where the probability distribution of z being generated based on a given input x is represented using the conditional probability $p(z|x)$.

Generative AI systems like VAE, FM, DM, and GAN will create new content x' based on the input x by using its generated latent variable z and the corresponding probability distribution $p(z)$ of the latent variable z. The new content that is generated can be in the form of data, information, images, sounds, music, computer program, design, etc. The probability distribution of the newly generated content x' is based on the given latent variable z using the conditional probability $p(x'|z)$.

Input variable x can be 1 dimensional (1D) or multi-dimensional. For example, the input x will be 1 dimensional (1D) if it is a sequence of data representing text, numbers, acoustic signal, etc. In this case, x is a sequence of scalar values, in which each value has a 1D form.

The input x can also be 2 dimensional (2D) or 3 dimensional (3D), in which the corresponding data will be represented in the computer program using a 2D vector (e.g., X_{2D} in (2.3)) or 3D vector (e.g., X_{3D} in (2.4)) in matrix form.

$$X_{2D} = \begin{bmatrix} x_1 \\ x_2 \end{bmatrix} \tag{2.3}$$

$$X_{3D} = \begin{bmatrix} x_1 \\ x_2 \\ x_3 \end{bmatrix} \tag{2.4}$$

This expansion in dimension can continue to grow to 4D, 5D, etc. In addition, a data point can be formed from a combination of multiple dimensional components, which is called a tensor (e.g., $X_{Tensor}^{a,b,c}$ in (2.5)).

$$X_{Tensor}^{a,b,c} = \begin{bmatrix} \vec{X}_a \\ \vec{X}_b \\ \vec{X}_c \end{bmatrix} = \begin{bmatrix} \begin{bmatrix} a_{0,0} & a_{0,1} \\ a_{1,0} & a_{1,1} \end{bmatrix} \\ \begin{bmatrix} b_{0,0} & b_{0,1} \\ b_{1,0} & b_{1,1} \end{bmatrix} \\ \begin{bmatrix} c_{0,0} & c_{0,1} \\ c_{1,0} & c_{1,1} \end{bmatrix} \end{bmatrix} \tag{2.5}$$

The latent variable z can be the same dimension as the input x or may have a smaller dimension.

2.2.1 Variational Auto-Encoder (VAE)

VAE systems are powerful in creating new content that is based on the input, but can diversify in many ways, resulting in input related content creativity. Due to this reason, VAE is used in the OpenAI text-to-image generation system DALL-E and various other generative AI systems [2, 9].

In order to describe the characteristics of a VAE, first the details of an autoencoder (AE) need to be provided.

In very simplistic terms, the Encoder of an AE will receive an input x and will generate a latent variable z based on a conditional probability $p(z|x)$. The creation of latent z values will form a probability distribution $p(z)$. After the encoding process of creating the probability distribution $p(z)$, a decoding process is used to recreate a new x' from the latent variable z using the conditional probability $p(x'|z)$, where x' has some properties of x but is different in a creative way.

As presented in Fig. 2.16(a), an AE system has two major parts, a Generative Network (GN) and an Invert Generative Network (IGN). The Encoder uses the IGN to encode the input x into a latent variable z using the conditional probability distribution $q(z|x)$ based DNN encoder. The second part of the VAE system is the Decoder which uses a GN to generate the new content x'. The GN decodes the latent variable z using the conditional probability distribution $p(x'|z)$ based DNN, which generates the new content x'.

Decoders of an AE directly generate output x' values from the latent variable z. However, VAE encoders will form a probability distribution of x to generate samples that are used to form the latent variable z and its probability distribution, as presented in Fig. 2.16(b). Then the probability distribution of z is used by the decoder to generate new output x' values. The probability distribution generated by the VAE encoder (IGN) commonly has a Gaussian normal (**N**) distribution with mean $\boldsymbol{\mu_x}$, standard deviation $\boldsymbol{\rho_x}$, where $\boldsymbol{\phi}$ represents the set of parameters of the encoder DNN (i.e., $\mathbf{N}_{\boldsymbol{\phi}}(\boldsymbol{\mu_x},\boldsymbol{\rho_x})$). The probability distribution generated by the VAE decoder (GN) commonly has a Gaussian normal (**N**) distribution with mean $\boldsymbol{\mu_z}$, standard deviation $\boldsymbol{\rho_z}$, where $\boldsymbol{\theta}$ represents the set of parameters of the decoder DNN (i.e., $\mathbf{N}_{\boldsymbol{\theta}}(\boldsymbol{\mu_z},\boldsymbol{\rho_z})$).

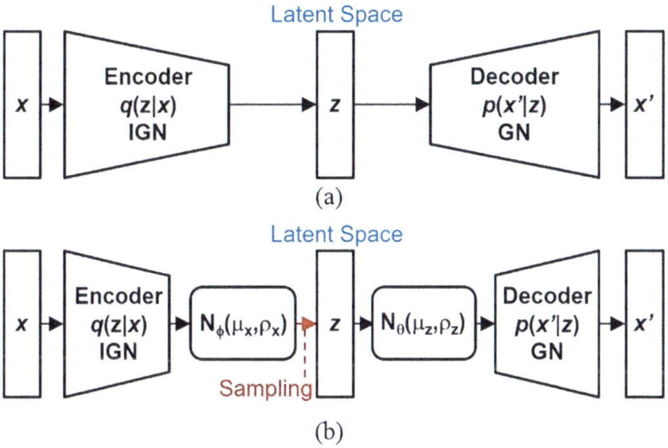

Fig. 2.16 Comparison of an AE to a VAE, where **a** is the AE and **b** is the VAE

The narrowing trapezoid of the encoder is used to represent that the dimension of the latent space may be smaller than the dimension of the input space. On the other hand, the widening trapezoid of the decoder is used to represent that the dimension of the latent space may be smaller than the dimension of the output space. The use of trapezoids in the other generative AI models have the same meaning [2].

VAE systems are based on the Deep Latent Variable Model (DLVM), where the latent variable z is generated from $p(z)$, and then z is used as the input to a DNN decoder to generate an output $x' = d(z)$ that is based on $p(x'|z)$. If the latent z variable's probability distribution $p(z)$ is a Gaussian distribution, then the DLVM is called a Deep Latent Gaussian Model (DLGM). DLGMs are frequently used in VAE models.

The VAE decoder needs to derive $p(x')$ from $p(z)$ and $p(x'|z)$, which requires the integration of (2.6), which is called posterior inference.

$$\int p(x'|z)p(z)dz = \int \frac{p(x',z)}{p(z)}p(z)dz = p(x') \tag{2.6}$$

The computation of (2.6) is burdensome as the probability distributions have to be multiplied and integrated, which is a complex process. To simplify this computation, the integration process of (2.6) can be divided into smaller parts that can be computed in a sequence. This will result in a closely approximated value of $p(x')$. However, the value can be made to be sufficient by controlling the generation resolution of the approximated computations. This process can be conducted using an inference network based on $q(z|x)$. These inference networks are commonly designed to have a Gaussian distribution with parameters provided from a DNN encoder.

A VAE system can be made to converge to locally optimized maximum likelihood estimated parameters with proper setup tunning and process control. VAEs can provide good results when the system is able to optimize the variational lower bound of the log-likelihood probability.

2.2.2 *Autoregressive Model (ARM)*

ARM systems use a linear combination of current and past values in generating new outputs. This is why ARM systems have an architecture in which each computation node is influenced by all previous nodes in operation. The ARM process is conducted using a fully connected Directed Acyclic Graph (DAG) that connects the state probabilities, in which the joint and conditional probability distributions are computed and used at each state of the process. In addition, ARM uses Attention Layer and Transformer technologies, which are also described in the following sections along with the ARM process structure [2].

2.2.2.1 Conditional Probability Used in ARM

Assuming that there are k inputs to the ARM so far (which are listed as $x_1, x_2, \ldots, x_k$), then the initial state's distribution (based on x_1) will be $p(x_1)$. The second state's distribution (based on x_2) will be $p(x_2|x_1)$, as it will depend on the former state x_1. The third state's distribution (based on x_3) will be $p(x_3|x_1, x_2)$, as it will depend on the two former state x_1 and x_2. Extending from this relation, the kth state's probability distribution (based on x_k) will be $p(x_k|x_1, x_2, \ldots, x_{k-1})$, as it will depend on the former states $x_1, x_2, \ldots, x_{k-1}$.

For convenience, $x_1, x_2, \ldots, x_{k-1}$ will be represented as $x_{1:k-1}$, and therefore, $p(x_k|x_1, x_2, \ldots, x_{k-1}) = p(x_k|x_{1:k-1})$, which is the conditional probability of arriving at x_k based on the condition that we have experienced the former states $x_1, x_2, \ldots, x_{k-1}$.

When a DNN is used to estimate the conditional probability distribution $p(x_k|x_{1:k-1})$ the system is called a Neural Auto-Regressive Density Estimator (NADE). In addition, frequently $p(x_k|x_{1:k-1})$ is estimated using a combination of conditional Gaussian probability distributions. In this case, the system is called as Real-Valued NADE (RNADE). Among DNN types, a Convolutional Neural Network (CNN) is very effective in predicting the state changes and the conditional probability distribution $p(x_k|x_{1:k-1})$. When using a CNN, a causal mask convolution needs to be applied to make sure that only past states (i.e., $x_1, x_2, \ldots, x_{k-1}$) are used to estimate x_k and $p(x_k|x_{1:k-1})$, and no future state values (i.e., $x_{k+1}, x_{k+2}, \ldots$) are used. This type of causal mask operation is commonly applied in ARM processes [1, 2].

2.2.2.2 Joint Probability Used in ARM

ARM systems also use the joint probability distribution $p(x_1, x_2, \ldots, x_k) = p(x_{1:k})$, which is the probability that the states of $x_1, x_2, \ldots, x_k$ have occurred in the ARM system (where $x_{1:1} = x_1$ and $x_{1:0}$ means there is no input to the ARM, so $p(x_1|x_{1:0}) = p(x1)$). The joint probability can be derived using the conditional probabilities, Bayes theory, and the chain rule of probability, as presented in (2.7).

$$\begin{aligned} p(x_1, \ldots, x_k) = p(x_{1:k}) &= \prod_{i=1}^{k} p(x_i|x_{1:i-1}) \\ &= p(x_1)p(x_2|x_1)\cdots p(x_k|x_{1:k-1}) \end{aligned} \tag{2.7}$$

In ARM systems, the likelihood probability of the data sequence can be optimized and the computation process is easy to understand and model. However, as k increases, it becomes more difficult to compute (2.7), resulting in longer time delays as well as requiring large data storage and a significant amount of computing resources.

This is why the ARM uses hidden states to assist the estimation of the values in (2.7). For example, one way to overcome this issue is to use a hidden Markov model to estimate the hidden state z_k based on the former hidden states of $z_1, z_2, \ldots, z_{k-1}$ that have occurred. A more powerful deep learning method to overcome this is to use a Recurrent Neural Network (RNN) to make highly accurate approximations using a hidden state z_k which is based on the input states of $x_1, x_2, \ldots, x_{k-1}$ that have occurred [1, 2].

Figure 2.17 presents an ARM system structure, which learns the conditional probability of each variable given past information.

ARM systems (as well as many generative AI and deep learning systems) use attention layers and transformers, which are described in the following.

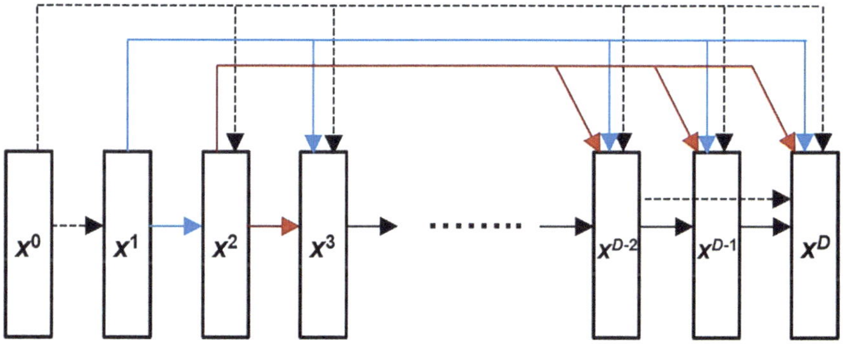

Fig. 2.17 ARM system structure

2.2.2.3 Attention

When attention is applied, the DNN attention weight vector A is computed based on the inputs, considering three types of essential parameters, which are the Keys, Values, and Query [2].

DNNs commonly process the input data vector using a linear combination of the input vector X to derive the hidden feature vector X^* applied in the deep learning system. Then the hidden vector X^* is multiplied with the weight vector W, which goes through a nonlinear process N to result in the hidden activations vector Z, which results in $Z = N(X^*W)$. The weight vector W can be obtained by training a DNN on a dataset to produce the desired Z.

Key vector K can be obtained from $K = X^*W$, in which $K = [k_1, k_2, \ldots, k_n]^T$. **Value** vector V can be obtained from $V = Y'W$, in which $V = [v_1, v_2, \ldots, v_n]^T$ and the stored examples of the output vector Y is replaced with the learned embedding output vector Y', which is multiplied to the weight vector W. Query value can be obtained from $q = Wx$, where the embedded input x triggers the query q.

A soft attention layer is used as the kernel function $Attn(q, (k_1, v_1), \ldots, (k_n, v_n))$ to generate the attention weight vector elements $a_1, a_2, \ldots, a_n$. Each element of the attention weight vector $A = [a_1, a_2, \ldots, a_n]^T$ has to be a real-value within the range of $0 \sim 1$ (i.e., $0 \leq a_i \leq 1$ for $i = 1,2, \ldots, n$) and the sum of all the attention weight vector elements must add up to 1 (i.e., $\{a_1 + a_2 + \cdots + a_n\} = 1$).

In summary, an attention system will use the input (x) to trigger a query (q) that will help assign a balanced amount of attention weights (a_i for $i = 1, \ldots, n$) to the key parameters such that more relevant content (which has already been embedded and learned by the system) will be proportionally combined when generating new content for the generative AI system.

2.2.2.4 Transformer

Transformers are autoregressive models that are used to create very powerful generative AI systems like BERT, GPT, GPT-2, GPT-3, ChatGPT, DALL-E, PARTI, etc. Among these, BERT, GPT, GPT-2, GPT-3, ChatGPT are generative AI based Large Language Models (LLMs), which have shown remarkable performances. However, still improvements need to be made, so problems like incorrect and biased answers can be avoided, and the amount of memory usage, computation resources, and required training needs to be significantly reduced [2].

Transformers are formed from multiple attention layers, where the output of one attention layer is used as an input to another attention layer.

Transformer types include Sequence Encoder, Sequence Decoder, Sequence Encoder-Decoder, as well as many variations. For a given context prompt c and input x, a transformer can be designed to provide joint probability $p(c, x)$ based sequence generation, conditional probability $p(x|c)$ based sequence generation, bidirectional encoder operations, and much more.

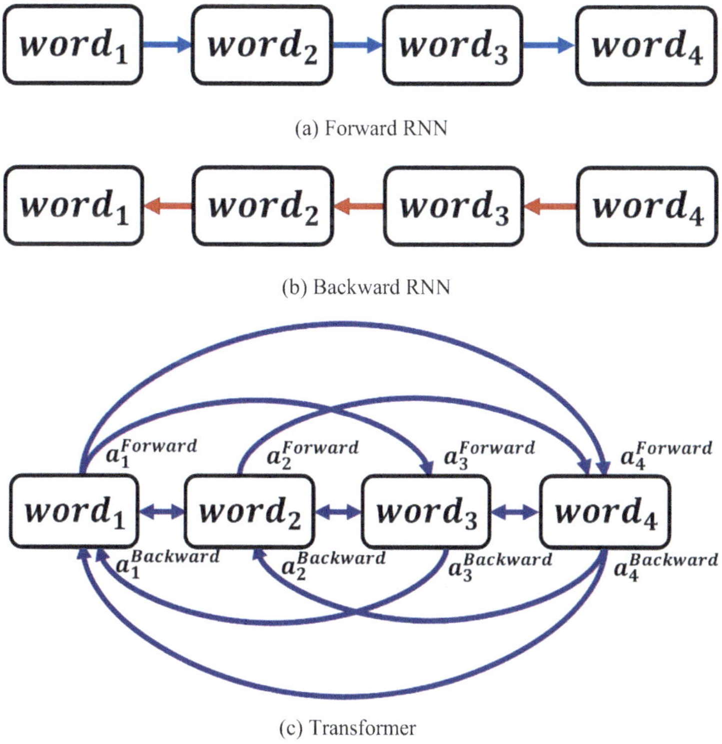

Fig. 2.18 Comparison of a transformer to a bidirectional RNN

The operation of a transformer can be more easily understood when compared to RNN forward and backward processing of a sequence of 4 words in a sentence (e.g., sentence "$word_1$ $word_2$ $word_3$ $word_4$.") while tagging syntax and semantic details to the vocabulary, as presented in Fig. 2.18 [1, 2].

In Fig. 2.18(c) Transformer, $a_i^{Forward}$ (for $i = 1, 2, 3, 4$) are the forward attention layer values, and $a_i^{Backward}$ (for $i = 1, 2, 3, 4$) are the backward attention layer values.

To analyze the syntax and semantic of a sentence, a bidirectional RNN (which is a combination of Fig. 2.18 parts (a) and (b)) will derive the relation and meaning of the vocabulary of the sentence by first applying an embedding process and then mapping the embedded data to the hidden states in a DNN.

On the other hand, a transformer can be designed to conduct embedding while using attention processes as a direct function of all the vocabulary in the sentence. This type of a system is called an **Encoder-Only Transformer**.

An alternative system to this would be a **Decoder-Only Transformer**. A decoder-only transformer (a.k.a. decoder transformer) will use the information of other previously obtained outputs to derive the current output. In case the decoder-only transformer will process data from a pre-stored dataset, it is important to make sure that the attention process is conducted only using earlier outputs in reference to the current

1. Repeat the following procedures for each step t.
   A. Compute the Attention weights $a_{1:t}$ from the causal masked self attention values of the inputs $y_{1:t}$.
   B. Compute the activation vector $z_t = \sum_{i=1}^{t} a_i y_i$.
   C. Repeat for all hidden layers of the DNN.
      i. Compute the DNN hidden layer weight values $h_t = \Xi(z_t)$, in which Ξ represents the DNN process.
   D. Predict the Decoder sequence's next categorical output (Cat) from $y_{t+1} \sim \Psi(S)$ using the following computation values.
      i. Ψ represents the Cat process
      ii. $S = Softmax(R)$
      iii. $R = Wh_t$ in which $R = [r_1\ r_2\ ...\ r_n]^T$
      iv. W represents the DNN's output layer weight values
      v. h_t values represent the hidden layer weights
2. Proceed to the next step $t + 1$ and repeat all of the above procedures.

Fig. 2.19 Transformer sequence decoder operation

output being processed (i.e., causality has to be maintained). In other words, when deriving a target output value, any future output value in reference to this target output should not be used by the transformer. This is an issue because decoder-only transformer's use pre-stored datasets.

A **sequence decoder** is one of the basic building blocks of a Transformer, and is also known as a **Generator**. Figure 2.19 presents an example of a transformer sequence decoder operation.

The Softmax function is used in Fig. 2.19 to obtain $S = Softmax(R)$, in which for input $R = [r_1 r_2 ... r_n]^T$ the elements of softmax vector $S = [s_1 s_2 \ldots s_n]^T$ are obtained from $s_i = \frac{e^{r_i}}{\sum_{k=1}^{n} e^{r_k}}$. Softmax element values are real-values within the range $0 \sim 1$ (i.e., $0 \leq s_i \leq 1$ for $i = 1, 2, \ldots, n$) and the sum of all the elements add up to 1 (i.e., $\{s_1 + s_2 + \cdots + s_n\} = 1$).

As presented in the Decoder procedures above, the transformer's output y_{t+1} is obtained using all former outputs $y_{1:t}$. Because transformers are formed from multiple attention layers, the output of one attention layer is used as an input to another attention layer. Although the Decoder description presented above is very simple, OpenAI's GPT uses only Decoders for its text generation, which demonstrates the power of a transformer decoder system.

In addition, sequence encoders are used in bidirectional encoders to implement systems like BERT. In addition, sequence encoder-decoder systems are used for conditional probability $p(y|c)$ based sequence generation.

Transformers based models are very powerful, as demonstrated in the BERT and GPT examples below.

2.2.2.5 BERT

Bidirectional Encoder Representations from Transformers (BERT) is a generative AI **Large Language Model (LLM)** based on an encoder-only transformer architecture (and does not use the decoder-only transformer ARM structure). BERT is considered as one of the most popular Natural Language Processing (NLP) reference models, based on its outstanding performance. Google introduced BERT in October of 2018 [2, 10].

BERT was originally implemented in two English language models based on size, which are BERTBASE and BERTLARGE, where these models were pre-trained using 800 million words in the Toronto Book Corpus and 2.5 billion words in the English Wikipedia. BERTBASE consists of 12 transformer encoders with 12 bidirectional self-attention heads processing a combination of 110 million parameters. BERTLARGE consists of 24 transformer encoders with 16 bidirectional self-attention heads processing a combination of 340 million parameters.

2.2.2.6 GPT

Generative Pre-trained Transformer (GPT) is a text generative AI transformer system developed by OpenAI. GPT uses a decoder-based transformer that uses causal masked attention values [2, 11].

GPT was released in 2018 by OpenAI, which was built using 4.5 GB (giga bytes of memory) of BookCorpus text to sustain the computation of 117 million parameters used in its transformer decoder system.

GPT-2 was released in 2019 by OpenAI, which is an enlarged version of GPT that applied maximum likelihood-based learning using 40 GB of WebText to train from the 8 million page corpus web data using 1.5 billion parameters.

GPT-3 was released in 2020 by OpenAI, which is an enlarged version of GPT-2 that also applies maximum likelihood-based learning using 175 billion parameters trained from a 300 billion word database. Text generation of GPT-3 was controllable through the conditioning prompt. GPT-3 can generate sophisticated text output based on the user requested prompt information received through its user interface, which is possible through the system's In-Context Learning (ICL) capability. The feature of In-Context Learning was observed for the first time by a computing system, which was both astonishing as well as alarming.

ChatGPT was released in 2022 by OpenAI, which is an enhanced version of GPT-3 with an interactive dialog-based conversation engine that applies **Reinforcement Learning from Human Feedback (RLHF)** technology. RLHF uses reinforcement learning to generate responses that are well aligned to the conversation objective based on an optimized reward ranking mechanism.

2.2.2.7 DALL-E

DALL-E is an image generative AI system that was released on 2021 by OpenAI. DALL-E was named based on the Spanish surrealist artist Salvador Dalí and the 2805 future robot WALL-E from the 2008 animated movie produced by Pixar for Walt Disney Pictures. DALL-E generates creative images using an autoregressive transformer and variational autoencoder system that consists of 12 billion parameters trained on 250 million text-based images collected from the Internet [2, 12].

2.2.3 Flow Model

Generative AI uses flow models in many ways, which include, content generation, inference, density modeling, etc. Flow modeling uses normalizing flows, which are probability density models that are easy to compute values, not too difficult to derive an inverse function, and can easily provide sample values [2, 13].

The flow modeling technology introduced in this section starts with a description of the inverse matrix determinant and Jacobian computations, as these are the fundamental functional tools that will be used. This is because flow models use multi-dimensional vectors as inputs, latent variables, internal parameters, and outputs, in which matrix or tensor operations are applied. Second, details on normalizing flows are introduced, which is a process of generating a latent variable z from the input x that will be used to compute the probability distribution of $p(z)$. Once $p(z)$ is obtained, random values of z will be used to generate the output x', which is the new generated content. This type of flow model that uses both a transformation and its inverse transformation together is called a **normalizing flow** model (Fig. 2.20). Third, from the basic affine flows and elementwise flows, coupling flow technology as well as affine autoregressive flow and reverse flow models are used to generate more complex generative AI flow models. This is because affine flows and elementwise flows have very limited expressive capabilities. More details are provided in the following sections.

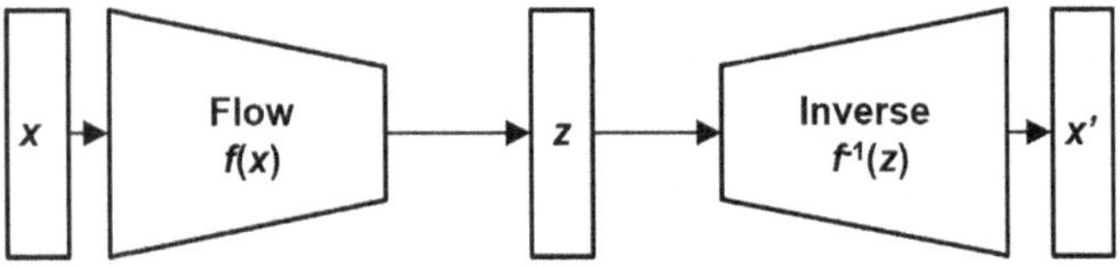

Fig. 2.20 Normalizing flow model

2.2.3.1 Inverse Matrix Determinant and Jacobian

In flow models, multi-dimensional vector values are used in the inputs, latent variables, internal parameters, and output values, so matrix or tensor operations are applied. In addition, flow models frequently use inverse functions in their normalizing flows or matrix operations. Inverse matrix operations require use of the matrix's determinant (det) and Jacobian matrix (J) computations.

The **determinant** of a matrix is a scalar value that is used to determine the mapping characteristic of the matrix and also used to check if the matrix can have an inverse matrix, which is possible when the determinant (det) of the matrix is not zero [2].

A matrix used in generative AI flow models needs to be invertible, so the matrix determinant has to be nonzero and the matrix needs to have an isomorphism structure. **Isomorphism** means that the matrix can map a dataset to another dataset preserving its base characteristics, which can also be inversed mapped back to the original dataset.

The **Jacobian matrix** (J) of a multi-variable vector (x) function (f) will include all possible first-order partial derivatives (values or functions) in matrix form based on the order of the variables. The Jacobian matrix (J) of function f evaluated at x will be commonly denoted in the form of $J(f(x))$ or $\left.\frac{df}{dx}\right|_x$.

2.2.3.2 Normalizing Flows

In generative AI, a '**flow**' is a process that can be transformed using a function f and can be reserve/inverse transformed using its subfunctions [2]. For example, in a flow, assume function f can be represented through its subfunctions $f_1, f_2, \ldots, f_N$ in the form of $f = f_1 \cdot f_2 \cdots f_{N-1} \cdot f_N$. In addition, each subfunction f_i has an inverse function $g_i = f_i^{-1}$ (for $i = 1, \ldots, N$). Then the inverse transformation function of f can be obtained from the inverse subfunctions, in which $f^{-1} = g = g_N \cdot g_{N-1} \cdots g_2 \cdot g_1$.

The input base probability distribution $p(x)$ is obtained from the input dataset values x. Gaussian or uniform base probability distributions have been frequently used for $p(z)$ in normalizing flow models. From the base probability distribution $p(z)$ a random variable z is selected. Nonlinear invertible transformation function f is used to transform x into z, which is represented as $z = f(x)$. Since the inverse of the transformation function f is represented as $f^{-1} = g$, we obtain $f^{-1}(z) = g(z) = x'$, in which x' is a newly generated content from the normalizing flow model. This inverse function has the effect of reversing (i.e., returning back to its original distribution) the transformation z back into the original base probability distribution $p(x)$.

Generation of z from $z = f(x)$ enables the computation of the probability distribution of $p(z)$ using (2.8), which involves matrix determinant (det) and Jacobian matrix (J) computations.

$$p(z) = p(x)|det(J(f(x)))|^{-1} \tag{2.8}$$

Once $p(z)$ is obtained, random values of z can be generated from sampling $p(z)$. When a flow model uses both a transformation (f) and inverse transformation ($f^{-1} = g$) together (for content creation in generative AI systems), it is called a normalizing flow, as presented in Fig. 2.20 [2, 13].

A transformation that is invertible is called a **Bijection**. A differentiable bijection that also has a differentiable inverse is called a **Diffeomorphism**.

In generative AI systems, normalizing flows are used in density estimation and variational inferences.

Density estimation is conducted over a dataset, which an estimate of the dataset's probability distribution can be obtained and transformed into a new probability distribution, in which samples can be extracted from and used for content generation. For this process, the likelihood function of (2.8) is maximized in the density estimation process.

Variational inference is the process of finding surrogate posterior probability functions that can be used instead of complex posterior probability functions, which are difficult to control. For this purpose, the latent variable model's approximate posterior probability function is derived. The evidence lower bound can be maximized through training of the variational parameters using a DNN.

2.2.3.3 Constructing Flows

Two basic types of flows are Affine flows and Elementwise flows, which are described in the following [2, 13].

Affine flow uses the transformation formula of $z = f(x) = Ax + b$, in which A is a square matrix that has an inverse A^{-1}. If $b = 0$ then this is called a linear flow.

Elementwise flow is created using a 1 dimensional (1D) scalar bijection h. A k dimensional vector bijection function f can be created from $f(x) = (h(x_1), \ldots, h(x_k))$, which applies the 1D scalar bijection h to each element of the k dimensional vector. The transformation function f has to be invertible to qualify as an elementwise flow.

Although affine flows and elementwise flows have very limited expressive functional capabilities, they can be used in coupling and autoregressive models as building blocks to create much more complex generative AI flow models.

Figure 2.21 presents a coupling flow example, where Φ represents the coupling layer function and Θ represents the conditioner that creates the dependency effect between the base flows (e.g., affine flows and/or elementwise flows). The coupling layer function Φ and conditioner Θ can be used to create many types of coupling flows. The coupling layer function Φ has to be invertible. In addition, DNNs are commonly used to form the conditioner Θ [2, 13].

An autoregressive flow process can be formed from the base flows (e.g., affine flows and/or elementwise flows), where an example is presented in the following.

Figure 2.22 shows an example of an affine autoregressive flow process and inverse flow process. From the base probability distribution $p(x)$ the sampled input values are $x_1, x_2, \ldots, x_k$. From the output probability distribution $p(z)$ the sampled output

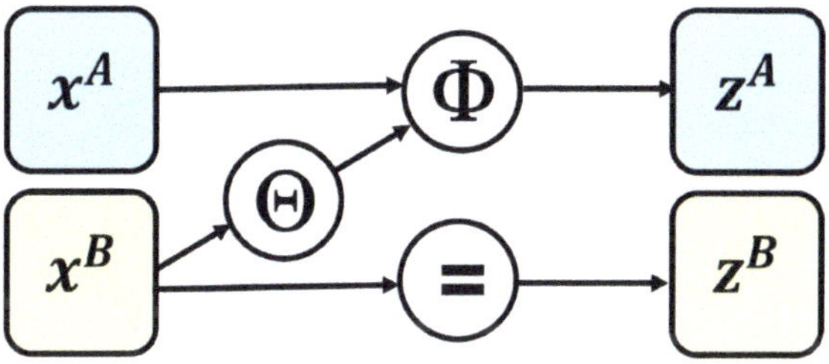

Fig. 2.21 Coupling flow example with the coupling layer function Φ and conditioner Θ

values are $z_1, z_2, \ldots, z_k$. The affine scalar bijection is h, where α represents the log scale and β is the bias. In addition, α_i (i.e., the ith component of α) is obtained from $z_{1:i-1}$ (in which $z_{1:i-1}$ is same as $z_1, z_2, \ldots, z_{i-1}$) based on the function $\alpha_i(z_{1:i-1})$. Using α_i and β_i, z_i can be obtained from (2.9) and x_i can be obtained from (2.10) as presented in Fig. 2.22 [2, 14].

$$z_i = x_i e^{\alpha_i(z_{1:i-1})} + \beta_i(z_{1:i-1}) \tag{2.9}$$

$$x_i = (z_i - \beta_i(z_{1:i-1}))e^{-\alpha_i(z_{1:i-1})} \tag{2.10}$$

The autoregressive bijection has to use causal output values only. So, a masked autoregressive flow needs to be applied, such that the conditioner Θ parameter θ_i (for $i = 1, \ldots, k$) is obtained from the causal outputs $z_1, z_2, \ldots, z_{i-1}$ only. In other words, when computing θ_i, the noncausal outputs $z_i, z_{i+1}, \ldots, z_k$ need to be masked out.

2.2.4 *Diffusion Model*

Diffusion models have played an important role in generative AI, especially in creative image generation. OpenAI's DALL-E 2, Google's Imagen, and Stable Diffusion from Stability AI all use diffusion technology [2].

Figure 2.23 illustrates a diffusion model with $2T$ stages. In each stage of the diffusion process, the diffusion model will gradually add noise until the image reaches a certain noise property, then the reverse process of gradually removing the noise is conducted. This process enables an image to turn into a new image with several creative aspects through the noise adding and removal process. Directly creating a creative structured image from noise is very difficult. But because the diffusion model's generation process starts from a source image(s), the newly created image will have some of the characteristics of the original image(s).

Diffusion models are built on similar generative AI techniques used in flow models and variational autoencoders as well as energy models. Diffusion models need to be trained with extensively large datasets, but due to their simple training process,

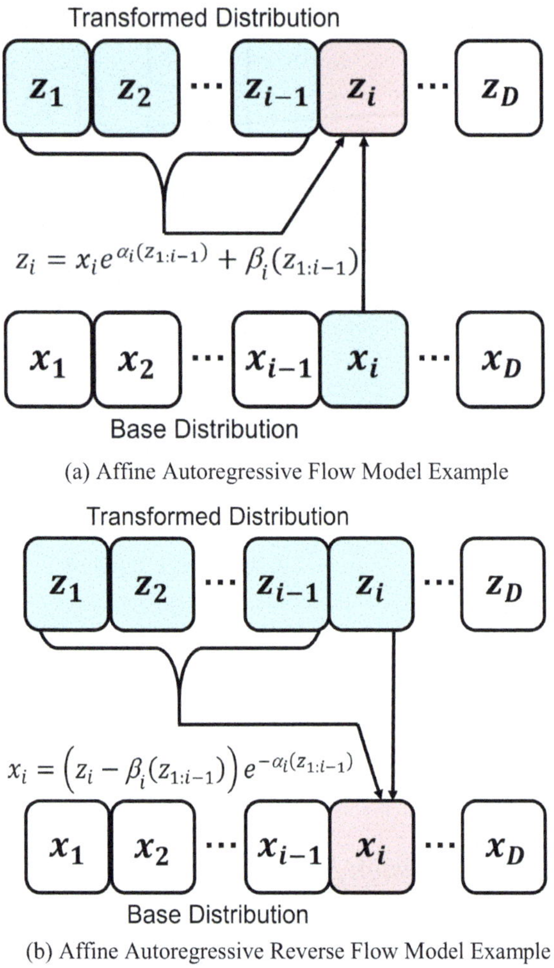

Fig. 2.22 Affine autoregressive **a** flow process and **b** reverse/inverse flow process

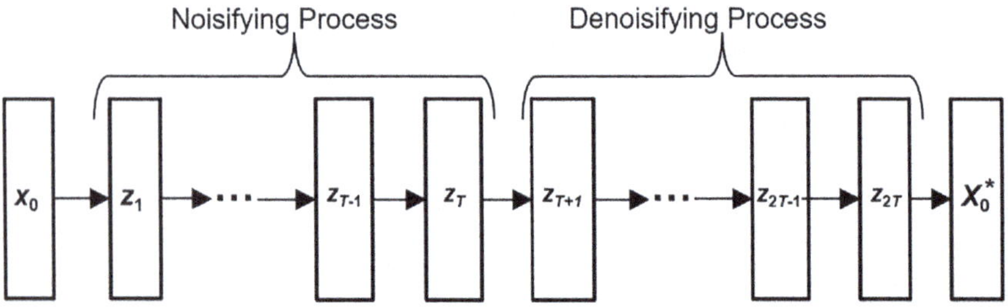

Fig. 2.23 Diffusion model with 2T stages, where stages z_1 to z_T are for noisifying, and stages z_{T+1} to z_{2T} are for denoisifying

they are scalable and can effectively utilize the information learned for new content creation in efficient ways.

A key model used in diffusion generative AI systems is the Denoising Diffusion Probabilistic Model (DDPM), which is described in the following section.

2.2.4.1 Denoising Diffusion Probabilistic Model (DDPM)

Among the $2T$ stages of a DDPM diffusion model, the first half of stages (i.e., stages 1 to T) are the noisifying process and the remaining half of the stages (i.e., stages $T + 1$ to $2T$) are the denoisifying process [2].

In each stage of the noisifying process, the diffusion model will first gradually add noise until the image reaches a certain noise model. A popular noise model to reach in the noisifying process is a **Gaussian normal probability distribution** with zero mean and a standard deviation of one (i.e., $N(0,1)$, where N represents the normal distribution). These stages are called the **diffusion process**, where other terms such as, forward process, noise encoding process, noisifying process have also been used. Based on the source image x_0, the diffusion process uses an encoder network Q, which is a linear Gaussian model or a DNN, in which the noisifying encoding process is gradually applied (where this encoder network Q is different from the reinforcement learning Q in Q-learning).

At stage 1 the source image x_0 is transformed into z_1 using $Q(z_1|x_0)$. At stage 2 the image z_1 is transformed into z_2 using $Q(z_2|z_1)$. Following this sequence, at stage T the image z_{T-1} is transformed into z_T using $Q(z_T|z_{T-1})$. The diffusion process would be scheduled to reach the target noisifyed level (e.g., $N(0,1)$) at the middle stage z_T of the diffusion model.

The remaining second part of a diffusion model will be the **Reverse Process** that will denoisify the noisifyed z_T image through multiple stages until x_0^* is reached, in which x_0^* is a new image generated from x_0. The diffusion process uses a decoder network P, which can be repeatedly used at each step of the reverse process.

The reverse process starts with stage $T + 1$, where a part of the noise in the image z_T is removed using the $P(z_{T+1}|z_T)$ process of the P network, turning it into z_{T+1}. This reverse process is repeated until the $2T$ stage in which z_{2T} is obtained from z_{2T-1} using the $P(z_{2T}|z_{2T-1})$ process. The output of the final z_{2T} stage will be the x_0^* new image generated from the source image(s) x_0.

A DDPM is similar to a hierarchical VAE with the differences of having a hierarchical diffusion stage structure where the latent states dimension used in the diffusion model are not reduced at the internal stages, but remain the same (which is why trapezoids are not used in Fig. 2.23), keeping all dimension information through the noisifying and denoisifying diffusion process. The DDPM is also similar to the flow model, however, diffusion models do not need any inverse transformations like flow models do. In addition, a DDPM will use stochastic processes in the weight updates of its DNN hidden layers.

2.2.4.2 Diffusion Model Content Generation Control

Conditional diffusion models are used to control the content generation directions [2]. A conditional diffusion model first needs to be trained on an extensive labeled dataset, which has data elements in the form of (L, X), in which L is the label of the data/image X. After sufficient training, the diffusion model will attempt to produce a new data/image X^* that will maximize the conditional probability of $p(X^*|R)$, in which R is the Request made by the user.

During the conditional diffusion model training process, the labeled dataset's data elements are mapped to an embedded vector in the latent space of the diffusion model, in which a DNN is commonly used. Latent spaces commonly have lower dimensions than the original input dataset's dimension. However, the latent space has to be able to sufficiently preserve the essential characteristics of the original input dataset. The embedding process converts source images, videos, signals, data into computer processible forms. This is why latent space mapping commonly requires the embedding process.

2.2.4.3 Diffusion Model Process Acceleration

Because diffusion models have many internal steps, the processing speed can be slow and require significant computation resources. Therefore, some diffusion models need process acceleration techniques to be applied for them to be more useful and efficient. Diffusion model processing requires a series of solving ordinary differential equations (ODEs), which is a complex memory and time-consuming process [2].

To enhance the processing speed of diffusion models, a deterministic generation technology that uses a non-Markovian forwarding diffusion model can be applied. This model uses the conditional information of the former step as well as the input together at each internal step. In addition, using given prior samples, a fully deterministic reverse processing model can be obtained. This type of a generative AI diffusion model is called a **Denoising Diffusion Implicit Model (DDIM)**.

Using a DNN, the diffusion model can be further compressed and the process can be accelerated using the **Distillation method**, which is described in the following. First, a Teacher model is created using a trained DDPM that is sampled using a DDIM. Second, the Teacher model is used to generate the intermediate latent states. Third, for every two steps of the Teachers output, a Student model learns the output characteristics and is trained to make accurate output predictions. Fourth, after the Student models are sufficiently trained, the Student models can be used to skip the intermediate Teacher models and reduce the diffusion steps by half. Fifth, this process can be repeated by treating the Student models as the Teacher models, and then generate new Student models to reduce the diffusion steps by another half. The Distillation method is known to be very effective in accelerating generative AI diffusion models.

Another generative AI diffusion model acceleration method is to use a **Latent Space Embedding** approach. In this approach, first the source images are embedded into a latent space. Secondly, the embeddings in the latent space are learned by the diffusion model. In diffusion-based latent space embedding methods, the embedded latent space may have the same or a lesser dimension compared to the dimension of the source images.

A representative latent space embedding diffusion model is the **Latent Diffusion Model (LDM)**, which uses a perceptual loss augmented VAE to embed the source images into the latent space, and then the embeddings of the latent space are fitted into the diffusion model through a learning process. LDM technology is used in the famous Stable Diffusion generative AI image creation system that was invented by Stability AI.

2.2.5 Generative Adversarial Network (GAN)

GAN systems use a Generator and Discriminator to generate new realistic content that is (almost) undistinguishable from the actual real content dataset [2, 15]. The Generator will learn to create new content that closely resembles the real existing content dataset. The Discriminator will learn to distinguish the Generator's created content from the actual real content dataset. The Discriminator will inform the Generator if the content was distinguishable from the original real content dataset. The Generator will improve its content generation skills to make its content less distinguishable from the original real content dataset. As the iterations of the GAN process continue, the Generator and Discriminator will learn more and more to become highly sophisticated. At some point, the Discriminator will not be able to distinguish the real existing content from the content newly generated from the Generator, which will complete the process. The GAN process is presented in Fig. 2.24.

In a more advanced GAN model, the Discriminator could inform the Generator how its content was distinguishable from the original real content, instead of just making a single number decision (in the range of 0–1) of the generated content being distinguishable or not.

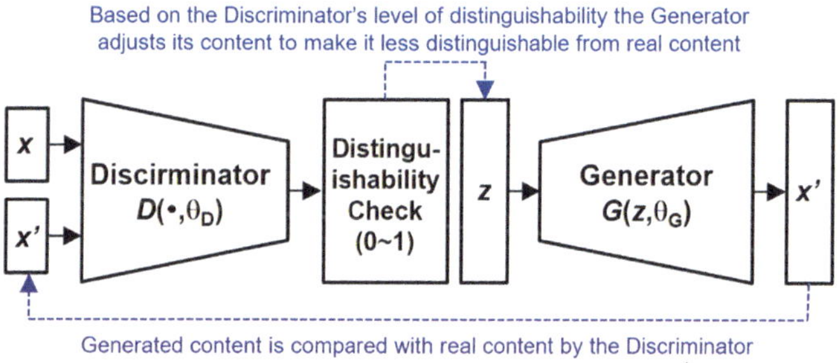

Fig. 2.24 GAN system model

GAN systems commonly use two separate DNNs, one for the Generator and one for the Discriminator [2, 15].

Based on data x from the input dataset, the Generators probability distribution p_G is learned using a (latent) input noise random variable z that has a prior probability distribution of $p_z(z)$. The Generator's mapping to the data space function is defined as $G(z, \theta_G)$, where this function is implmented using a DNN that has the parameter set θ_G.

The Discriminator's DNN implements the function $D(\cdot, \theta_D)$ that compares x and x', and outputs a decision value (that is used in the V computation) in the range of 0–1. The decision value will be closer to 0 if the tested content seems like it was created by the Generator (i.e., generated from the Generator's p_G distribution). On the other hand, the decision value will be closer to 1 if the content seems like it came from the original reference dataset (i.e., it was not created by the Generator). So the value function V will be influenced by both the Discriminator and Generator. The Discriminator's DNN parameter set θ_D is trained to maximize the probability of the function $D(\cdot, \theta_D)$ to correctly produce a decision value. So, the Discriminator's DNN parameter set θ_D has to be trained using the input from the input dataset (i.e., x) and the dataset of content generated from the Generator (i.e., $x' = G(z, \theta_G)$).

The GAN system conducts a **min–max game** using the value function V, which has the form of (2.11).

$$\min_G \max_D V(D, G) = E_{x\sim p(x)}\left[\log D(x, \theta_D)\right] + E_{z\sim p(z)}\left[\log\left(1 - D\left(x', \theta_D\right)\right)\right] \tag{2.11}$$

The three terms in (2.11) have the following meaning. The first term "$\min_G \max_D V(D, G)$" in (2.11) means that the value function (V) (which is based on the Discriminator (D) and Generator (G) outputs) will have two objectives, where the Discriminator (D) will try to maximize ($\max_D$) the value function (V), and the Generator (G) will try to minimize ($\min_G$) the value function (V).

The second term "$E_{x\sim p(x)}\left[logD(x, \theta_D)\right]$" in (2.11) means that the output of the Discriminator $D(x, \theta_D)$ (which is based on input x and the DNN parametersθ_D) will be converted into log scale (log), and an expectation (E) will be computed based on the random variable x, which has a probability distribution function of $p(x)$ (i.e., $E_{x\sim p(x)}$). This second term evaluates how accurately the Discriminator can discriminate (i.e., classify) the actual data.

The third term "$E_{z\sim p(z)}\left[\log\left(1 - D\left(x', \theta_D\right)\right)\right]$" in (2.11) means that the one minus the output of the Discriminator $D\left(x', \theta_D\right)$ (which is based on the Generator's output x' and the DNN parameters θ_D) will be converted into log scale (log), and an expectation (E) will be computed based on the latent random variable z, which has a probability distribution function of $p(z)$ (i.e., $E_{z\sim p(z)}$). This third term evaluates how accurately the Discriminator can discriminate (i.e., identify) the data generated from the Generator.

For the evaluation of a GAN's performance, the **inception score** and/or the **Fréchet inception distance** can be used, where other measures are also useable. The inception score is based on the amount of variation the Generator's output shows. Fréchet inception distance is based on the level of similarity between the reference dataset and the output of the Generator.

Using GAN technologies, generative AI systems can create new content and make adjustments to make it seem as original realistic content. Based on the type of data the Generator was used to train, the Generator's direction of content creativity will be influenced. In addition, based on how the Discriminator is setup and how it learns, the level of creativity and limitations placed on the Generator can diversly change.

2.2.5.1 GAN Types

There are various GAN types, where some that are related to the generative AI technologies that were described above are introduced in the following [2, 15].

When GAN is used as an image generative AI system, a Deconvolutional Neural Network (DCNN) can be used for the Generator's DNN model and a Convolutional Neural Network (CNN) can be used for the Discriminator's DNN model. This type of a GAN is called a **Deep Convolutional GAN (DCGAN)**, as the Generator and Discriminator both use deep convolutional technology (e.g., DCNN and CNN).

A **Flow-GAN (FGAN)** system will use a flow model for the Generator to make the likelihood function computations faster and more efficient.

Self-attention GAN (SAGAN) adds residually-connected standard self-attention modules to the DCGAN model's Generator and Discriminator to enhance the performance.

Variational AutoEncoder GAN (VAEGAN) model uses a VAE system for the GAN Generator.

Transformer GAN (TransGAN) systems are different to DCGAN systems, as a TransGAN system will not use a DCNN or CNN at all. Instead, a TransGAN system will use two transformers, one for the Generator and one for the Discriminator.

Many other GAN models do exist, so continuing research is recommended.

2.3 Federated Learning

Federated Learning is a deep learning technology that enables multiple AI systems to learn together without having any training data exchanged. This is why federated learning is also called **Collaborative Learning**. Due to this special characteristic, federated learning can be used in many systems that require data security and privacy protection such that data exchange is prohibited or limited [16].

A similar technology that needs to be compared to federated learning is **Distributed Learning**. Distributed learning systems were originally focused on distribution of computing resources, where a single AI system is trained on multiple independent computers with local datasets that were approximately the same size and have independent and identically distributed (*iid*) statistical characteristics. Although the supporting technologies of distributed learning have extended over time, a few of the main differences with federated learning are that federated learning systems are focused on using heterogeneous datasets that are not approximately the size and do not have *iid* statistical requirements.

How to collect and combine the AI system configurations is based on the federated learning type, which include Centralized federated learning, Decentralized federated learning, and Heterogeneous federated learning. More details on these federated learning types will be described in the following section.

2.3.1 Federated Learning Types

In this section, a description of centralized federated learning, decentralized federated learning, and heterogeneous federated learning is provided.

Centralized federated learning uses a central server to control all participating AI systems on the federated learning network. The central server will decide the operational parameters and the program(s) that the participating AI systems need to have installed and have the AI systems authorized and authenticated before the operation begins. Details of the federated learning operational parameters are described in the following section. The central server will collect the trained weights of the distributed AI systems (based on the processed local objective functions and databases) and combine the AI systems configurations into one weight set and objective function. Then the central server will distribute the combined weight set and objective function to the AI systems of the federated learning network. The central server will repeat the federated learning process and will terminate it when needed.

Decentralized federated learning is used when there is no defined central server to coordinate the federated learning process, and the participating AI systems form a pattern that enables collection of the trained weights based on the local objective functions and databases. In addition, the objective functions and databases of the distributed AI systems need to be combined into one weight set and objective function in an agreeable manner. Then the combined weight set and objective function needs to be accessible to the AI systems of the federated learning network. One method to conduct a secure and reliable decentralized federated learning network is to use a blockchain. The blockchain's consensus algorithm will be used to determine the most acceptable combination of the weight set and objective function. Details of blockchain consensus algorithms and reliability (i.e., fault tolerance) are provided in Chap. 8 of this book.

Heterogeneous federated learning considers the difference of the AI systems participating in the federated learning process. Centralized and decentralized federated learning technologies commonly accept systems that have the same or similar specs for their AI systems in order to participate in their federated learning process.

Heterogeneous federated learning enables the AI systems participating in the federated learning process to have much more diversity in both hardware and software. The hardware aspects include processor types (CPU and GPUs), processing capability, memory, and the network type, where the software aspects include the computing kernel, operating system, and application software. New types of heterogeneous federated learning technologies are being invented, which will promote more intelligent networks to use heterogeneous federated learning in the future.

2.3.2 Federated Learning Operation

The operation of federated learning is described below and formulated in Fig. 2.25. Federated learning is conducted over a network consisting of multiple AI systems that have a preinstalled program. Each AI system that will participate in federated learning will have the required application program or Software as a Service (SaaS) installed.

1. Select the type of federated learning network (e.g., Centralized, Decentralized, Heterogeneous, etc.)
2. Each AI system serving the federated learning process will have the required application program or SaaS installed (authentication and authorization may be needed).
3. Among the N AI systems in the federated learning network, a total of K AI systems will be chosen.
4. Federated learning operational parameters are selected (e.g., round (R_j), local dataset training iterations (Φ), learning rate (Δ), etc.) and sent to (or retrieved by) the K distributed AI systems over the network.
5. For each round conduct the following procedures (where R_j represents the jth round for $j = 1,2,\ldots$).
   A. Each AI system will train its weights and local objective function (i.e., $O_i(w_i)$ for $i = 1,2,\ldots,K$).
   B. The federated learning network will collect the trained weights (i.e., $w_1, w_2, \ldots, w_K$) of the local objective functions (i.e., $O_1, O_2, \ldots, O_K$) and combine the AI systems configurations into one weight set using the federated learning objective function for the jth round (e.g., $F_j(w_1, w_2, \ldots, w_K) = \frac{1}{K}\sum_{i=1}^{K} O_i(w_i)$, where other models do exist).
   C. The combined weight set and objective function (i.e., $F_j(w_1, w_2, \ldots, w_K)$) will be distributed to the AI systems in the federated learning network.
6. Repeat for the $(j + 1)$th round (until the limit in rounds is reached, if defined).

Fig. 2.25 Federated learning operational procedures

If centralized federated learning is used, then the central server may check if the required application program or SaaS has been installed by the federated learning participating AI systems, and then conduct authentication and authorization [16].

For a federated learning network consisting of N AI systems, a total of K AI systems will be chosen in the jth round R_j, which will conduct training on their local dataset for a duration of Φ iterations using a learning rate of Δ. The total number of rounds that will be conducted needs to be decided based on the application type and the characteristics of the local datasets and the federated learning network. If the local datasets are continuously being updated, then the total number of rounds may not be set at the beginning of the federated learning operation.

For the K chosen AI systems, the ith AI system will have the trained weights w_i and local objective function O_i. The federated learning network will need to collect the trained weights (i.e., $w_1, w_2, \ldots, w_K$) of the local objective functions (i.e., $O_1, O_2, \ldots, O_K$) and combine the AI systems configurations into one weight set using the objective function (e.g., $F_j(w_1, w_2, \ldots, w_K) = \frac{1}{K}\sum_{i=1}^{K} O_i(w_i)$ for the jth round, where other combining algorithms exist) and distribute the combined weight set to the AI systems of the federated learning network.

2.4 Digital Twin Technology

Actual systems can be modeled into a virtual digital form that has all the same essential properties such that testing, analysis, and modeling can be accurately conducted. Such systems are called Digital Twins, which are an essential part of modern designing as accurate performance estimations and predictions can be made. Using advanced digital twin technology, testing, controlling, monitoring, integration, decomposition, maintenance, and modifying can be possible. Based on the level of design details and the specifications of a digital twin, all or some of these features can be possible [17, 18].

Digital twin technology evolved from computer simulation and digital modeling, where the first use of the terminology "Digital Twin" was used in the "2010 Roadmap Report" from the National Aeronautics and Space Administration (NASA). Related terminology had appeared in earlier documents using slightly different names like "virtual twin" et al.

2.4.1 Digital Twin Types

One way to classify the types of digital twins is Digital Twin Prototype (DTP), Digital Twin Instance (DTI), and Digital Twin Aggregate (DTA).

Digital Twin Prototype (DTP) is used to help design and test a prototype of a physical system or service before it is released for official/commercial use by customers/users. DTPs need to include design features and behavioral characteristics that are sufficiently accurate such that the testing and analysis results of the DTP are reliable. DTPs enable corrections to be made before the product or service is released. A great benefit of a DTP is to be able to see how the physical system or service would perform and appear to its customers/users before it is officially made public for sale or service. Evidently, the value of a good DTP is significant, which is why almost all high-tech and high-value companies use DTPs to prepare their products and services.

Digital Twin Instance (DTI) is used to check a product or service status once the product or services has been manufactured, released to the public, and/or in operation. A DTI will be used by the manufacturing company to check problems in a product that has been reported by its users, or to predict a failure due to upcoming events in which the product or service may be used. A manufacturer may not be able to predict all environments in which its products or services could be used in. So, as new situations arise, the manufacture or retailer would need a DTI to determine the level of performance its products or service would be able to provide in the new environment. In addition, even if the environment has not changed, as the product or service ages, component failures or module deterioration can result in different characteristics and performance levels, in which a DTI would be able to significantly help to inform what parts need to be replaced or upgraded in advance to help prevent a failure. As such failures of a system or service can result in critical damage to property as well as result in human casualties, the value of an accurate DTI is significant. This is why DTIs are used by automobile and airplane companies, as they help to predict failures, which may lead to recalls for parts replacement so accidents can be prevented.

Digital Twin Aggregate (DTA) is needed when it is not be possible to create a DTI for a large system or a network of systems. A DTA is an aggregation of multiple DTIs in which the interconnected behavioral characteristics are integrated into the DTA such that the comprehensive virtual functionalities are accurately expressed. Evidently, a DTA takes much more effort to design and build than a DTI, due to the complex interconnected effects between the DTIs. A complex system prototype would also need a DTA to be created from the combination of its DTP and DTI modules, in which a network that characterizes the integrated interconnection of the DTPs and DTIs must be included. Although a DTA is more difficult and expensive to design and build, a properly designed DTA would be able to help fix malfunctions, prevent large scale accidents, and predict methods to maximize the performance. This would result in prevention of large-scale system failures and security breaches, as well as help increase the performance of the networked systems in which savings and profits could be significant. In many cases, investments into accurate DTAs are worthwhile as causalities and/or major property loss can be prevented.

2.4.2 Digital Twin Technology

A digital twin will need to include a virtual model of all essential parts of the actual system such that an accurate behavior model can be created. Since actual physical systems commonly include modules and parts made from various companies, digital twins commonly need its parts to be also designed by multiple programmers and designers, which may be working for different companies/organizations. To make the virtual modules and parts that were designed from other companies/organizations work effectively together, digital twin modeling requires standardized guidelines [18].

Digital twin technologies need to be developed to fit these guidelines, so they can be integrated together in an efficient and reliable way. For this reason, the International Organization for Standardization (ISO) has created the ISO 23247 standard digital twin framework for manufacturing, which focuses on automation systems and integration, which is presented in Fig. 2.26. The ISO 23247 standard series are not the only standards for digital twin modeling. For example, the ISO 15926 series can be used for digital twin modeling for the oil and gas industry, and the ISO 16739 series can be used for digital twin modeling for building and constructions, et al.

There are four parts to the ISO 23247 standards, which are listed below and presented in Fig. 2.26.

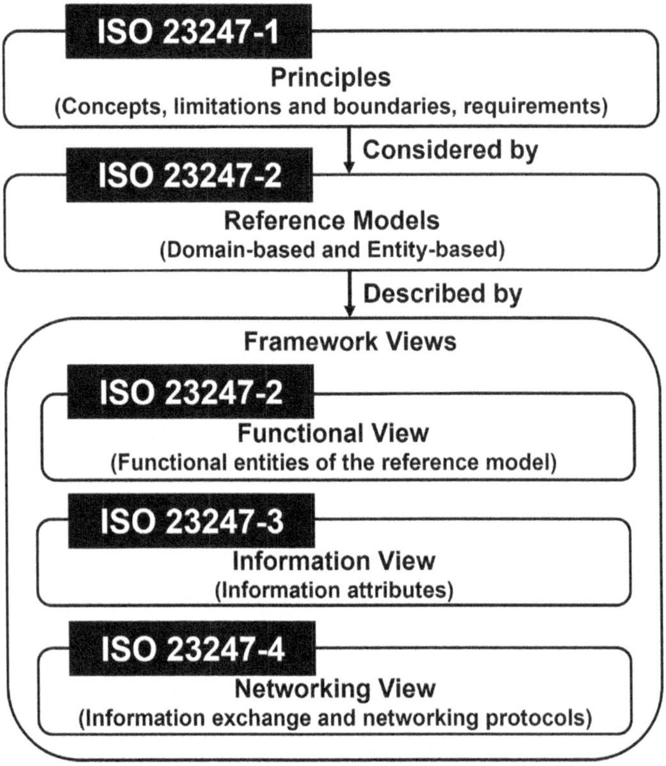

Fig. 2.26 Digital twin framework for manufacturing ISO 23247 series structure

The first part ISO 23247-1 is titled "General principles and requirements for developing digital twins in manufacturing," which covers the concept and requirements as well as the limitations and boundaries.

The second part ISO 23247-2 is titled "Reference architecture with functional views," which include domain and entity references to be used for digital twin models.

In addition, the second (ISO 23247-2), third (ISO 23247-3), and fourth part (ISO 23247-4) of the ISO 23247 series provide digital twin framework views for automation systems and integration manufacturing. The functional view of ISO 23247-2 includes details of how the functional entities of the reference digital twin can be modeled. The third part ISO 23247-3 is titled "List of basic information attributes for the observable manufacturing elements," which provide details on how the information attributes of the digital twin's information view can be modeled. The fourth part ISO 23247-4 is titled "Technical requirements for information exchange between entities within the reference architecture," which describes the information exchange and networking protocols that need to be designed to support the digital twin's networking view for automation systems and integration manufacturing.

Figure 2.27 illustrates the ISO 23247 digital twin framework for manufacturing. The **Digital Twin Entity** is the center block, which is the center of the modeling process. The digital twin entity includes three main units, which are the **Operation and Management Sub-Entity**, **Application and Service Sub-Entity**, and the **Resource Access and Interchange Sub-Entity**. The three sub-entities and their functional elements (FEs) must be interconnected so information exchange and functional collaboration is possible at all sub-layer and sub-domains in the digital twin model [18–20].

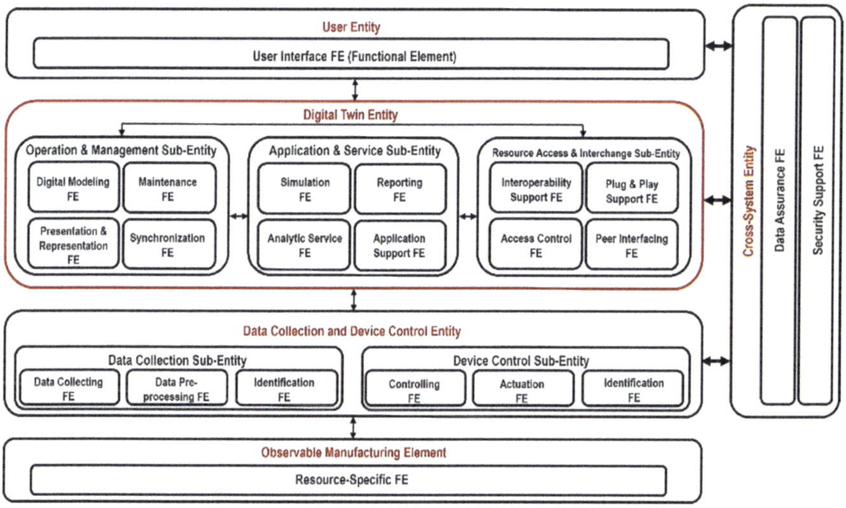

Fig. 2.27 Structure of the ISO 23247 digital twin framework for manufacturing

The sub-entities of the Digital Twin Entity are described in the following. The Operation and Management Sub-Entity includes functional elements for digital modeling, maintenance, presentation and representation, and synchronization. The Application and Service Sub-Entity includes the functional elements for simulation, reporting, analytic services, and application support. The Resource Access and Interchange Sub-Entity includes the functional elements for interoperability support, plug-and-play support, access control, and peer interfacing. These sub-entities and functional elements are the core of the digital twin model.

On one side, the Digital Twin Entity interfaces with the **User Entity** through the user interface functional element. On the other side, the Digital Twin Entity interfaces with the **Data Collection and Device Control Entity**, which is consists of two sub-entities, which are the Data Collection Sub-Entity and the Device Control Sub-Entity. The Data Collection Sub-Entity includes the functional elements for data collecting, data pre-processing, and identification. The Device Control Sub-Entity includes the functional elements for controlling, actuation, and identification.

The User Entity, Digital Twin Entity, and the Data Collection and Device Control Entity interfaces with the **Cross-System Entity** to enable efficient, reliable, and secure collaboration and coordination. The Cross-System Entity includes the functional elements for data assurance and security support. The data assurance functional element includes support for in-time delivery of important information as well as data source reliability support, which include data access authorization as well as data authentication. The security support functional element includes various cyber-attack protection functions, anomaly detection, as well as counter defense measures to protect the digital twin network and operational entities.

The Data Collection and Device Control Entity interfaces with the **Observable Manufacturing Element**, which include the resource-specific functional elements.

2.5 Summary

This chapter introduces the three most powerful state-of-the-art artificial intelligence (AI) technologies along with the most advanced emulation/simulation technology, which include Deep Reinforcement Learning, Generative AI, Federated Learning, and Digital Twin. All of these technologies have the potential to strengthen the security level of networks and blockchains. On the other hand, these advanced AI technologies can be used in conducting cyber-attacks to networks and blockchains. More details and examples of security issues in networks and blockchains will be described in Chap. 11 of this book.

References

1. J.-M. Chung, *Emerging Metaverse XR and Video Multimedia Technologies* (Springer Nature, Apress, 2023)
2. K. Murphy, *Probabilistic Machine Learning: Advanced Topics* (MIT Press, 2023)
3. V. Mnih et al., Human-level control through deep reinforcement learning. Nature **518**, 529–533 (2015)
4. H.V. Hasselt, A. Guez, D. Silver, Deep Reinforcement Learning with Double Q-learning (2015). arXiv:1509.06461
5. Wikipedia, Reinforcement learning. https://en.wikipedia.org/wiki/Reinforcement_learning
6. Wikipedia, Model-free (reinforcement learning). https://en.wikipedia.org/wiki/Model-free_(reinforcement_learning)
7. S. Kim, Lecture 3: Dummy Q-learning (table), YouTube. https://www.youtube.com/watch?v=Vd-gmo-qO5E&t=9s
8. T.P. Lillicrap, et al., Continuous control with deep reinforcement learning (2015). arXiv:1509.02971
9. L. Girin, S. Leglaive, X. Bie, J. Diard, T. Hueber, X. Alameda-Pineda, Dynamical variational autoencoders: a comprehensive review. Found. Trends® Machine Learn. **15**(1–2), 1–175 (2021). https://doi.org/10.1561/2200000089
10. Wikipedia, BERT (language model). https://en.wikipedia.org/wiki/BERT_(language_model)
11. Wikipedia, Generative pre-trained transformer. https://en.wikipedia.org/wiki/Generative_pre-trained_transformer
12. Wikipedia, DALL-E. https://en.wikipedia.org/wiki/DALL-E
13. I. Kobyzev, S. Prince, M.A. Brubaker, *Normalizing Flows: An Introduction and Review of Current Methods* (2019) arXiv:1908.09257
14. E. Jang, Normalizing Flows Tutorial (2018). https://blog.evjang.com/2018/01/nf1.html
15. I. Goodfellow, J. Pouget-Abadie, M. Mirza, B. Xu, D. Warde-Farley, S. Ozair, A. Courville, Y. Bengio, Generative adversarial nets. Proc. NIPS **2014**, 2672–2680 (2014)
16. Wikipedia, Federated learning. https://en.wikipedia.org/wiki/Federated_learning
17. Wikipedia, Digital twin. https://en.wikipedia.org/wiki/Digital_twin
18. ISO 23247-1:2021, Automation systems and integration Digital twin framework for manufacturing Part 1: Overview and general principles. ISO, https://www.iso.org/standard/75066.html
19. STEP Tools, Inc., Demonstration of Three ISO 23247 Digital Twin Use Cases, YouTube. https://www.youtube.com/watch?v=wbsC_qzB8us
20. Y. Goh, D. Jung, G. Hwang, J.-M. Chung, Consumer electronics product manufacturing time reduction and optimization using AI-based PCB and VLSI circuit designing. IEEE Trans. Consum. Electron. **69**(3), 240–249 (2023)

Chapter 3
Internet Access for My PC and Smartphone

In this chapter, details of how your personal computer (PC) and smartphone are setup to enable you to use the Internet is explained. The Internet settings information explained in this chapter applies to all PCs, laptops, tablet computers, pad computers, and smartphones [1].

This chapter has three sections, which all use simple and practical examples to explain the IP network setup and gateway/router operation principles. The examples are based on the IPv4 protocol, but the same rules apply to the IPv6 protocol, where a difference to consider is that IPv4 addresses and subnet masks are 32 bits while IPv6 addresses and subnet masks are 128 bits.

First, in Sect. 3.1 the IPv4 network connection setup is described based on a PC connection to an Ethernet network. Both the manual TCP/IP network connection setup method is introduced followed by a description of the automatic setup method. In addition, the technical relevance of a wired connected PC is compared to a wireless connected smartphone or laptop computer is explained.

In Sect. 3.2, the IPv4 address and subnet configuration method is described using a project example. The method of how to set the subnet mask size and how to assign IPv4 addresses in a subnet is described. This includes setting the IP addresses for the gateway/router interface(s), all PCs as well as the broadcast IPv4 address of the subnets.

Using the IPv4 address and subnet configuration setup in Sect. 3.2, in Sect. 3.3 the project continues to setup the gateway/router's IPv4 routing table. In addition, detailed descriptions of the routing operations conducted when an IPv4 packet is received are provided.

In summary, Sect. 3.1 provides you with the information of what a device (i.e., PC, smartphone, or laptop) needs to establish a connection with the Internet. In addition, Sects. 3.2 and 3.3 provide the information of what needs to be prepared on the network side and how the network equipment (i.e., gateways and routers) actually operates.

J.-M. Chung, *Emerging Secure Networks, Blockchains and Smart Contract Technologies*, https://doi.org/10.1007/978-3-031-65866-2_3

Using the practical examples of this chapter, the project descriptions can help in understanding realistic TCP/IP network setup and operations.

3.1 IPv4 Network Connection Setup

In your PC, assuming you are using Microsoft Windows 11, if you enter the 'Control Panel' within the main menu, under 'Network & Internet' you can enter 'Network Connections' in which you can select the network connection type to be able to check your PC's settings. The example in Fig. 3.1 is for a computer using Microsoft Windows connected to an Ethernet Local Area Network (LAN). Unix, Linux, and Apple iOS computers have similar but slightly different control panel forms, but the Internet setup information is the same as described in this chapter.

Figure 3.1 presents the Control Panel of a PC using Microsoft Windows 11 that is connected to an Ethernet LAN. To access the Ethernet properties (as shown in Fig. 3.2), first enter the (a) "Control Panel" and click "Network and Sharing Center," then (b) select "Ethernet," and (c) in the "Ethernet Status" select "Properties."

In Fig. 3.2, when scrolling down the selectable items, if the 'Internet Protocol Version 4 (TCP/IPv4)' is selected then the 'Internet Protocol Version 4 (TCP/IPv4) Properties' window will open, as presented in Fig. 3.3.

In Fig. 3.3, the button (○) of '**Use the following IP address**' is selected (◉) and the windows for 'IP address,' 'Subnet mask,' and 'Default gateway' have information inserted. If the button (○) of '**Obtain an IP address automatically**' were selected, then the **Dynamic Host Configuration Protocol (DHCP)** would have been activated and all information for 'IP address,' 'Subnet mask,' and 'Default gateway' would be automatically assigned, and therefore, these three windows would be empty.

In the lower part of Fig. 3.3, the Domain Name System (DNS) option can be selected. The button (○) of '**Use the following DNS server address**' is selected (◉) and the windows for 'Preferred DNS server' and 'Alternate DNS server' have IPv4 addresses inserted. If the button (○) of '**Obtain DNS server address automatically**' were selected, then the IPv4 addresses of 'Preferred DNS server' and 'Alternate DNS server' would be automatically assigned, and therefore, these two windows would be empty.

First let's study about how the 'IP address,' 'Subnet mask,' 'Default gateway,' and 'DNS server address' is setup and later a description of the DHCP automatic assignment methods will be provided. This is because DHCP does the same setup operation using an automatic assignment method.

DHCP enables the network administrator to dynamically control who and when a PC/smartphone can connect to the network and be very efficient in network resource management, which includes control of network bandwidth usage, data traffic usage amount, user authorization, network access management, and IP address usage within the subnet. This is why laptop and tablet computers as well as smartphones commonly use DHCP (i.e., the '**Obtain an IP address automatically**' option) when connecting to use Wi-Fi or the mobile wireless network.

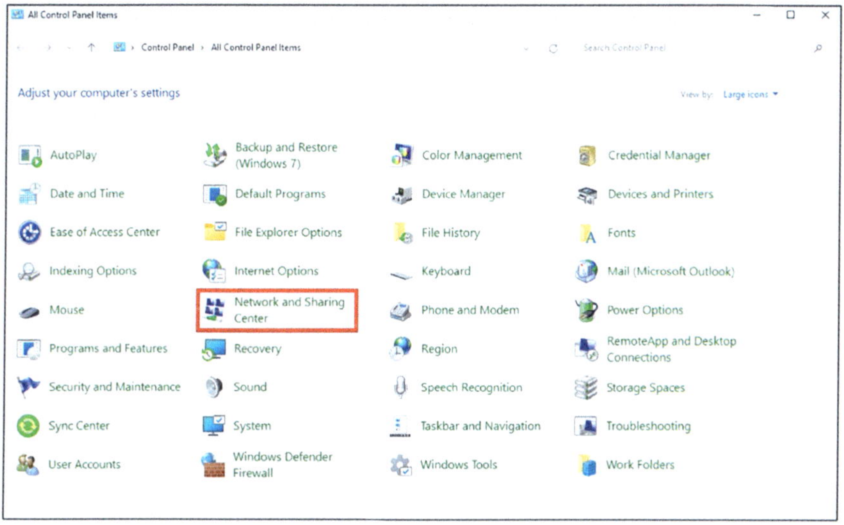

(a)

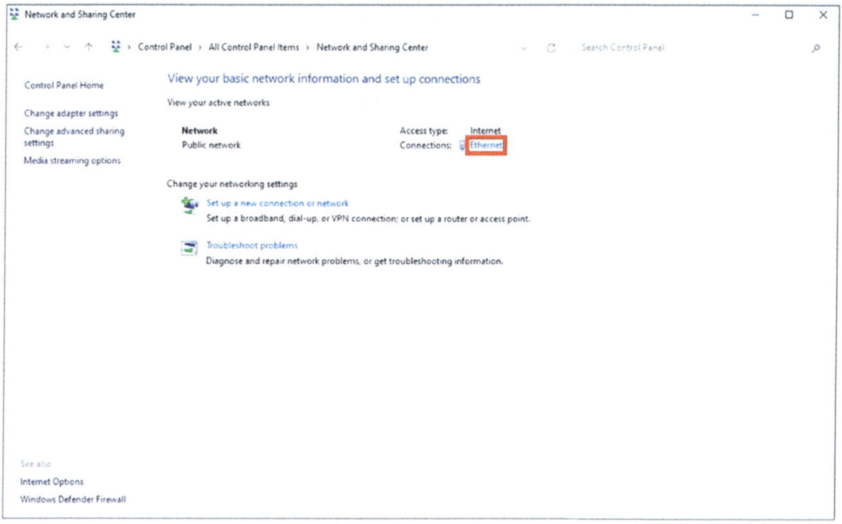

(b)

Fig. 3.1 Control panel of a PC using Microsoft Windows 11 that is connected to an Ethernet LAN. First, enter the **a** "Control Panel" and click "Network and Sharing Center," then **b** select "Ethernet," and **c** in the "Ethernet Status" select "Properties" to enter the Ethernet properties as shown in Fig. 3.2

Fig. 3.1 (continued)

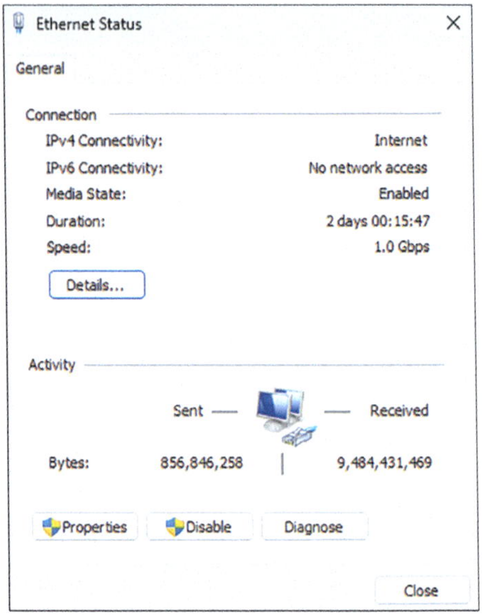

(c)

The IPv4 address is a 32 bit (i.e., 4 bytes) address where the format of the address is presented with periods between every 8 bits (i.e., for each byte) to make it easier for us to read. However, the actual 32 bit IPv4 address does not have any periods, as it is a sequence of only 32 bits. The formula to convert 8 bits into a decimal number is described in the following.

For example, 8 bits in a sequence are represented as **ABCDEFGH** where **A**, **B**, **C**, **D**, **E**, **F**, **G**, and **H** are each a 0 or 1 bit. Then the decimal number of the binary number **ABCDEFGH** would be obtained from $\{(\mathbf{A} \times 2^7) + (\mathbf{B} \times 2^6) + (\mathbf{C} \times 2^5) + (\mathbf{D} \times 2^4) + (\mathbf{E} \times 2^3) + (\mathbf{F} \times 2^2) + (\mathbf{G} \times 2^1) + (\mathbf{H} \times 2^0)\}$. For example, the binary number 10110001 would be equal to $\{(1 \times 2^7) + (0 \times 2^6) + (1 \times 2^5) + (1 \times 2^4) + (0 \times 2^3) + (0 \times 2^2) + (0 \times 2^1) + (1 \times 2^0)\}$ which is same as $\{(1 \times 128) + (0 \times 64) + (1 \times 32) + (1 \times 16) + (0 \times 8) + (0 \times 4) + (0 \times 2) + (1 \times 1)\} = \{128 + 32 + 16 + 1\} = 177$ as a decimal number. The largest 8 bit binary number is 11111111 which is equal to the decimal number 255. In addition, the smallest 8 bit binary number is 00000000 which is equal to the decimal number 0. So, an 8 bit binary number can be a decimal number in the range of 0–255, as presented in Fig. 3.4. In addition, the IPv4 address 165.132.126.159 converted into binary is 10100101.10000100.01111110.10011111, which is 32 bits divided into four 8 bit segments, divided with a period in between, as presented in Fig. 3.5. The actual IPv4 binary address is 10100101100001000111111010011111 without the periods in between.

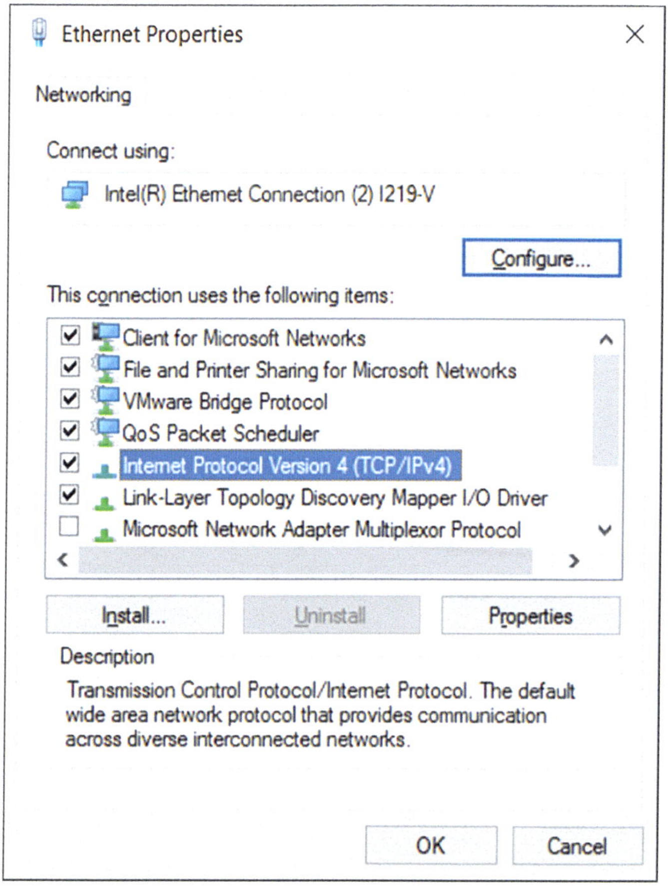

Fig. 3.2 Ethernet properties control panel

3.1.1 Subnet Mask

The Internet is divided into subnets, and subnets are divided into smaller subnets. Subnets are used to simplify the process of network setup, management, IP address assignment, routing, etc. The Internet Assigned Numbers Authority (IANA) defines use of IP address blocks [2] and major subnet divisions. In addition, the IANA manages the IPv4 Special-Purpose Address Registry [3, 4].

Subnet masks are based on the size of the subnet that the Client (PC or smartphone) is connected to. An IPv4 subnet mask is formed by 32 bits, with a format of ones or zeros in a sequence. The left-most bit is the Most Significant Bit (MSB) and the right-most bit is the Least Significant Bit (LSB) of the subnet mask.

The subnet mask is used to mask (filter) the IP address (in the received IPv4 packet) to easily determine if the packet belongs to this subnet or not. This process

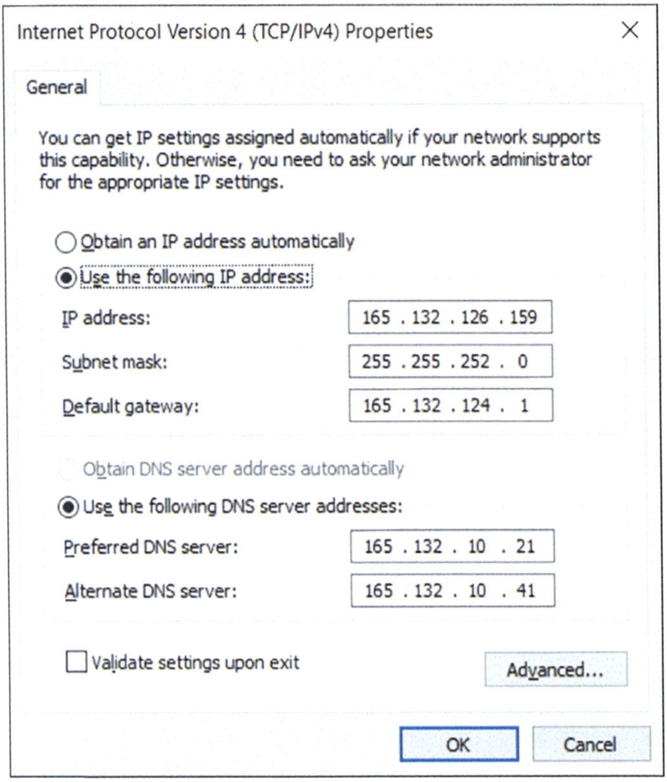

Fig. 3.3 TCP/IPv4 properties control panel

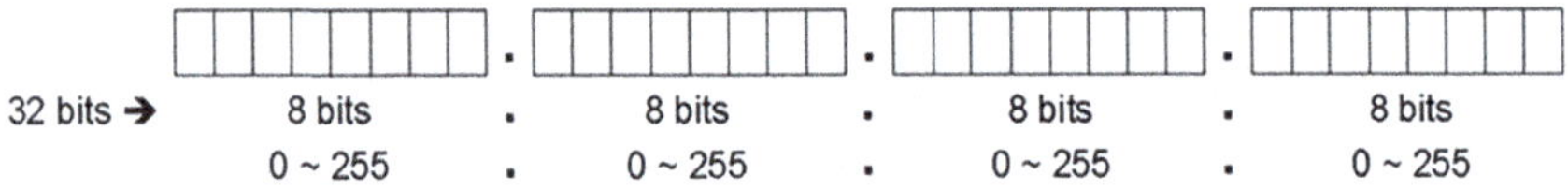

Fig. 3.4 A 32 bit IPv4 address format divided into four 8 bit segments with dividing periods between the segments

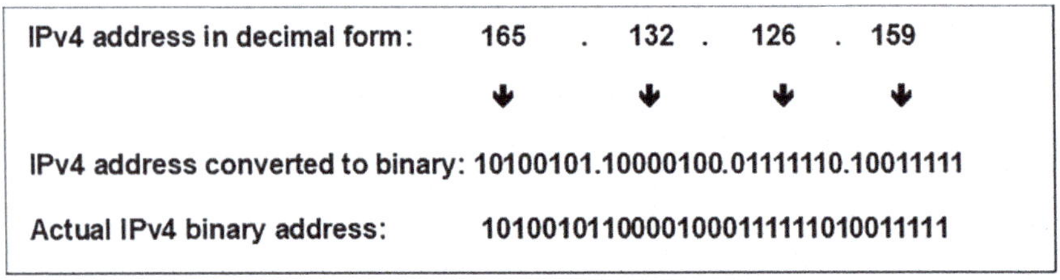

Fig. 3.5 IPv4 address in decimal form converted to binary form

Subnet mask decimal to binary conversion example
255.255.252.0 → 1111111.11111111.11111100.00000000
10 zeros
Due to the 10 consecutive zeros in the subnet mask, the size of this subnet is 2^{10} = 1,024 IP addresses.

Fig. 3.6 Subnet mask converted into binary to reveal subnet size

is easier to understand by first converting the subnet mask into binary, as presented in Fig. 3.6.

As presented in Fig. 3.6, the subnet mask 255.255.252.0 is converted into binary, which becomes 11111111.11111111.11111100.00000000, where there are 10 zeros at the end. Due to the 10 consecutive zeros in the subnet mask, the size of this subnet is $2^{10} = 1{,}024$ IP addresses.

Subnet masks commonly have the form of a sequence of ones starting from the MSB that changes into a sequence of zeros in the middle and continues until the end LSB. If the subnet mask is all ones (i.e., 255.255.255.255 = 1111111.11111111.11111111.11111111) then the subnet mask is closed and all network connections will be closed, which will block network access to all devices in the subnet. On the other hand, if the subnet mask is all zeros (i.e., 0.0.0.0 = 0000000.0000000.0000000.00000000) then the subnet mask is all open and every network connection will be accepted.

Subnet masks are determined by the size of the subnet. A subnet mask is also called a 'netmask' in various network documents.

In the following Sect. 3.2 (IP Address and Subnet Configuration (Project Example)), the method of forming a subnet, how IP addresses are assigned, and how a subnet is used in the routing table of gateways and routers is explained.

3.1.2 *Default Gateway*

A default gateway is the dedicated Internet router that will send and receive all Internet IP packets for a PC/smartphone. In other words, a PC/smartphone will access the Internet (i.e., send and receive all IP packets) through its default gateway. Therefore, the default gateway is very important to the PC/smartphone. In a following section of this chapter, the operations of gateways/routers are explained.

3.1.3 Domain Name Server (DNS)

DNS is a server that converts hostnames (like 'google.com') into IP addresses. Hostname website addresses are commonly in Uniform Resource Locator (URL) form, which contain information about the location, subfolders, and files of a webpage. DNS makes the Internet and email much easier to use, because we can use names and words that are easy to remember. The IPv4 or IPv6 addresses that are converted by the DNS will be used to send IP packets to the destination IP address.

For example, the Google public DNS IPv4 address is 8.8.8.8 or 8.8.4.4 and IPv6 addresses are 2001:4860:4860::8888 and 2001:4860:4860::8844 [5]. So, IP packets sent to Google will commonly have one of these IPv4 or IPv6 addresses included in the IP packet header (as its destination IP address) with the help of the DNS. More details on the IPv4 protocol are provided in Sect. 4.3 and details of the IPv6 protocol and IPv6 address is provided in Sect. 4.4 of this book.

In addition, for popular social network website addresses like www.facebook.com the DNS will change it into 173.252.110.27. As you can see, the DNS is very important for any type of Internet connection. This is why two DNS servers are included (i.e., 'Preferred DNS server' and 'Alternate DNS server') in the IPv4 properties control panel, as shown in Fig. 3.3. However, even if the 'Alternate DNS server' information is not recorded, the 'Preferred DNS server' can do the job, but sometimes a longer waiting time or a connection failure could occur.

3.1.4 Dynamic Host Configuration Protocol (DHCP)

As can be noticed based on the descriptions above, setting up an IP connection to a network to use the Internet is difficult. In addition, all of the above information (i.e., IP address, subnet mask, default gateway, preferred DNS server, and alternate DNS server) are not self-selectable, but must be setup and provided by the network administrator. Therefore, in many networks (like your smartphone), the 'Obtain an IP address automatically' option is used and the 'Obtain DNS server address automatically' is also used. When these automatic setup options are used, the Dynamic Host Configuration Protocol (DHCP) will be activated and all network connection information will be automatically assigned by the DHCP server, and therefore, all of the network connection values (i.e., IP address, subnet mask, default gateway, preferred DNS server, and alternate DNS server) would be empty in the 'Internet Protocol Version 4 (TCP/IPv4) Properties' windows of Fig. 3.3.

DHCP enables a PC/smartphone to automatically contact its local DHCP server and request for an IP address and networking parameters to connect to the Internet. When DHCP is used, the network setup is automatic, so there is no need to contact the local network administrator to have the Internet connection setup manually. This is why smartphones, laptops, and mobile devices commonly use DHCP. DHCP is used for both IPv4 and IPv6 connection setups. But, in order to be able to use DHCP, an active local DHCP server must exist.

The device's local DHCP server will provide the device with the IP address, subnet mask, default gateway, preferred DNS server, and alternate DNS server (and other Internet configuration parameters) to use. The IP address can be either an IPv4 or IPv6 address, based on the requirements of the network that the device is connecting to.

3.1.4.1 DHCP Enables Reuse of IP Addresses

When DHCP is used, only a PC/smartphone that needs an Internet connection at that time is assigned an IP address to use. So, after the Internet connection ends, that IP address can be reused by another device. This is very effective for subnets, as they can reuse IP addresses, which is especially important for IPv4 networks (due to shortage in IPv4 addresses).

Considering that IPv4 addresses are 32 bits long, a total of 2^{32} IPv4 addresses can be created, but considering that $2^{32} < 4.3$ billion, there are less than 4.3 billion IPv4 addresses that can exist.

Currently there is a lack of IPv4 address because there are many unused IP addresses in various subnets (which is explained later in Sect. 3.2.2. Subnet setup for Room A) and also because there are many active devices using multiple IP connections to the Internet simultaneously. For example, one smartphone may establish multiple network connections (through 3G, 4G, 5G, Wi-Fi, Bluetooth networks) and therefore use multiple IP addresses simultaneously, which is explained in further detail in the following section.

3.1.4.2 Multiple IP Addresses Used Per Device (Smartphone Example)

The IPv4 properties control panel setup presented in Fig. 3.3 (either manually or automatically setup) is needed for each and every (wired or wireless) network connection, as a separate IP address (IPv4 or IPv6 or both) needs to be assigned to each interface port, because each Internet interface connection to a network needs to be setup and operated independently. Therefore, one smartphone may have many interface ports in use at the same time, based on the number of wireless applications the device is accessing simultaneously.

For example, my **Samsung Galaxy A54** smartphone (this product was released on March 24, 2023) supports 2G Global System for Mobile Communications (GSM), 3G High Speed Packet Access (HSPA), 4G Long Term Evolution (LTE), and 5G mobile communications by **AT&T** as well as Wi-Fi 6 (which is equipped with IEEE 802.11 a/b/g/n/ac modes with dual-band and Wi-Fi Direct support) and Bluetooth 5.3 (equipped with Advanced Audio Distribution Profile (A2DP) and Low Energy (LE)) [6]. Each connection to the Internet would need a separate TCP/IP interface to be setup (like the one presented in Fig. 3.3), which will each require a different source IP address (for the same smartphone) to connect to a different network type.

For example, if the smartphone was using a Voice over IP (VoIP) connection over the 5G network, while being connected to an app using 4G LTE and using a website through the Wi-Fi 6 network, then these three Internet connections from the same smartphone would all use different IPv4 or IPv6 interface connections (and different IPv4 or IPv6 source addresses).

In other words, network connections are established independently for each network interface (e.g., 5G, 4G LTE, and Wi-Fi 6 in this example). This is why there is a lack of IPv4 addresses and therefore DHCP and IPv6 are needed.

3.1.4.3 DHCP Operations

DHCP operates on a Client–Server model, where the Client would be your PC or smartphone. The Server would be the application, website, media portal, cloud, or database you are accessing with your PC/smartphone.

The DHCP server manages a group of IP addresses, where the client (e.g., PC, smartphone) information is kept on the DHCP server. The client information includes the IP address of the default gateway, domain name servers, time servers, and more. The DHCP setup messages and operation are illustrated in Fig. 3.7 and described in the following.

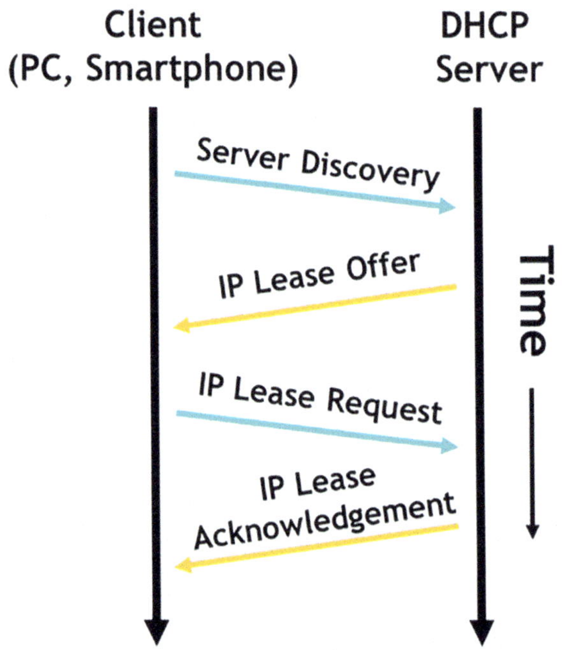

Fig. 3.7 DHCP setup messages and operation

DHCP Setup Messages and Operation

1. Client (e.g., PC, Smartphone, etc.) connects to the network.
2. DHCP uses User Datagram Protocol (UDP) with Client port 68 and DHCP Server port 67.
3. Client's DHCP program broadcasts a 'Server Discovery' message requesting for network information.
4. Any DHCP Server on the network can provide service by replying an 'IP Lease Offer' message to the Client.
5. Client will send an 'IP Lease Request' back to the DHCP Server.
6. DHCP Server will send back an 'IP Lease Acknowledgement' enabling use of an IP Address and network parameters for a limited time duration.

When a client (e.g., PC or smartphone) tries to reconnect to the Internet, the DHCP server may give the same IP address that was used before by that PC/smartphone. However, a different IP address may be assigned if that IP address is being used by some other device or due to the Network Administrator's IP address assignment policies, which are based on the organization's network security policies.

3.2 IP Address and Subnet Configuration (Project Example)

Gateways and routers both have IP routing tables inside, which are used to determine the routing operation for IP packets that pass through. Gateways and routers are different based on the following aspects. A router is a device that provides shortest path routing in a network that uses the same network protocols throughout the network. This type of a homogeneous protocol network is commonly operated by a single Internet Service Provider (ISP) and is called an Autonomous System (AS). For example, within an Ethernet network supporting IPv4 packets, a router would be used because all packets in this network would use the same protocols, which are the Ethernet protocol and IPv4 protocol. So, in this Ethernet network, Ethernet frames with an IPv4 packet inside will be sent and received with the help of the router. Routers have IP routing tables inside to support the IP packet routing operations [1, 7, 8].

The details of how IP routing tables are used to support the IP packet routing operations are explained in Sect. 3.3 "IPv4 Routing Table Setup and Operations (Project Example)" later in this chapter.

A gateway is used to provide routing between different networks (i.e., heterogeneous networks). For example, a gateway would be used as an interface between a 5G mobile communication network and an Ethernet network, because the 5G network protocol is different to the Ethernet network protocol. Because each 5G frame or Ethernet frame will have an IPv4 or IPv6 packet inside, gateways need to have IP routing tables to support the IP packet routing operations between the 5G and Ethernet networks.

Because gateways and routers both use IP routing tables and provide routing connection of IPv4 and IPv6 packets (based on the destination IP address in the IP packets), we will study the IP routing operation of a gateway and router together in the project example below. More details on how IPv4 packets are included inside Ethernet frames will be described in detail in Chap. 4 of this book.

3.2.1 IP Gateway/Router Configuration Project Example

In this section, the method of how TCP/IP is used to access the Internet and how to setup a gateway/router is described based on a project example. In this example, assume that we have been assigned a project to setup a network connecting a total of 175 personal computers (PCs) that are located in four rooms (A, B, C, and D) that need to have access to the Internet through a gateway/router. In this project, we need to setup the IPv4 address of the 175 computers and configure the gateway's IPv4 routing table. The project details are described in the following.

3.2.1.1 IP Network Setup Project Requirements

Configure the gateway/router routing table and assign all PCs IPv4 addresses to support the following network setup, which is illustrated in Fig. 3.8.

1. Subnet the IPv4 network such that IPv4 addresses for all PCs in each of the 4 rooms A, B, C, D can be assigned individual IPv4 addresses.
2. Subnet off of 165.132.9.0 using 256 IPv4 addresses.
3. Gateway connects rooms A, B, C, D, and the Internet.
4. Room A has 100 PCs that need to be connected to interface (port) 1 of the gateway/router.

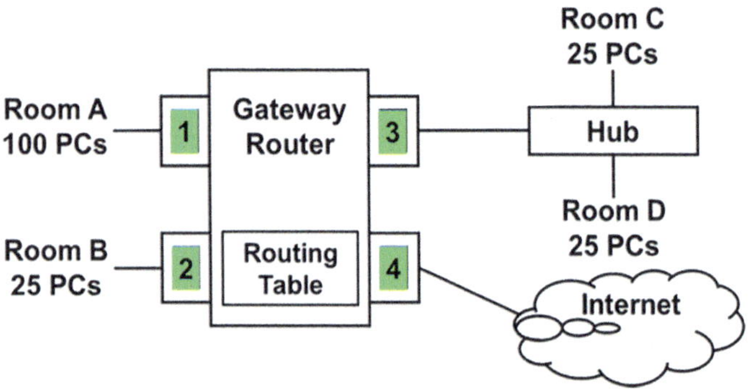

Fig. 3.8 IPv4 gateway/router configuration network

5. Room B has 25 PCs that need to be connected to interface (port) 2 of the gateway/router.
6. Rooms C and D each have 25 PCs connected to a Hub/Switch that is connected to interface (port) 3 of the gateway/router.
7. All PCs need to have connection to the Internet, where the Internet is connected to interface (port) 4 of the gateway/router.
8. The Internet port connection is assigned by the network administrator/manager as follows.
   - IPv4 Address: 165.132.15.56
   - Subnet Mask: 255.255.255.252.

Next, we need to setup the subnet for each room. First, we start with the IPv4 addresses and subnet setup of room A.

3.2.2 Subnet Setup for Room A

The subnet for room A needs to be prepared. Room A has 100 PCs that each will need an IPv4 address. In addition, one IPv4 address will be needed for the network destination, which will serve as the IPv4 address assigned to the subnet of room A (which will be used like a name for room A's subnet). One IPv4 address needs to be assigned for broadcast purposes (which will be used to send a message to all PCs in the subnet of room A). For example, if a control and management IPv4 packet (i.e., Internet Control Message Protocol (ICMP) IPv4 packet (ICMPv4)) would need to be quickly sent to all interfaces (i.e., the interfaces on the 100 PCs and 1 on the gateway/router) on this subnet, then the broadcast IPv4 address would be used in the destination IP address field of that ICMPv4 packet. In addition, one IP address will be needed for interface (port) 1 of the gateway/router. So overall, at least 103 = (100 + 1 + 1 + 1) IP addresses would be needed for the subnet of room A.

Subnet sizes can only be in sizes of $2^{\mathbf{X}}$ where **X** is a positive integer. So, subnet sizes can only be 2, 4, 8, 16, 32, 64, 128, 256, 512, 1024, 2048, 4096, etc. Since room A needs 103 IP addresses in its subnet, the smallest subnet size that is large enough to include 103 IP addresses is 128. So, a 128 size subnet will have to be used for room A's subnet.

Since $128 = 2^7$, room A's subnet will have 7 consecutive zeros from the LSB and up. Considering that the subnet mask is 32 bits constructed with a series of ones (starting from the MSB position and down) and zeros, room A's subnet mask will be 11111111.11111111.11111111.10000000 in binary form, which is equivalent to 255.255.255.128 in decimal form. The 7 consecutive zeros in room A's subnet means that this subnet has opened up 128 IPv4 addresses that can be used.

In this project example, we will assign IPv4 addresses using the following rule (however, other IP assignment methods do exist). The network destination IPv4 address will be assigned the lowest IPv4 address in the subnet. Following the network

IPv4 Addresses Required = 103 addresses ≤ 128 = 2^7

Subnet mask: 11111111.11111111.11111111.10000000
= 255.255.255.128 (7 zeros)

Broadcast IP: 1 IPv4 address ← Highest IPv4 address in the subnet

Gateway Interface 1: 1 IPv4 address ← Second highest IPv4 address in the subnet

Room A (100 PCs): 100 IPv4 addresses ← Assign each PC an IPv4 addresses higher than the Network Destination's IPv4 address

Network Destination: 1 IPv4 address ← Lowest IPv4 address in the subnet

Fig. 3.9 IPv4 address requirements of the gateway/router interface and 100 PCs for room A's subnet (other IPv4 address assignment methods do exist)

destination IPv4 address, in increasing IPv4 address order, each PC will be assigned an IPv4 address. The broadcast IPv4 address will be assigned the highest IPv4 address in the subnet and the gateway interface IPv4 address will be assigned the second highest IPv4 address in the subnet. In Fig. 3.9, this IPv4 gateway/router configuration network example is presented.

It is important to note that the network destination IPv4 address should always be assigned the lowest IPv4 address in the subnet and the broadcast IPv4 address should always be assigned the highest IPv4 address in the subnet. But the IPv4 address assignments for the gateway/router interface and PCs can be changed as long as they are larger than the network destination IPv4 address (which is the lowest IP address in the subnet) but smaller than the broadcast IPv4 address of the subnet (which is the highest IPv4 address in the subnet). Therefore, **there are other IPv4 address assignment methods that do exist.** For example, instead of assigning the gateway/router interface IPv4 address the second highest IPv4 address in the subnet, it can be assigned the second lowest IPv4 address in the subnet, one IPv4 address larger than the network destination IPv4 address, and in increasing IPv4 address order, each PC in room A can be assigned an IPv4 address. Various other gateway/router and PC IPv4 address assignment methods do exist. However, in the following, we will continue to explain the gateway/router operations using the method that was described above in Fig. 3.9. In addition, the basic subnet setup method for IPv6 networks using IPv6 addresses and subnets is the same (except that IPv6 addresses and subnet masks are 128 bits long).

Figure 3.10 presents the IPv4 addresses and subnet setup for room A, based on the IPv4 address ordering described in Fig. 3.9. Using the IPv4 network requirements item 2 given above in Sect. 3.2.1.1, we need to subnet off of 165.132.9.0 using 256 IPv4 addresses to support all the PCs in the rooms A, B, C, and D.

0	Network Destination
1~100	IP Addresses for 100 PCs
101~125	IP Addresses not used (25 reserved)
126	Gateway Interface 1 IP address
127	Broadcast IP Address

(★ marks rows 0–127)

- **IPv4 address for Room A's PCs: 165.132.9.★**
- **We needed 103 ≤ 128 = 2^7 = Subnet Size**
  Subnet mask: 255.255.255.128 ← (256-2^7) = 128
- **CIDR (Classless Inter-Domain Routing)**
  - **Network Destination / Subnet Mask 1's (32-7=25)**
  - **CIDR Network ID: 165.132.9.0/25**

Fig. 3.10 IPv4 addresses and subnet setup for room A

Therefore, we can use the IPv4 addresses starting from 165.132.9.0 to 165.132.9.255, in which the front part '165.132.9.' is all the same (which is the prefix part) and only the last part changes from 0 to 255, resulting in a total of 256 IP addresses we can use (i.e., we can use the 256 IPv4 addresses from 165.132.9.0 to 165.132.9.255 to solve this project). In Fig. 3.10, the ★ represents the position of the last number to include in the IP address 165.132.9.★.

Among these 256 IPv4 addresses, since the subnet of room A was set to a size of 128 IPv4 address, we can use the IPv4 addresses starting from 165.132.9.0 to 165.132.9.127.

Figure 3.10 presents the IPv4 address assignments (based on Fig. 3.9), where the network destination IPv4 address will be assigned the lowest IPv4 address in the subnet, which is 165.132.9.0. Next, in increasing IPv4 address order, each PC in room A will be assigned an IPv4 address. As a result, the IPv4 addresses assigned to the 100 PCs in room A will be from 165.132.9.1 to 165.132.9.100. The broadcast IP address will be assigned the highest IPv4 address in the subnet, which is 165.132.9.127. The gateway interface IPv4 address will be assigned the second highest IPv4 address in the subnet, which is 165.132.9.126.

Among the 128 IPv4 addresses in the subnet of room A, 25 of these IPv4 addresses (i.e., from 165.132.9.101 to 165.132.9.125) are not assigned to anything.

These 25 unused IPv4 addresses resulted from needing to select the subnet size based on $2^{\mathbf{X}}$ where **X** is a positive integer. So, subnet sizes can only be 2, 4, 8, 16, 32, 64, 128, 256, 512, 1024, 2048, 4096, etc. Since room A needed 103 IPv4 addresses in its subnet, and the smallest subnet size that was large enough to include 103 IPv4 addresses is 128, as a result 128 became the subnet size for room A, which resulted in 25 leftover IPv4 addresses in room A.

These 25 unused IPv4 addresses can be used when adding more PCs to room A, or will be wasted if left unused. This is why there are many unused (wasted) IP addresses in various subnets, resulting in a premature lack of IPv4 addresses.

Room A's subnet is represented using the Classless Inter-Domain Routing (CIDR) network identification (ID), which has the notation of the 'IPv4 address of the network destination' followed by a '/' followed by the 'number of ones in the subnet mask.' For room A, the IPv4 address of the network destination is 165.132.9.0. In addition, room A's subnet mask is 11111111.11111111.11111111.10000000 in binary form (i.e., a sequence of 25 ones followed by 7 zeros), which is equivalent to 255.255.255.128 in decimal form. So, room A's CIDR Network ID is 165.132.9.0/25, as presented in Fig. 3.10.

In other words, from the CIDR network ID '165.132.9.0/25' the network administrator and managers can easily see that 165.132.9.0 is the IPv4 address of room A's network destination and the subnet mask has 25 ones (i.e., 11111111.11111111.11111111.10000000) so it will have a size of 128 IP address included, therefore, room A's subnet has the IP addresses in the range of 165.132.9.0 to 165.132.9.127.

The positions that are 1 in a subnet mask are the IP address 'Prefix' part, which applies to both IPv4 and IPv6 addresses. All IP addresses used in a subnet should have the same prefix. The subnet mask is used to check the prefix part of all received IP packets, to identify if the IP packet belongs to this subnet. Because the prefix part of an IP address is fixed (i.e., the 'fix' part) in advance (i.e., the 'pre' part) when the subnet is setup, it is called a 'prefix.' Therefore, all IP addresses used in subnet A should have the same prefix 25 bits as 165.132.9.0.

3.2.2.1 Subnet Mask Operations

In addition, another reason that the subnet mask uses a form of a sequence of ones followed by a sequence of zeros, is due to the way it is used when masking the IPv4 address when an IPv4 packet is newly received. When a new IPv4 packet is received at the gateway/router, this packet's destination IPv4 address is masked by the subnet mask, and all the bit positions of the IPv4 address that are in the sequence of the ones position in the subnet mask are compared to the network destination IP address of the subnet. In other words, the first 25 bits of the IPv4 destination address (in a newly received IPv4 packet) must match the first 25 bits of the network destination IPv4 address. If these 25 bits do match, then this IPv4 packet's destination address is within room A's subnet, in which the gateway/router will deliver this IPv4 packet to the proper PC in room A.

Figure 3.11 presents an example of the binary logic operation of how the subnet mask is used to check if the destination IPv4 address of a newly received IPv4 packet belongs to room A's subnet. This operation uses the subnet mask and the network destination IPv4 address of room A, which are provided in room A's CIDR network ID 165.132.9.0/25.

As illustrated in Fig. 3.11, in this example, the received IPv4 packet's destination IPv4 address is 165.132.9.12, which is 10100101.10000100.00001001.00001100 in binary form. The subnet mask is 255.255.255.128 which is 11111111.11111111.11111111.10000000 in binary form. The IP address

1. 10100101.10000100.00001001.00001100 ← Received packet's destination IPv4 address 165.132.9.12 in binary

2. 11111111.11111111.11111111.10000000 ← Room A's subnet mask (25 ones)

3. 10100101.10000100.00001001.0 ← First 25 bits of the received packet's destination IPv4 address

4. 10100101.10000100.00001001.0 ← First 25 bits of the network destination IP address 165.132.9.0

5. Since the first 25 bits of the received IPv4 packet's destination IPv4 address matches the first 25 bits of the network destination IPv4 address, this packet should be delivered to room A's subnet.

Fig. 3.11 Example of the binary logic operation of how the subnet mask is used to check if the destination IPv4 address of a newly received IPv4 packet belongs to room A's subnet, using the subnet mask and the network destination IPv4 address of room A (i.e., prefix checking)

of the network destination is 165.132.9.0, which is equivalent to 10100101.10000100.00001001.00000000 in binary. When a new IP packet is received, the gateway/router will check the destination IP address in this new IP packet and compare it to the subnet mask. The 25 bit positions of ones in the subnet mask are the comparing positions to check if the 25 bits in the destination IPv4 address of the new packet matches the 25 bits of the IPv4 address of the network destination (i.e., prefix part comparison).

As presented in Fig. 3.11, the first 25 bits of the received IPv4 packet's destination IPv4 address '10100101.10000100.00001001.0' matches the first 25 bits of the network destination IPv4 address '10100101.10000100.00001001.0' in this example. Because the first 25 bits of the received IP packet's destination IPv4 address matches the first 25 bits of the network destination IPv4 address, this packet should be delivered to room A's subnet.

This process is equivalent to checking if the received IPv4 packet's destination IPv4 address is in the range of 165.132.9.0 to 165.132.9.127, which is room A's subnet range of IPv4 addresses. Since computers have to conduct this process using a simple software program, the binary logic operation (example) presented in Fig. 3.11 is used.

3.2.3 Subnet Setup for Room B

Next, the subnet for room B has to be prepared. Room B has 25 PCs that each will need an IPv4 address. In addition, one IPv4 address will be needed for the network destination IPv4 address of room B's subnet. One IPv4 address needs to be assigned as the broadcast IPv4 address and one IPv4 address will be needed for interface (port) 2 of the gateway/router. So overall, at least $28 = (25 + 1 + 1 + 1)$ IPv4 addresses are needed for room B's subnet.

Since subnet sizes are selectable among 2^X numbers (where **X** is a positive integer), for the 28 IPv4 addresses needed for room B's subnet, the smallest subnet size that is large enough to include 28 IP addresses is 32. As $32 = 2^5$, room B's subnet will have 27 ones from the MSB position and 5 consecutive zeros from the LSB and up, resulting in a subnet mask of 11111111.11111111.11111111.11100000 in binary, which is equivalent to 255.255.255.224 in decimal form, as presented in Fig. 3.12.

Figure 3.13 presents the IPv4 addresses and subnet setup for room B, based on the IPv4 address ordering described in Fig. 3.12. In Fig. 3.13, the ★ represents the position of the last number to include in the IP address 165.132.9.★.

IPv4 Addresses Required =28 addresses ≤ 32=2^5

Subnet mask: 11111111.11111111.11111111.11100000 =
= 255.255.255.224 (5 zeros)

Broadcast IPv4: 1 IPv4 address ← Highest IPv4 address in the subnet

Gateway Interface 2: 1 IPv4 address ← Second highest IPv4 address in the subnet

Room B (25 PCs): 25 IPv4 addresses ← Assign each PC an IPv4 addresses higher than the Network Destination's IPv4 address

Network Destination: 1 IPv4 address ← Lowest IPv4 address in the subnet

Fig. 3.12 IPv4 gateway/router configuration network example for room B's subnet (other IPv4 address assignment methods do exist)

- **IPv4 address for Room B: 165.132.9.★**

★	
128	Network Destination
129~153	IP Addresses for 25 PCs
154~157	IP Addresses not used (4 reserved)
158	Gateway Interface 2 IP address
159	Broadcast IP Address

- **We needed 28 ≤ 32 = 2^5 = Subnet Size**
  Subnet mask: 255.255.255.224 ← $(256-2^5) = 224$
  - **Subnet Mask 1's (32-5=27)**
  - **CIDR Network ID: 165.132.9.128/27**

Fig. 3.13 IPv4 addresses and subnet setup for room B

Because the subnet of room A used 128 IPv4 addresses 165.132.9.0 to 165.132.9.127, room B will use the next 32 IPv4 addresses starting from 165.132.9.128 to 165.132.9.159.

The network destination IPv4 address will be assigned the lowest IP address in the subnet, which is 165.132.9.128. Next, in increasing IPv4 address order, each PC in room B will be assigned an IPv4 address. As a result, the IPv4 addresses assigned to the 25 PCs in room B will be from 165.132.9.129 to 165.132.9.153. The broadcast IPv4 address will be assigned the highest IPv4 address in the subnet, which is 165.132.9.159. The gateway interface IPv4 address will be assigned the second highest IPv4 address in the subnet, which is 165.132.9.158. Among the 32 IPv4 addresses in the subnet of room B, 4 of these IPv4 addresses (i.e., from 165.132.9.154 to 165.132.9.157) are not assigned to anything and are reserved for future use. Room B's subnet is represented using the CIDR network ID of 165.132.9.128/27.

3.2.4 Subnet Setup for Room C and D

Next, the subnet for rooms C and D have to be prepared. Room C has 25 PCs and room D also has 25 PCs, where each PC needs an IPv4 address. In addition, one IPv4 address will be needed for the network destination IPv4 address of the subnet shared by rooms C and D. One IPv4 address needs to be assigned as the broadcast IPv4 address and one IPv4 address will be needed for interface (port) 3 of the gateway/router. So, overall, $53 = (25 + 25 + 1 + 1 + 1)$ IPv4 addresses are needed for the subnet to support rooms C and D.

Since subnet sizes are selectable among $2^{\mathbf{X}}$ numbers (where **X** is a positive integer), for the 53 IPv4 addresses needed for the subnet supporting rooms C and D, the smallest subnet size that is large enough to include 53 IP addresses is 64. As $64 = 2^6$, the subnet supporting rooms C and D will have 26 ones from the MSB position and 6 consecutive zeros from the LSB and up in the subnet mask used for rooms C and D. Resulting in a subnet mask of 11111111.11111111.11111111.11000000 in binary, which is equivalent to 255.255.255.192 in decimal form, as presented in Fig. 3.14.

Figure 3.15 presents the IPv4 addresses and subnet setup that supports rooms C and D, based on the IPv4 address ordering described in Fig. 3.14. In Fig. 3.15, the ★ represents the position of the last number to include in the IPv4 address 165.132.9.★.

Because the subnets supporting rooms A and B have already used 160 IPv4 addresses from 165.132.9.0 to 165.132.9.159, the subnet for rooms C and D will use the next 64 IPv4 addresses starting from 165.132.9.160 to 165.132.9.223.

The network destination IP address will be assigned the lowest IPv4 address in the subnet, which is 165.132.9.160. Next, in increasing IPv4 address order, each PC in rooms C and D will be assigned an IP address. As a result, the IPv4 addresses assigned to the 50 PCs in rooms C and D will be from 165.132.9.161 to 165.132.9.210. The broadcast IPv4 address will be assigned the highest IPv4 address in the subnet, which is 165.132.9.223. The gateway interface IPv4 address will be assigned the

IPv4 Addresses Required = 53 addresses ≤ 64 =2^6

Subnet mask: 11111111.11111111.11111111.11000000
= 255.255.255.192
6 zeros

Broadcast IP: 1 IPv4 address ← Highest IPv4 address in the subnet

Gateway Interface 3: 1 IPv4 address ← Second highest IPv4 address in the subnet

Room C+D (50 PCs): 50 IPv4 addresses ← Assign each PC an IPv4 addresses higher than the Network Destination's IPv4 address

Network Destination: 1 IPv4 address ← Lowest IPv4 address in the subnet

Fig. 3.14 IPv4 gateway/router configuration network example for the subnet of rooms C and D (other IPv4 address assignment methods do exist)

- **IPv4 for Room C & D: 165.132.9.★**

★	
160	Network Destination
161~210	IP Addresses for 50 PCs
211~221	IP Addresses not used (11 reserved)
222	Gateway Interface 3 IP address
223	Broadcast IP Address

- **We needed 53 ≤ 64 = 2^6 = Subnet Size**
  Subnet mask: 255.255.255.192 ← ($256-2^6$) = 192
- **CIDR Network ID: 165.132.9.160/26**

Fig. 3.15 IPv4 addresses and subnet setup for rooms C and D

second highest IPv4 address in the subnet, which is 165.132.9.222. Among the 64 IPv4 addresses in the subnet of rooms C and D, 11 of these IPv4 addresses (i.e., from 165.132.9.211 to 165.132.9.221) are not assigned to anything and are reserved for future use. The subnet supporting rooms C and D has a CIDR network ID of 165.132.9.160/26.

In case the Hub/Switch needs to have one or two IPv4 addresses assigned (to its interface ports), then the IPv4 addresses for the Hub/Switch can be assigned from the unused (reserved) IPv4 addresses. Therefore, we can assign IPv4 addresses for the Hub/Switch using the IPv4 addresses among 165.132.9.211 to 165.132.9.221.

Or the IPv4 addresses of the gateway/router interface, PCs, and Hub/Switch can all be reassigned using the IPv4 addresses from 165.132.9.161 to 165.132.9.222. However, for the subnet supporting rooms C and D, the IPv4 address of the network destination has to be the lowest IPv4 address of the subnet (which is 165.132.9.160) and the broadcast IPv4 address has to be the highest IPv4 address of the subnet (which is 165.132.9.223).

3.2.5 Subnet Setup for Internet Connection

Based on the IPv4 network requirements assigned above, (in item 8) the Internet port connection is assigned to the IPv4 address 165.132.15.56 using a subnet mask of 255.255.255.252. In Fig. 3.8, the Internet port is connected to the IPv4 gateway/router interface 4. The subnet mask is 255.255.255.252 which is equivalent to 11111111.11111111.11111111.11111100 in binary. Since there are 2 zeros at the end of the subnet mask, this subnet has $4 = 2^2$ IPv4 addresses.

As mentioned above, other IP address assignment methods do exist. Therefore, when setting up a network subnet or when trying to make changes to an existing network or gateway/router, the exact IPv4 or IPv6 address assignment rules that have been used previously need to be checked in advance and followed.

3.3 IPv4 Routing Table Setup and Operations (Project Example)

In the previous section, the IPv4 routing table setup and operations for this project were described. We have already assigned the IPv4 network destination, subnet mask, and gateway/router interface for all subnets and PCs supporting the rooms A, B, C, D and the Internet connection. Figure 3.16 presents the IPv4 routing table configuration for the gateway/router illustrated in Fig. 3.8. The IPv4 routing table consist of the room, network destination, subnet mask, gateway/router interface, and metric value.

The ‘metric’ in the routing table indicates the associated cost (efficiency) of the route. The metric in the routing table is used in the multihop route selection process (which is explained in Chap. 4 of this book), where the minimum cost routing path is selected as the path to send IP packets towards a destination IP address. The sum of metric values (cost) of each path is computed and compared for all paths from the source IP address to the destination IP address. The routing path with the minimum cost is selected, and all packets sent from the source IP address will follow this routing path to reach the destination IP address. Minimum cost routing paths used in a network are updated either when metric value (cost) changes in the network are detected or on a periodic basis. More details on the minimum cost routing path selection process will be explained in Chap. 4 of this book.

Since the minimum cost routing path is desired, selection of routes with smaller metric values are preferred. The metric value (cost) can be based on various characteristics of the network, where network companies may use a different metric value

Room	Destination (Network Destination)	Subnet Mask (Netmask)	Gateway	Interface (GW Port)	Metric
A	165.132.9.0	255.255.255.128	165.132.9.126	1	10
B	165.132.9.128	255.255.255.224	165.132.9.158	2	10
C, D	165.132.9.160	255.255.255.192	165.132.9.222	3	1
Internet	165.132.15.56	255.255.255.252	165.132.15.58	4	10
Default Route	0.0.0.0	0.0.0.0	165.132.15.58	4	10

Fig. 3.16 IPv4 routing table configuration for the gateway/router

(cost) computation method. Some of the basic parameters used in the metric value (cost) are time delay, hop count (i.e., number of gateway/routers between source to destination), data traffic amount, link utilization, level of stability, or a combination of these values.

Any IPv4 packet that is received by this gateway/router, will have its IPv4 destination address compared to each row of the routing table, starting from the top row. If the IPv4 destination address does not belong to the subnet of the comparing row, then it will be checked if it belongs to the next row of the IPv4 routing table.

For example, the IPv4 destination address of a newly received IP packet will be checked in the following order starting from the top row, going down row-by-row of the routing table.

1. First, corresponding to the row for room A, the following status is checked for the IPv4 destination address of the newly received IPv4 packet.

   A. Is the IPv4 destination address in the subnet for room A?
      a. This needs to be answered considering that the subnet for room A includes the IPv4 addresses from 165.132.9.0 to 165.132.9.127 (the method described in Sect. 3.2.2.1 'Subnet Mask Operations' is used here).
      b. If the answer is 'yes' then the IP packet is sent to the subnet of room A through the gateway/router interface (port) 1.
      c. If the answer is 'no' then the IP packet is sent to the next row to be checked.

2. Second, corresponding to the row for room B, the following status is checked for the IPv4 destination address of the newly received IPv4 packet.

   A. Is the IPv4 destination address in the subnet for room B?
      a. This needs to be answered considering that the subnet for room B includes the IPv4 addresses from 165.132.9.128 to 165.132.9.159.

   b. If the answer is 'yes' then the IP packet is sent to the subnet of room B through the gateway/router interface (port) 2.
   c. If the answer is 'no' then the IP packet is sent to the next row to be checked.

3. Third, corresponding to the row for rooms C and D, the following status is checked for the IPv4 destination address of the newly received IPv4 packet.

   A. Is the IPv4 destination address in the subnet for rooms C and D?
      a. This needs to be answered considering that the subnet for rooms C and D includes the IPv4 addresses from 165.132.9.160 to 165.132.9.223.
      b. If the answer is 'yes' then the IP packet is sent to the subnet of rooms C and D through the gateway/router interface (port) 3.
      c. If the answer is 'no' then the IP packet is sent to the next row to be checked.

4. Fourth, corresponding to the row for the Internet, the following status is checked for the IPv4 destination address of the newly received IPv4 packet.

   A. Is the IPv4 destination address in the subnet for the Internet connection?
      a. If the answer is 'yes' then the IP packet is sent to the Internet through the gateway/router interface (port) 4.
      b. If the answer is 'no' then the IP packet is sent to the next row to be checked.

5. Fifth, if the IPv4 destination address does not belong to any of the above subnets, then it will be sent out the Default Route, which is the last row of the routing table, which will send the newly received IPv4 packet out to the Internet through the gateway/router interface (port) 4.

The last row of the routing table is the 'Default Route' row, where the network destination IPv4 address is 0.0.0.0, which is the lowest IPv4 address possible, and the subnet mask is all zeros (there are no ones in it), so it is a fully open subnet mask. As a result, the IPv4 address range of the 'Default Route' is from the IPv4 address 0.0.0.0 to 255.255.255.255, which includes all IPv4 addresses that exist. Therefore, if the IPv4 destination address of the received packet does not belong to any of the above subnets of the rooms A, B, C + D, and Internet, then it will be sent out the Default Route, which is the last row of the routing table, because all packets have an IPv4 destination address within the Default Route's IPv4 address range.

Because the last two rows of the routing table result in the same operation (i.e., the packet will be sent to the **Internet subnet through the gateway/router interface (port) 4**) and the 'Default Route' includes the subnet of the Internet connection (because the 'Default Route' includes all IPv4 addresses), the 'Internet' row is combined with the 'Default Route' row in the routing table.

As a result, the routing table in Fig. 3.17 will be used to check which subnet the newly arriving packet's destination IPv4 address belongs to. This process starts with room A, followed by room B, followed by room C and D, and if the IPv4 destination address does not belong to any of these rooms, then it will be sent out to the Internet via gateway/router interface (port) 4.

Room	Destination (Network Destination)	Subnet Mask (Netmask)	Gateway	Interface (GW Port)	Metric
A	165.132.9.0	255.255.255.128	165.132.9.126	1	10
B	165.132.9.128	255.255.255.224	165.132.9.158	2	10
C, D	165.132.9.160	255.255.255.192	165.132.9.222	3	1
Internet	0.0.0.0	0.0.0.0	165.132.15.58	4	10

Fig. 3.17 The revised IPv4 routing table configuration for the gateway/router, where the 'Internet' row was combined with the 'Default Route' row

3.4 Summary

Most of the Internet connections we make from our smartphone, laptop, or PC will use the 'Obtain an IP address automatically' option and the 'Obtain DNS server address automatically' option in which the Dynamic Host Configuration Protocol (DHCP) will help setup the network connection for us. However, knowing exactly what your smartphone, laptop, or PC uses to make an Internet connection enables you to understand the exact procedures and information needed for a device to connect to the network, which is the objective of Sect. 3.1. In addition, Sects. 3.2 and 3.3 provide the setup and operation details of the network and network equipment (i.e., gateway and routers) so you can setup or reconfigure a network properly. Further details on the IPv4, IPv6, TCP, and UDP protocol as well as routing techniques are introduced in Chap. 4 of this book. With knowledge of this chapter and Chap. 4, new techniques can be created to improve the network performance and make networks safer.

References

1. J.-M. Chung, Introduction to TCP/IP. Coursera. https://www.coursera.org/learn/tcpip
2. Wikipedia, List of assigned/8 IPv4 address blocks. https://en.wikipedia.org/wiki/List_of_assigned_/8_IPv4_address_blocks
3. IANA, IANA IPv4 Special-Purpose Address Registry. https://www.iana.org/assignments/iana-ipv4-special-registry/iana-ipv4-special-registry.xhtml
4. Internet Engineering Task Force. www.ietf.org
5. Google, Configure your network settings to use Google Public DNS. https://developers.google.com/speed/public-dns/docs/using
6. GSMAREAN.com, Samsung Galaxy A54. https://www.gsmarena.com/samsung_galaxy_a54-12070.php
7. W. Stallings, *Data and Computer Communications*, 10th edn. (Pearson, 2014)
8. L.L. Peterson, B.S. Davie, *Computer Networks: A Systems Approach*, 5th edn. (Morgan Kaufmann, 2011)

Chapter 4
TCP/IP Protocols and IP Routing

This chapter starts with an introduction of the OSI model and the TCP/IP model. In addition, the operations of TCP/IP are explained using an example of two computers on an Ethernet network transferring an application file. Next, the IPv4, IPv6, TCP, and UDP protocol descriptions are provided followed by details of route setup and routing operations [1].

In the TCP/IP standard Internet Engineering Task Force (IETF) documents, the terminology 'octet' is used to represent a set of 8 bits, which is same as a 'byte' which is commonly used. Historically, a byte was originally used to describe a character size (in bit units) to be stored in a computer's memory. So, there were several bit group sizes using the terminology of 'byte' in different documents. The most popular one was 8 bits equals 1 byte. For example, the popular American Standard Code for Information Interchange (ASCII) code uses 8 bits to represent each character (including the error detection parity bit). As the term 'octet' refers to an 8 bit character in the ITEF standards that defines IPv4, IPv6, TCP, UDP, as well as many other Internet standards, it will also be used in this book [1, 2].

4.1 OSI 7 Layer Model and TCP/IP 5 Layer Model

In this chapter, first the two representative network protocol models are compared, which are the OSI model and the TCP/IP model [2]. OSI stands for the 'Open Systems Interconnection' which is a 7 layer model that is used to characterize and standardize communication functions of a telecommunication or computing system. TCP/IP stands for the 'Transmission Control Protocol (TCP) and Internet Protocol (IP),' which is a 5 layer model that is simpler than the OSI 7 layer model, and is the most widely used network protocol in the world. A structural comparison of the network protocol architectures based on the OSI 7 layer model and the TCP/IP 5 layer model is provided in Fig. 4.1.

J.-M. Chung, *Emerging Secure Networks, Blockchains and Smart Contract Technologies*, https://doi.org/10.1007/978-3-031-65866-2_4

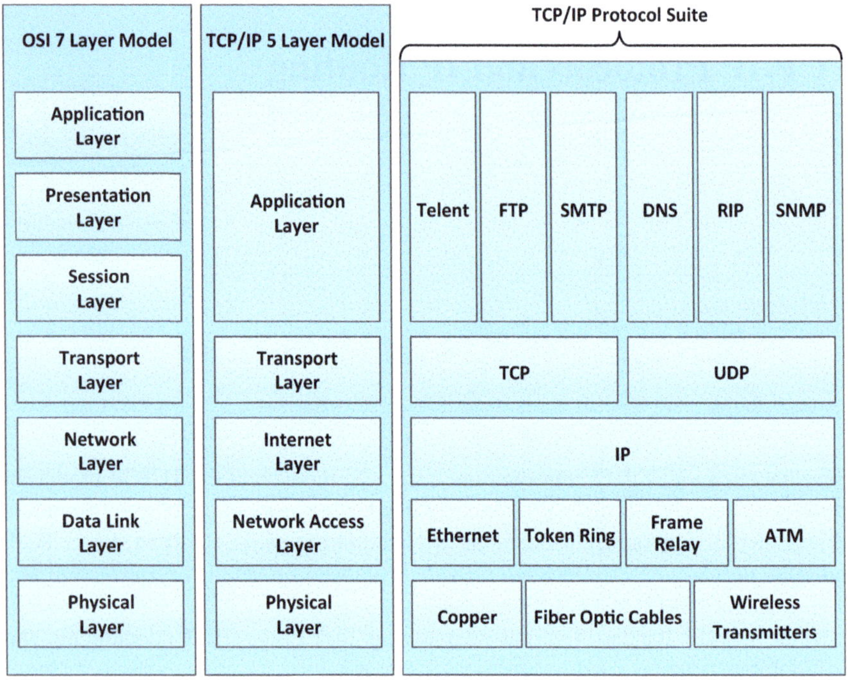

Fig. 4.1 Comparison of the network protocol architectures of the OSI 7 layer model and the TCP/IP 5 layer model

4.1.1 OSI 7 Layer Model

The OSI model was made to characterize and standardize communication functions of a telecommunication or computing system. The OSI Basic Reference Model was released in 1983. The International Organization for Standardization (ISO) 7498 standard for the OSI model was published in 1984, and later updated to the ISO/IEC 7498-1 standards, which is currently the International Telecommunication Union-Telecommunications Standardization Sector (ITU-T) standard X.200 (ISO/IEC 7498-1). A drawback of the OSI 7 layer model is that it is more complex and has some redundant network functionalities in its layers, making it less efficient than the TCP/IP model. As a result, the TCP/IP 5 layer model is considered the standard model for networks.

4.1.2 TCP/IP 5 Layer Model

The TCP/IP model was developed for the United States (US) Department of Defense (DoD) based on a packet switching network project named 'Advanced Research

Projects Agency Network (ARPANET)' which was also called the DoD model. ARPANET planning initiated in 1968, and was declared as the US DoD's standard for all military computer networking in 1982. ARPANET was decommissioned in 1990, where its upgraded standards are currently maintained by the IETF [2, 3].

As presented in Fig. 4.1, the **Application Layer** of the TCP/IP protocol suite is where various applications are controlled (e.g., Telnet, FTP, SMTP, DNS, RIP, Simple Network Management Protocol (SNMP), etc.). The **Transport Layer** is in charge of providing data segments to the Application Layer. Since there are many types of applications a computer may be using at the same time, the Port ID is used to distinguish exactly what application the data segment should be delivered to. The Transport Layer has two representative protocol types that are frequently used for this purpose, which are TCP and UDP, where TCP controls the data segment transfer flow speed and retransmission of missing segments, so it is more complicated and reliable compared to UDP. The **Internet Layer** has two options, IPv4 and IPv6. Because the Internet is a combination of various types of networks that use different protocols, the Internet layer forms IP packets and provides end-to-end routing support through the Internet. The **Network Access Layer** (a.k.a. **Data Link Layer**) changes based on network type, which includes Ethernet, Token Ring, Frame Relay, ATM, Synchronous Optical Network (SONET), Synchronous Digital Hierarchy (SDH), Wi-Fi, MPLS, as well as mobile communication protocols such as 3G, 4G LTE, 5G, and 6G [2]. Based on the Network Access Layer type, the **Physical Layer** will be decided. For example, Ethernet and Token Ring will send electrical signals over a copper wire, SONET and SDH will send optical signals over fiber optical cables, and Wi-Fi and mobile communication will send RF signals through wireless transmitters.

4.1.3 TCP/IP Network Operations

In the following, the TCP/IP network operation is described through an application file transfer example, presented in Fig. 4.2. In this example, Application X (which we will call APP X for short) running on computer Host 1 (H1) created an application data file that needs to be sent to the Application X (APP X) running on computer Host 5 (H5) through the Internet using the TCP/IP protocol. Computers H1 and H5 are connected to an Ethernet network where routers R1, R2, and R3 are on the routing path between the computers H1 and H5. Above the Ethernet layer lies the IP layer, TCP layer, and the Application layer is located at the top of both computers H1 and H5. In this example, APP X is operating within the Application layer of computers H1 and H5.

Figure 4.2 presents the TCP/IP networking operation details to enable APP X file transfer from computer H1 to computer H5. In this example, it is assumed that the Ethernet network is large and divided into multiple subnets to enhance the efficiency and scalability of the network. There are three Ethernet switches on the routing path from H1 to H5, where these three Ethernet switches have IP router functionalities inside.

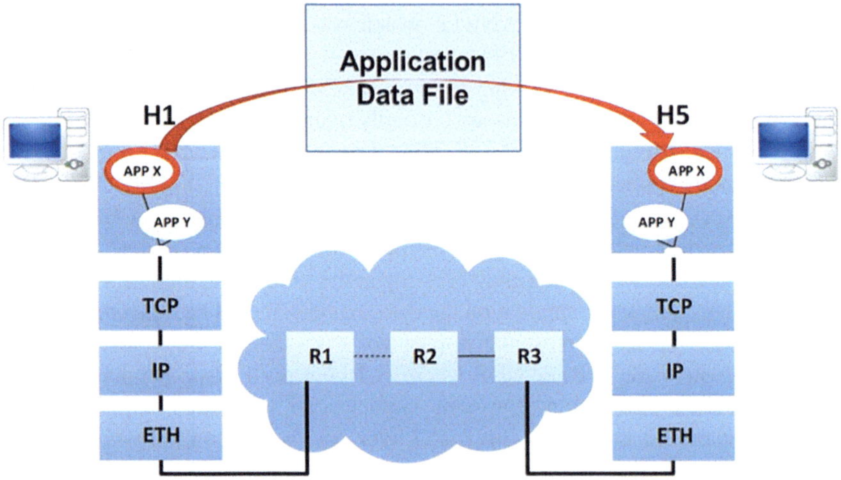

Fig. 4.2 TCP/IP network application file transfer example

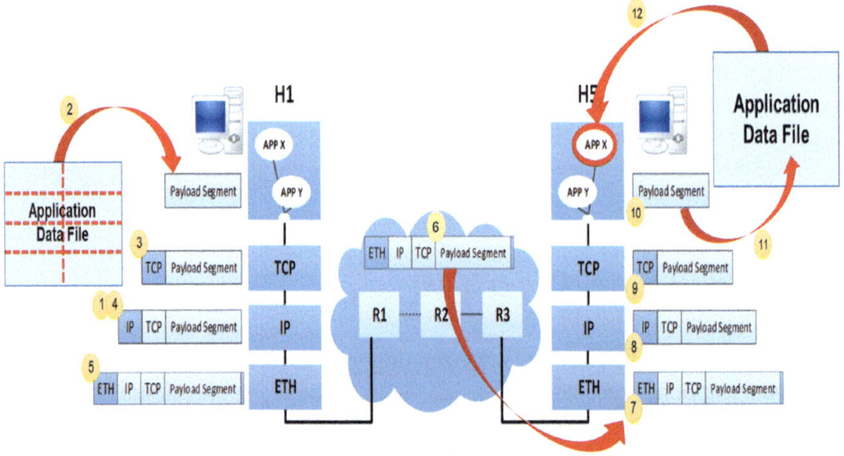

Fig. 4.3 TCP/IP networking operation supporting the APP X application file transfer from computer H1 to computer H5

Figure 4.3 presents the numbered sequence of the TCP/IP protocol networking procedures used by H1 and H5 as well as the routers R1, R2, and R3, which are each in Ethernet switches in a large Ethernet Local Area Network (LAN). Details of the TCP/IP protocol networking procedure sequence are described below.

① The APP X application data file needs to be divided into segments because the segments must fit into the payload area of the Ethernet frame that will be used to deliver the segment of the APP X file to H5. The maximum size a divided segment can be is based on the Ethernet Maximum Transfer Unit (MTU). Ethernet's MTU is 1,500 octets. In this Ethernet MTU, an IP packet must fit inside. An IP packet will include an IPv4 or IPv6 header followed by a TCP or UDP header followed by the segment of the APP X data. When no options are added, the IPv4 header size is 20 octets, the IPv6 header size is 40 octets, the TCP header size is 20 octets, and the UDP header size is 8 octets. So, (with no options added) if an IPv4 packet with TCP is used then the segment size would be 1460 = (1500 – 20 – 20) octets, if an IPv4 packet with UDP is used then the segment size would be 1472 = (1500 – 20 – 8) octets, if an IPv6 packet with TCP is used then the segment size would be 1440 = (1500 – 40 – 20) octets, and if an IPv6 packet with UDP is used then the segment size would be 1452 = (1500 – 40 – 8) octets. If options are added to the IPv4, IPv6, or TCP header, then the APP X segment would need to be that much smaller. More details on the Ethernet MTU are provided in the following Sect. 4.1.4 and Fig. 4.4.

② Using the MTU defined in step ①, the APP X data file will be divided into segments of the MTU size and the last segment may be smaller [1, 4].

③ A TCP header is added to the segment to help control the data transmission speed (i.e., flow control using Window size adaptation) and assist erroneous data segment recovery (i.e., error control using repeated acknowledgments (ACKs)). The TCP session of segments transfer (flow speed and error recovery) is controlled end-to-end by H1 and H5 through the TCP session operation. The TCP session operation is setup between computers H1 and H5 using a TCP three-way handshake. The TCP header attached in front of each segment includes a sequence number to make sure the segments are combined in the correct order (at the destination H5) to construct the original APP X data file properly. More details on TCP (including the three-way handshake) are provided in the following Sect. 4.5.

④ An IP header is added in front of the TCP header to form an IP packet. The IP header needs to be prepared considering the version of the network, which can be either IPv4 or IPv6. IPv4 and IPv6 have different IP header formats, in which details are explained in the following section. IP packet headers include a Source IP address and a Destination IP address, in which IPv4 addresses are 32 bits and IPv6 addresses are 128 bits. More details on IPv4 and IPv6 are provided in the following section.

⑤ An Ethernet header and trailer are respectively added to the front and back of the IP packet to form an Ethernet frame, in which details are explained in the following section and in Fig. 4.4. The Ethernet header and trailer are used by Ethernet Switches in the Ethernet network to help the flow control and error control. Flow control is the act of controlling the speed of frames that are transmitted (i.e., controlling the Ethernet frames/s sent). Error control is the act of detecting if a frame has an error, and if so, then the erroneous frame will be retransmitted. Other protocols like Wi-Fi use their own protocol header and trailer to form a frame (e.g., Fig. 4.5), which are different from the Ethernet frame.

⑥ There are three Ethernet switches on the routing path from H1 to H5, where each of these three Ethernet switches have an IP router inside. Each Ethernet switch

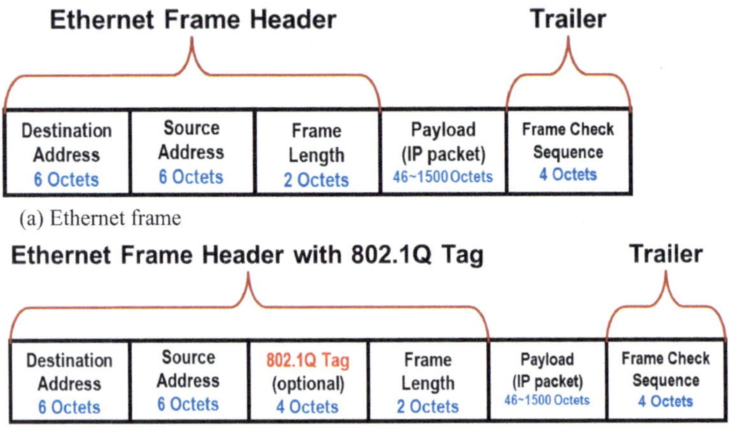

(a) Ethernet frame

(b) Ethernet frame with a 802.1Q tag

Fig. 4.4 Ethernet frame structure based on the IEEE 802.3 LAN standards

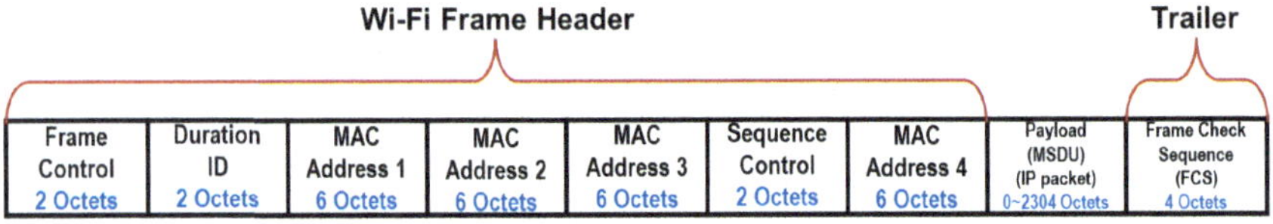

Fig. 4.5 Wi-Fi frame structure based on the IEEE 802.11 WLAN standards

will check the Ethernet frame's header and trailer to detect errors in the Ethernet frame. Using the Frame Check Sequence (FCS) field in Fig. 4.4, if an error is detected, then the Ethernet frame is discarded, and a frame retransmission is requested. If there are no errors in the Ethernet frame, then the Ethernet frame's header and trailer will be removed to reveal the IP packet. The destination IP address in the IP packet header will be read. The three IP routers R1, R2, and R3 use the destination IP address in the IP packet to route the IP packet to the destination IP address, which is the IP address of computer H5. As described in Chap. 3, routers (and gateways) use routing tables to determine where to send each IP packet. Then this IP packet will have an Ethernet header and tailer added to form an Ethernet frame, which will be sent out to the network. This process is repeated by R1, R2, and R3 until the Ethernet frame arrives at the H5 destination.

⑦ The last router R3 on the routing path delivers the Ethernet frame to computer H5. After the Network Interface Card (NIC) of computer H5 receives this frame, the Ethernet header and trailer are used to detect errors in the Ethernet frame. If an error is detected, then the Ethernet frame is discarded, and a frame retransmission will be requested to R3.

⑧ When H5 receives an errorless Ethernet frame from R3, the Ethernet header and trailer will be removed to reveal the IP packet. Using the IP packet header, several

IP network control functions can be conducted (e.g., IPv4 Explicit Congestion Notification (ECN)) to support the network.

⑨ After the IP operations are completed, the IP header is removed to reveal the TCP header. The segment sequence number in the TCP header is checked, and if there are no errors then the TCP header is removed and the segment of data is saved in the memory of the H5 computer, corresponding to a location based on the TCP segment's sequence number. In addition, the control information in the TCP header will be used to control the transmission Window size to increase or decrease the data transfer rate. If a segment is missing (or if segments arrive not following the proper sequence number order) then duplicated acknowledgments (ACKs) are sent back to H1 informing H1 of the missing segment's sequence number. If three of the same ACKs are received at H1, then H1 will retransmit the missing segment to H5. The TCP header also includes the Port address that connects the received segment of data to H5's APP X.

⑩ After the TCP operation is completed, the segment of data will be sent to the Port address in the TCP header, which is an application specific location in the memory of computer H5 to store data segments for APP X.

⑪ In the application specific location in the memory of computer H5, the data segments for APP X will be reconstructed as more and more segments of APP X are received.

⑫ When all segments that computer H1 sent are received at H5 and the application data file for APP X is fully reconstructed, then it is sent to the APP X processor to be used for APP X running on computer H5.

Through these procedures, APP X on computer H1 has successfully transferred its data file to computer H5, which can now be used by APP X in H5.

4.1.4 Ethernet and Wi-Fi Frames and MTU

Figure 4.4 presents the Ethernet frame structure based on the IEEE 802.3 Local Area Network (LAN) standards. The MTU is defined as the maximum size of an IP packet that can be transmitted without fragmentation over a medium. The MTU of an Ethernet (IEEE 802.3 standard) frame is 1,500 octets, meaning that the largest IP packet size that can be included in an Ethernet frame is 1,500 octets, which is the maximum size of the 'Payload' field of an Ethernet frame, as presented in Fig. 4.4. So, the data segment size (of APP X) has to be determined considering the IP header and TCP header size.

When forming an Ethernet frame (based on the IEEE 802.3 standards) the Ethernet header and trailer have to be added respectively to the front and back of the IP packet as presented in Fig. 4.4. The Ethernet frame consists of a Destination Address field (6 octets), Source Address field (6 octets), Frame Length field (2 octets), Payload Medium Access Control (MAC) Service Data Unit (MSDU) field (46–1500 octets, the IP packet goes in here), and Frame Check Sequence (FCS) field (4 octets).

When an Ethernet frame is transmitted, a Preamble (7 octets) and Start Frame Delimiter (SFD) field (1 octet) are added to the front of the frame to help the detection and synchronization of the frame at the receiving system. The bit pattern of the Preamble (7 octets) is a sequence of 10101010.... for 56 bits and the SFD is 10101011 to indicate that the Ethernet frame will start (using the 11 at the end). The bit pattern of the Preamble and SFD were designed to help frame detection and bit synchronization as much as possible, considering Ethernet's asynchronous communication characteristics.

Adding the Ethernet header and trailer to an IP packet adds on a total of 18 octets when forming an Ethernet frame, as presented in Fig. 4.4(a). If the 4 octet IEEE 802.1Q tag is included in the header of the Ethernet frame, then the Ethernet header becomes 4 octets longer, as presented in Fig. 4.4(b). The IEEE 802.1Q tag is used to assist Virtual Local Area Network (VLAN) operations within the Ethernet network, which supports enhanced network connection services. A VLAN is a logically formed subnetwork inside a larger network that has a group of devices specially configured such that the devices operate like a coordinated isolated network.

It is important to note that the MTU is different for other protocols. For example, Wi-Fi frames are based on the IEEE 802.11 Wireless Local Area Network (WLAN) standards. As presented in Fig. 4.5, Wi-Fi frames have a MTU of 2304 octets, which is called the maximum size Medium Access Control (MAC) Service Data Unit (MSDU), based on the IEEE 802.11 WLAN standards. The Wi-Fi frame is formed by adding the header and trailer to the MSDU. The Wi-Fi frame is called the MAC Protocol Data Unit (MPDU), which includes the Frame Control (FC) field (2 octets), Duration ID field (2 octets), four Address fields (6 octets each), Sequence Control field (2 octets), MSDU field (0–2312 octets, the IP packet goes in here), and Frame Check Sequence (FCS) field (4 octets).

4.2 TCP/IP 5 Layers

In this section, the operations of the TCP/IP layer stack are explained through an example. As presented in Fig. 4.6, the TCP/IP protocol has 5 layers.

Layer 1 is the Physical (PHY) layer that deals with the actual frame signal transmission and reception through the network.

Layer 2 is the Network Access Layer (NAL) that forms the frame to transmit and conduct flow and error control of the frames. The layer 2 NAL protocol commonly becomes the name of the network type, such as, Ethernet network, Wi-Fi network, LTE network, 5G network, 6G network, satellite network, etc.

Since there are many NAL protocols, a higher-level common protocol needs to be used to internetwork, which is the operation of being able to pass through different (heterogeneous) layer 2 NAL networks. This is the role of the **layer 3** IP protocols (i.e., IPv4 and IPv6 protocols). IP addresses are globally recognized unique addresses, which are used to identify a sender (i.e., source IP address) and a receiver (i.e., destination IP address) in the Internet. IP packets are also used as Internet Control

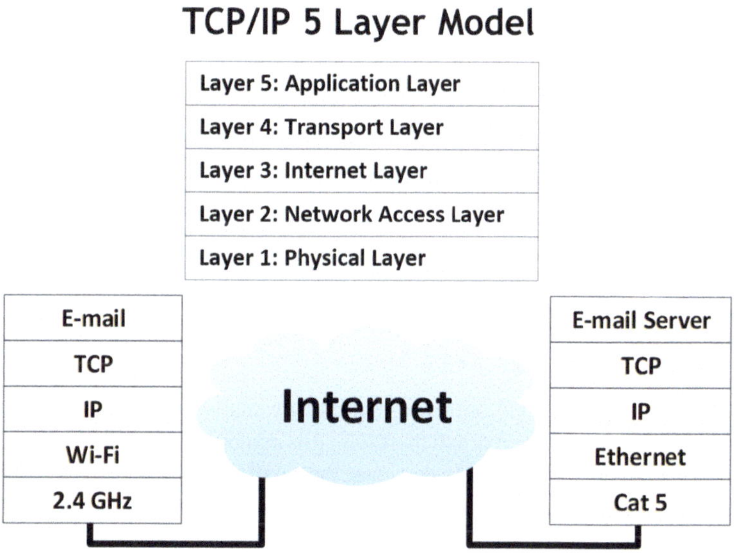

Fig. 4.6 Example to describe the operations of the TCP/IP layers

Message Protocol (ICMP) packets, which help control and manage the functionality of network devices, routers, and gateways.

In a Local Area Network (LAN) or a Wireless LAN (WLAN), the layer 2 NAL is divided into two sublayers called the Logical Link Control (LLC) layer and the Medium Access Control (MAC) layer. The LLC layer is above the MAC layer, where both are sublayers of the NAL layer in the TCP/IP 5 layer model. The LLC layer provides support of the logical connection (e.g., multiplexing) as well as frame loss detection and retransmission operations. The MAC layer is in charge of the frame formation, source and destination MAC addresses, as well as the frame transmission timing and control.

The **4th layer** conducts transport operations commonly supported by either TCP or UDP. When TCP is used, the layer 4 operation adds a TCP header to the data segment to control the data transmission flow speed and conduct data segment retransmissions when a segment does not properly arrive at the destination. The TCP session is controlled by the end devices only (e.g., computers H1 and H5 in Fig. 4.2), where the end devices setup the TCP session using a TCP three-way handshake (described in Fig. 4.13). Instead of TCP, layer 4 can use User Datagram Protocol (UDP) instead. UDP is much simpler compared to TCP as UDP does not control the data transmission speed (flow) nor does it assist missing data segment retransmission. UDP simply provides Port information once the IP packet is delivered to the proper destination so the proper application (e.g., APP X in H5 of Fig. 4.2) can receive the data file. Due to the simplified operations of UDP, it does not need a three-way handshake to be setup. More details on TCP and UDP will be provided in the following sections.

The highest **5th layer** of the TCP/IP protocol stack is the Application (APP) layer. This is the layer that the applications (e.g., APP X in Fig. 4.2) are executed in.

4.2.1 Analogy of the TCP/IP Layer Operations

Figure 4.7 presents an analogy of mailing books to the TCP/IP layer operations. This example is presented because the TCP/IP 5 layer operation was modeled after the common way we mail packages to each other.

Layer 5: Application Layer (APP)

Assume you want to send 7 books to a friend by mail. This can be seen as the application you want to achieve, and therefore, this can be considered as an operation of the application layer (layer 5). In this case, the set of books that needs to be mailed are like a data file that needs to be sent over the network.

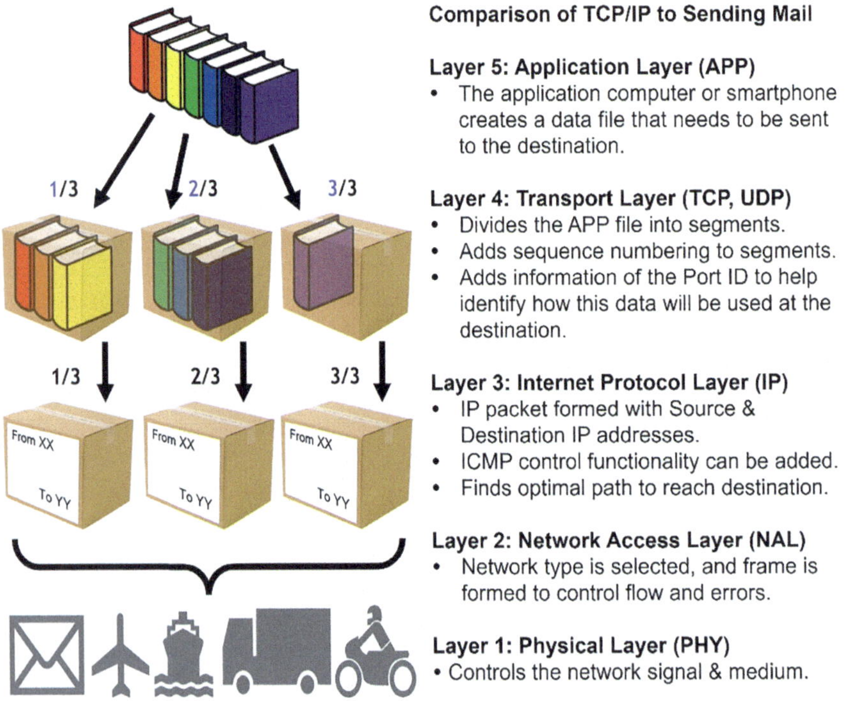

Fig. 4.7 Analogy of mailing books to the TCP/IP layer operations

Layer 4: Transport Layer (TCP, UDP)

Since there are 7 books to mail, you would need to divide the books into groups so they can be put into boxes, in which the boxes would be numbered like 'Box 1 out of 3' and 'Box 2 out of 3' and 'Box 3 out of 3.' The grouping of books is like dividing the data file into segments such that they fit in to the frame payload field (i.e., the MSDU field) to be sent. The segment size needs to consider the layer 2 protocol frame size as well as the TCP header and IP header size. Likewise, when deciding how many books to include in a box depends on the box size as well as the amount of protection wrapping you plan to use in the box to protect the books.

The numbering of the boxes helps to identify if a box is missing and informs the receiver how many boxes will arrive. In addition, the box numbering helps in reordering the boxes at the destination. This is what the TCP layer (layer 4) does, as the box numbering is what the TCP sequence numbers were modeled after.

In this example, 3 books fit into a box. So, three boxes will be used to send the 7 books, where 3 books will be put into the first box, the next 3 books will be put into the second box, and the remaining 1 book will be put into the third box. For the third box a smaller box can be used since there is only one book to include. Likewise, the last data segment may be smaller compared to the former segments, resulting in a smaller IP packet.

Layer 3: Internet Protocol (IP) Layer

On each box, the mailing address of the sender and receiver needs to be recorded. In addition, the postal office will decide what route to deliver your mailed boxes. This is what the IP layer (layer 3) does for IP packets. The sender's mailing address written on each box is equivalent to the source IP address. The receiver's mailing address written on each box is equivalent to the destination IP address. The sender and receiver mailing addresses on the boxes have to be globally unique addresses for proper mail delivery. Likewise, the source IP address and the receiver IP address in an IP packet have to be globally unique IP addresses (the address pair will be either IPv4 or IPv6 based).

The postal office will decide the most efficient and reliable way to send the boxes. Likewise, the IP routers/gateways will use the metric values in the routing tables to decide the best way to send the IP packets to the destination IP address across the Internet. Details of the optimal (least cost) routing path setup are described in the following Sect. 4.7.

In addition, mailing addresses are divided into regions like, nation, state, city, district, and local address. Likewise, the Internet is divided into subnets based on nation, state, city, district, and local address. A district's subnet will include the IP addresses for local computers. A city subnet will include multiple district subnets. A state subnet will include the subnets of several cities and surrounding rural regions. A nation's subnet will include the subnet of all states in the nation.

Layer 2: Network Access Layer (NAL)

Based on available options, the postal office will mail your boxes by air mail, surface mail, priority mail, etc. Likewise, based on the available options of networks, a network type will be selected. For example, when you send an email from your smartphone, you can use Wi-Fi, Bluetooth, 3G, 4G LTE, 5G, which are your smartphone's NAL (layer 2) options.

Layer 1: Physical Layer (PHY)

Based on the mail type (e.g., air mail, surface mail, priority mail, etc.) the physical vehicle to transport your mailed boxes will be different (e.g., airplane, ship, truck, bicycle, motorcycle, walking, etc.). Likewise, based on what NAL is selected to be used, the transmission signal type, signal frequency, signal bandwidth, frame format will be different.

In the following sections, details of the IPv4, IPv6, TCP, and UDP protocols are provided, followed by the routing technologies in the next section.

4.3 IPv4 Protocol

The IP protocol was designed to be a connectionless network protocol equipped as a 'datagram.' The term 'datagram' is defined in the Internet standards RFC 1594 as "A self-contained, independent entity of data carrying sufficient information to be routed from the source to the destination computer without reliance on earlier exchanges between this source and destination computer and the transporting network" [5]. These design objectives were applied to make the IP protocol networking very reliable and robust, good enough to be used in war situations, as it was created to be used by the U.S. Department of Defense (DoD). A connectionless network protocol is a protocol that does not require end-to-end network connection setup and control when it is used.

During 1973–1978, the Internet Protocol version 1 (IPv1), version 2 (IPv2), and version 3 (IPv3) were designed and used in various network experiments. These network experiments and protocol models resulted in the Internet Protocol version 4 (IPv4). In 1982, IPv4 was first used on the Atlantic Packet Satellite Network (SATNET). In 1983, IPv4 was used to support the Advanced Research Projects Agency Network (ARPANET), which is the world's first Packet-Switched (PS) Wide-Area Network (WAN) that was created by the Advanced Research Projects Agency (ARPA) of the US DoD. ARPANET was the world's first network that used TCP/IPv4 to enable computers to communicate, which is the origin of the Internet [6].

IPv4 was initially defined in the Internet Engineering Task Force (IETF) document RFC 760, which was released in January of 1980. In September of 1981, RFC 791 was released (replacing RFC 760) to include the new upgrades of the IPv4 standards. In March of 1982, the US DoD decided to use IPv4 for the TCP/IP suite to support all computer data communication over their military networks.

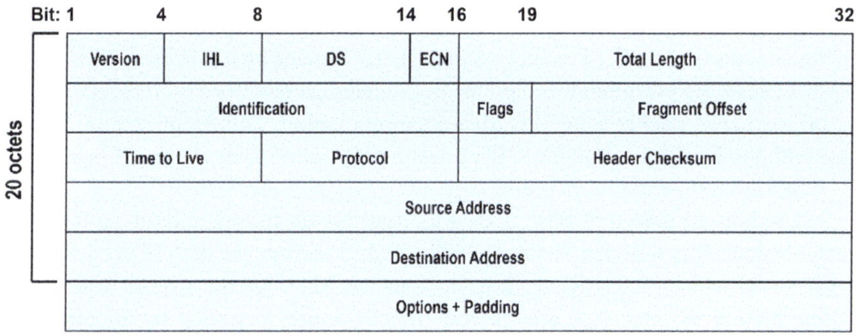

Fig. 4.8 IPv4 packet header structure

The IPv4 packet can be divided into a header and payload part. The IPv4 essential functions are inter-networking and routing, where the IPv4 packet header contains all required functionalities to deliver an IPv4 packet to its destination. This is why IP4 (and IPv6) packets are called datagrams. The IPv4 payload part includes the TCP or UDP header and the data segment payload.

The IPv4 packet header structure is presented in Fig. 4.8. Details of the IPv4 packet header are described in the following.

Version field (4 bits) is used to distinguish different IP packet versions, where the IPv4 packets will have the version value 4, which is 0100 in binary. Currently, in the Internet, only IPv4 and IPv6 are used.

Internet Header Length (IHL) field (4 bits) is used to represent the length of an IPv4 header in units of 32 bit words (i.e., 4 octet words). The minimum IHL value is five, so the minimum header length is 20 octets (since $5 \times 4 = 20$ octets).

Differentiated Services (DS) field (6 bits) is used to distinguish service priority assignments.

Explicit Congestion Notification (ECN) field (2 bits) is used for explicit signaling of congestion in the network to slow down the speed of packets being transmitted when delay or congestion is detected. The DS and ECN bits are used together to control and manage data services in a network.

Total Length field (16 bits) represents the total length of the IPv4 packet in units of octets. The Total Length field includes the whole packet including the IPv4 packet header and payload. This is obtained by using the 16 bits in the Total Length field in the exponent of 2. The maximum IPv4 packet length would be obtained when all 16 bits of the Total Length field are ones (i.e., 1111111111111111), which is equal to 65,535 in decimal, since $(2^{16} - 1) = 65{,}535$. Therefore, the maximum IPv4 packet length is 65,535 octets, which is equivalent to 524,280 bits. However, as described above in Sect. 4.1.4, the maximum IP packet size is based on the layer 2 (NAL) frame's MTU size. For example, based on the IEEE 802.3 standards for Ethernet, the largest IP packet in an Ethernet frame is 1,500 octets (presented in Fig. 4.4). In addition, based on the IEEE 802.11 standards for WLANs, the largest IP packet in a Wi-Fi frame is 2,304 octets (presented in Fig. 4.5).

Identification field (16 bits) is a sequence number used to uniquely identify fragmented parts of an IPv4 packet when packet fragmenting has been applied. So, if an IPv4 packet was fragmented, then the fragmented parts will be used to reconstruct the original IPv4 packet using this Identification number. This Identification field is used with the IPv4 source address, IPv4 destination address, and protocol field in the IPv4 packet reconstruction process.

It is worthy to note that IPv6 does not allow packet fragmentation, so the IPv6 packet header does not include this field. Instead, IPv6 uses the path MTU discovery process (defined in IETF RFC 8201) to find the MTU of the overall end-to-end routing path in advance, and uses it from the beginning, so packet fragmentation is not needed [1, 7, 8].

Flags field (3 bits) consists of three bits (i.e., Bit 0, Bit 1, and Bit 2) in which each flag bit is used to control a specific functionality. Only two (Bit 1 and Bit 2) of the three bits are used (based on the RFC 791 standards).

Bit 0 is reserved (i.e., not used) and is always set to 0.

Bit 1 is used to indicate 'Don't Fragment (DF)' in which if the DF flag is set (i.e., DF bit equals 1) then this IPv4 packet should not be fragmented. However, if the DF flag is clear (i.e., DF bit equals 0) then this IPv4 packet can be fragmented. For a flag bit, the term 'clear' and 'set' are used in the standards. Fragmenting is used when a part of the network uses a smaller frame size with a smaller MTU, so the IPv4 packet needs to be divided (fragmented) to fit into multiple smaller NAL (layer 2) frames. In case frame fragmenting is needed but the DF flag is set, then the IPv4 packet cannot be fragmented, so an alternate route to the destination needs to be used (to avoid passing through the subnetwork that uses the smaller MTU), or the sending computer needs to create smaller IPv4 packets, so fragmenting would not be needed when passing the subnetwork that uses the smaller MTU.

Bit 2 is used to indicate 'More Fragments (MF)' when fragmenting is permitted (i.e., when Bit 1 equals 0). If the MF flag is set (i.e., MF bit equals 1), then there are more fragments to this IPv4 packet following. However, if the MF flag is clear (i.e., MF bit equals 0), then this is the last fragment of a fragmented IPv4 packet. If fragmenting is not permitted (i.e., when Bit 1 equals 1), then the MF bit will be zero.

Fragment Offset field (13 bits) is used to indicate where this fragment belongs to (in 64 bit word units) when reconstructing the IPv4 packet. Fragment Offset is only used when fragmenting is permitted (i.e., when flag Bit 1 equals 0). If fragmenting is not permitted (i.e., when flag Bit 1 equals 1), then this field will be all zeros. Fragments (excluding the last fragment) must contain a data field that is a multiple of 64 bits in length.

Time to Live (TTL) field (8 bits) is used to specify the time length (in seconds) that a datagram is allowed to remain in the Internet. Every router that the IP packet passes through should decrease the TTL by at least one. When the TTL reaches 0, the IPv4 packet should be discarded (regardless of its arrival at the destination IPv4 address or not). TTL is used to discard old outdated IPv4 packets from showing up too late, which helps to prevent confusion at the destination computer. TTL is similar to the hop count used in IPv6 packets.

Protocol field (8 bits) is used to identify the type of the next header that directly follows the IPv4 header, which is the 'Options + Padding' field shown in Fig. 4.8. Table 4.1 presents a list of the representative Protocol IDs used in the IPv4 packet header.

Header Checksum field (16 bits) is an error detection code added to protect the IPv4 packet header from errors. The header checksum does not detect errors in the IPv4 packet payload part (i.e., the TCP header and data segment are not protected by the IPv4 Header Checksum). Using the header checksum, errors in the IPv4 packet header are checked at each router. Because information in the IPv4 packet header can change at each router (i.e., the TTL, flags, fragmentation offset, etc. can change) the header checksum value needs to be updated at each router before the IPv4 packet is sent out.

Source and Destination IPv4 addresses are each 32 bits. In 1993, the Internet Engineering Task Force (IETF) switched from using Classful IPv4 addresses (i.e., Class A, B, C, D, E) to Classless Inter-Domain Routing (CIDR) IPv4 addresses. CIDR uses Variable-Length Subnet Masking (VLSM) and is used in all IPv4 and IPv6 address assignments and subnet masks. The Internet Assigned Numbers Authority (IANA) assigns Regional Internet Registry (RIR) CIDR blocks to countries or territories (e.g., Europe, North America, etc.). RIRs are subdivided into Local Internet Registry (LIR) subnets, which are further subdivided into subnets to fit local network sizes. All source and destination IPv4 addresses are assigned within the range of the smallest local subnet, which is a small part of the LIR subnet.

Options field is a variable length field that includes the options requested by the sending host computer (e.g., computer H1 in Fig. 4.2). Since the options field is a variable length field, padding may be used to make sure that the IPv4 packet header is a multiple of 32 bit word units in length. Table 4.1 provides a list of representative option protocol IDs used in the IPv4 packet header. There are many more options (and corresponding protocol IDs) that support various IPv4 networking functionalities.

4.4 IPv6 Protocol

In 1979, the Internet Stream Protocol (ST) was first defined in Internet Experiment Note IEN-119. The ST protocol was later upgraded to ST-II defined in RFC 1190 and upgraded again to ST2+ in the RFC 1819 IETF standards, which were to become IPv5. This Internet ST protocol was never used publicly, which is why there are no IPv5 packets used in the Internet. However, the design concept and objectives of the ST protocol influenced later protocols, such as, Asynchronous Transfer Mode (ATM), Multiprotocol Label Switching (MPLS), and Voice over IP (VoIP) [9].

Internet Protocol version 6 (IPv6) was designed to become a successor to the IPv4 protocol. IPv6 was designed to solve several of the limitations of the IPv4 protocol, in which having a larger address pool was one of the main objectives, as there was a lack of IPv4 addresses, as the number of Internet users had rapidly grown. The IPv4 addresses were 32 bits, and therefore less than 4.3 billion IPv4 addresses were

Table 4.1 List of representative protocol IDs used in the IPv4 packet header

Protocol Number		Protocol Name	Abbreviation
1	00000001	Internet Control Message Protocol	ICMP
2	00000010	Internet Group Management Protocol	IGMP
6	00000110	Transmission Control Protocol	TCP
17	00010001	User Datagram Protocol	UDP
41	00101001	IPv6 encapsulation	ENCAP
89	01011001	Open Shortest Path First	OSPF
132	10000100	Stream Control Transmission Protocol	SCTP

available, in which many addresses were not used because subnets had to be large enough to include all IPv4 addresses needed, while only being able to take sizes based on 2^{X} in which **X** is a positive integer (i.e., 2, 4, 8, 16, 32, 64, 128, 256, 512, 1024, 2048, 4096, etc.). Due to the limitations in subnet sizes, many subnets have unused IPv4 addresses. In addition, many advanced devices (e.g., smartphones) have multiple Internet access modems, where each network connection interface needs to have an independent IP address assigned. As the number of devices (e.g., smartphones) that use multiple IP addresses exponentially grew, and the number of Internet users significantly grew, and considering the large waste of leftover unused IPv4 addresses in subnets (as described in Sect. 3.2.2), a serious lack of IPv4 addresses occurred [10].

To solve this problem, IPv6 was designed to have 128 bit addresses. Since 2^{128} is larger than 3.4×10^{38}, the number of IPv6 addresses are truly abundant and there is little worry of running out of IPv6 addresses in the near future. However, because IPv6 address are 128 bits, a more efficient way to write and read binary numbers had to be used. Therefore, instead of using binary to decimal representations, IPv6 uses hexadecimal representations.

In binary to hexadecimal conversions, 4 bits are converted into a single hexadecimal digit using the mapping presented in Fig. 4.9. Due to this mapping structure, 128 bit IPv6 addresses can be represented with 32 hexadecimal digits.

Hexadecimal digits always use an '0x' in front to avoid confusion with binary or decimal numbers, which are the most common numbering mechanisms. The hexadecimal digits are 0, 1, 2, 3, 4, 5, 6, 7, 8, 9, a, b, c, d, e, and f, which are a total of 16 characters to match the 4 bits in binary form, as presented in Fig. 4.9.

Binary = Hexa = Decimal	Binary = Hexa = Decimal
0000 = 0x0 = 0	1000 = 0x8 = 8
0001 = 0x1 = 1	1001 = 0x9 = 9
0010 = 0x2 = 2	1010 = 0xa =10
0011 = 0x3 = 3	1011 = 0xb =11
0100 = 0x4 = 4	1100 = 0xc =12
0101 = 0x5 = 5	1101 = 0xd =13
0110 = 0x6 = 6	1110 = 0xe =14
0111 = 0x7 = 7	1111 = 0xf =15

Fig. 4.9 Binary to hexadecimal to decimal conversion

Figure 4.10 presents the IPv6 packet header structure. The IPv6 packet is a datagram with a header that contains all required functionalities to deliver an IPv6 packet to its destination.

In the following, the fields (and corresponding functionalities) in the IPv6 packet header are described [1, 2, 8, 11–14].

Version field (4 bits) is used to distinguish different IP packet versions, in which IPv6 packets have the version value of 6, which is '0110' in binary. Currently in the Internet only IPv4 and IPv6 packets are used.

Differentiated Services (DS) field (6 bits) is used to distinguish service priority assignments.

Explicit Congestion Notification (ECN) field (2 bits) is used for explicit signaling of congestion in the network. ECN can be used to slow down the speed of packets being transmitted when congestion is detected.

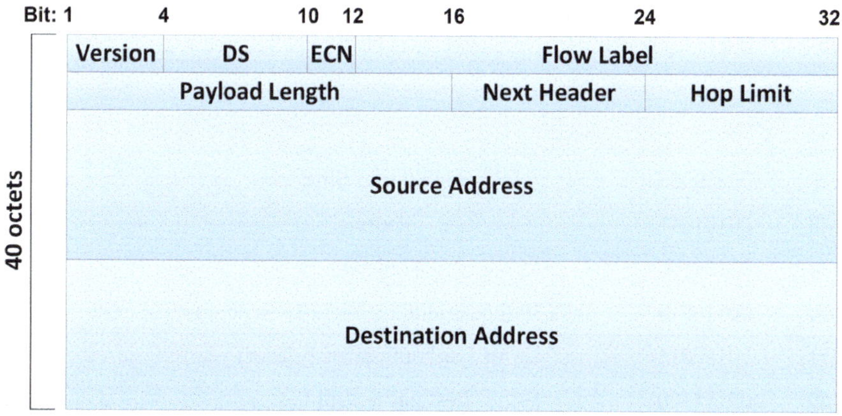

Fig. 4.10 IPv6 packet header structure

Flow Label field (20 bits) was originally created to provide special service support for real-time applications. However, the flow label is currently used to inform routers and switches (with multiple outbound paths) that routing path changes should not be applied to these packets, because these packets will not be reordered at the receiver.

Payload Length field (16 bits) is used to represent the size of the payload in octets, which includes extension headers. The maximum IPv6 packet payload size is 65,535 octets (since $(2^{16} - 1) = 65{,}535$).

For networks that have very low error rates and high throughput, an IPv6 **Jumbogram** can be used, which is an IPv6 packet with a payload size larger than 65,535 octets. An IPv6 Jumbogram can have a payload size up to $4{,}294{,}967{,}295 = (2^{32} - 1)$ octets, which is nearly a 4.3 Giga Byte size single IPv6 packet. When an IPv6 Jumbogram is used the Payload Length field is set to all zeros and an IPv6 Jumbo Payload Option extension header has to be used (attached to the IPv6 header). IPv6 Jumbo packets provide enhanced performance over networks that can support Jumbograms.

Next Header field (8 bits) specifies the type of the next header. The next header can be a TCP or UDP header, or an extension header (attached to the IPv6 header). Multiple extension headers can be added to the IPv6 header, where each extension header has a different format, but most follow the Type, Length, Value (TLV) format in which padding bits may be used to satisfy the basic header size. Table 4.2 presents some of the popular IPv6 next headers, where many more exist.

Hop Limit (HL) field (8 bits) is set when the IPv6 packet is transmitted. The HL is decreased by one at each intermediate router/gateway that the IPv6 packet passes through, where the IPv6 packet is discarded when the HL value reaches 0. The HL value will be set to be sufficient to reach the destination IP address. But the HL should not be set to be too large. This will help prevent a packet from staying alive in the network for too long, in case routing or delay problems occur and to prevent looping issues.

Looping is a phenomenon where a packet is routed in a cyclic loop, returning to the same router after being routed away. This can occur due to errors in the routing

Table 4.2 List of representative next headers used in the IPv6 packet header

Type		Extension Header
0	00000000	Hop-by-Hop Options
43	00101011	Routing
44	00101100	Fragment
50	00110010	ESP (Encapsulating Security Payload)
51	00110011	AH (Authentication Header)
60	00111100	Destination Options
135	10000111	Mobility

`2ac1:05b8:123e:0000:0000:0000:03a0:c234`

`1 : 2 : 3 : 4 : 5 : 6 : 7 : 8` ← **8 groups 16 bits**

Fig. 4.11 Example of an IPv6 address representation of 8 groups of 4 hexadecimal (0x) digits (4 bits/hexadecimal) divided by colons

`2ac1:05b8:123e:0000:0000:0000:03a0:c234` ← **IPv6 address**

`2ac1: 5b8:123e:                : 3a0:c234` ← **Zeros erased**

`2ac1:5b8:123e::3a0:c234` ← **Shortened form**

Fig. 4.12 Example of an IPv6 address representation in which zeros in front of the 4 hexadecimal digits can be omitted (i.e., "05b8"→"5b8" and "03a0"→"3a0") and one consecutive group of zeros (i.e., ":0000:0000:0000:"→"::") are replaced with two consecutive colons "::" in the middle.

process, an erroneous IPv6 address, or due to a cyber-attack. This is why all routing path setup algorithms prevent looping routes from forming.

Source and Destination IPv6 addresses are each 128 bits, where their representation follows the Classless Inter-Domain Routing (CIDR) notation. The 128 bits of the IPv6 address are represented in 8 groups of 16 bits, where 16 bits are equivalent to 4 hexadecimal digits. Between each group of 4 hexadecimal digits a colon ':' is used to help read the 128 bit long IPv6 address as presented in Fig. 4.11.

To make IPv6 addresses easier for us to read and write, the zeros in front of the 4 hexadecimal digits can be omitted, and one consecutive group of zeros in the middle may be replaced with two consecutive colons '::' in an IPv6 address (where the actual IPv6 address will always be 128 bits with no colons), as shown in the example presented in Fig. 4.12.

Some of the frequently used IPv6 addresses are described in the following.

In CIDR notation form, the unspecified address ::/128 is the all zero address. This is equal to the IPv4 all zero address 0.0.0.0/32.

The IPv6 Default Route CIDR notation ::/0 is used in the last row of almost all IPv6 routing tables. The all zero network address is the lowest IPv6 address that exists and the all zero subnet mask is a fully open subnet mask that accepts all IPv6 addresses. This is equivalent to the IPv4 Default Route CIDR notation 0.0.0.0/0 which is used in the last row of almost all IPv4 routing tables.

The IPv4 address mapped to an IPv6 address uses ::ffff:0:0/96 in the mapping process, where few exceptions do apply.

4.5 TCP Protocol

TCP is used by only the end systems (e.g., computer H1 and computer H5 in Fig. 3.1) to setup a session for flow and error control of the data segments. The other routers and switches in the network do not read the TCP header information. A **TCP session** is setup at the beginning using a three-way handshake. The **three-way handshake** uses the synchronize (SYN) flag, acknowledge (ACK) flag, and sequence numbers (SNs). The TCP session is closed using a finish (FIN) flag after all data segments have been received at the destination successfully and the source is informed of this successful reception at the destination. SYN, ACK, FIN, and SN are in the TCP header [1, 2].

Figure 4.13 presents an example of how a TCP session is setup using a three-way handshake between a PC and Server. In this example, the PC makes a request to connect to the Server by sending a SYN message. The Sever sends back an ACK to the PC to accept the connection from the PC, and also sends back a SYN to establish a reverse direction connection from the Server to the PC, so a bidirectional flow of data segments can be exchanged between the PC and Server. The PC sends back an ACK to the Server to accept the connection from the Server. In addition, along with each SYN that is sent (from the PC to Server and the Server to PC), a SN is included to be used as the starting sequence number that will be assigned to the data segments when communicating. This three-way handshake procedure completes the TCP session setup.

The TCP header structure is presented in Fig. 4.14.

The TCP header's minimum length is 20 octets. Various options can be added to the end of the TCP header to conduct special functionalities. Padding may be added to the end of the options field to satisfy length requirements. The TCP header contains various data segment flow control functionalities.

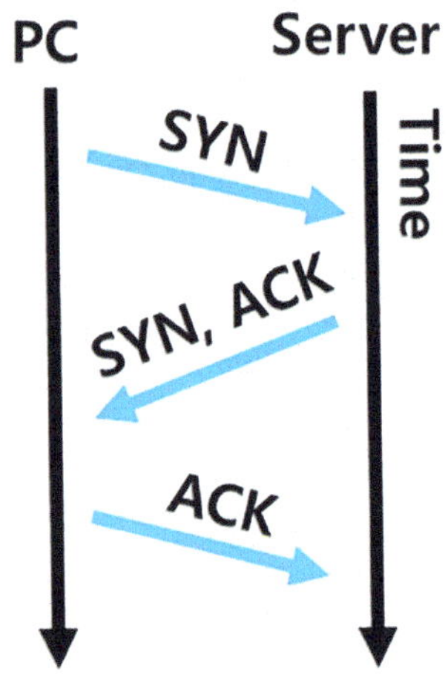

Fig. 4.13 TCP three-way handshake

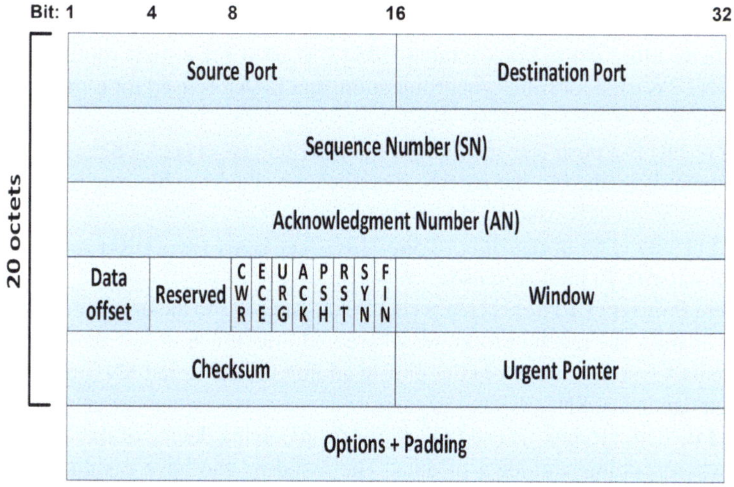

Fig. 4.14 TCP header structure

Checksum field (16 bits) is used for error detection to protect the TCP header from errors. As presented in Fig. 4.15, the TCP Checksum is computed using the TCP Pseudo Header, TCP Header, and the TCP data segments. The TCP Pseudo Header includes the IP Source Address, IP Destination Address, Reserved field, Protocol field, and the TCP Length field. When computing the Checksum, the Checksum bits in the TCP Header are preset to all zeros. Then the computed Checksum value will replace the all zero Checksum in the TCP Header.

Data Offset field (4 bits) is a number that represents the offset (in 32 bit word units) of the data segments (in this TCP frame) in reference to the original data file. The minimum TCP header length is 20 octets. When additional options are used, padding may be added to make the TCP header length a multiple of 32 bit words.

Reserved field (4 bits) is saved for future use and are not currently used, and therefore is normally set to all zeros.

Source and Destination Ports field (16 bits) is used to indicate what port the segments of data should be delivered to in the destination computer. Table 4.3

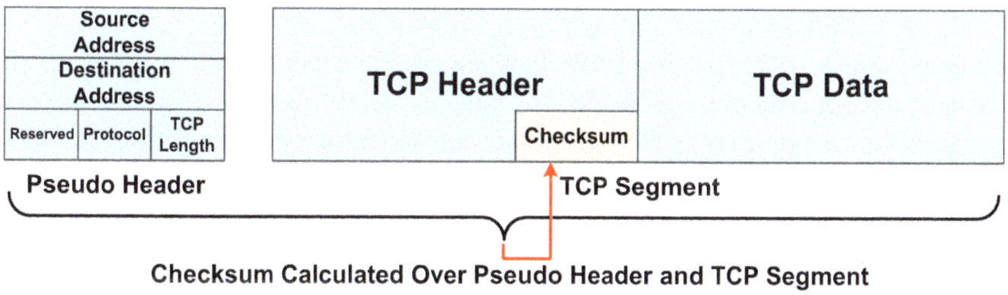

Fig. 4.15 TCP checksum computation method

provides a list of popular TCP and UDP port IDs and corresponding applications, where many more exist. For example, SMTP is used for e-mail routing between servers, NTP is used for time synchronization, and IMAP is used for e-mail message management.

PSH (1 bit) is a push flag in which if the PSH flag is set (i.e., PSH = 1) then the data segment is pushed to the receiving application. Push enables the received data segments to be quickly used by the application.

URG (1 bit) is an urgent flag. If the URG flag is set (i.e., URG = 1) then the Urgent Pointer (UP) field is in use.

Urgent Pointer (UP) field (16 bits) is a field that points to the urgent data location. The UP enables the receiver to know where and how much urgent data is coming. Urgent data's last SN can be obtained from adding the UP to the SN (i.e., SN + UP = Urgent data's last SN).

Window field (16 bits) is used to inform the receiving device of the window size to use when sending segments of data. The Window can have a value from 0 to 65,535 (i.e., $2^{16} - 1 = 65{,}535$). If the Window value is 0 then the Window is closed and this will stop the transmission of data segments. The larger the Window value, the faster segments will be sent to the destination. Therefore, the Window is used to control the flow of segments of data in which the destination and the routers of the routing path need to avoid delay and overflow. If problems occur on the routing path of the segments, then delays in arrival and missing segments will be detected at the destination computer. Then the destination computer will send back repeated ACKs to the source computer along with a smaller Window value in the TCP header. This way the sending computer will make its Window smaller, which will slow down the segment sending speed. As described, duplicated ACKs are used for recovery of erroneous and missing data segments and Window size control is used to control the sending speed of the data segments. When the receiving conditions improve, a larger window size will be requested by the receiving computer/server.

4.6 UDP Protocol

User Datagram Protocol (UDP) is like a reduced version of the TCP protocol, in which only the port information and check sum function is supported. UDP provides source and destination port information to support the application connection, as presented in Fig. 4.16. Because UDP does not establish any end-to-end connection manager to check on the received data segments, UDP is called a connectionless protocol. This is opposite to TCP, which sets up an end-to-end connection (through a three-way handshake) between the end computers to control the flow and recover missing data segments. This is why TCP is called a connection-oriented protocol.

Source Port field (16 bits) provides information of the port in the source computer that the data file was generated from.

Destination Port field (16 bits) provides information of the port in the destination computer that the data file needs to be delivered to.

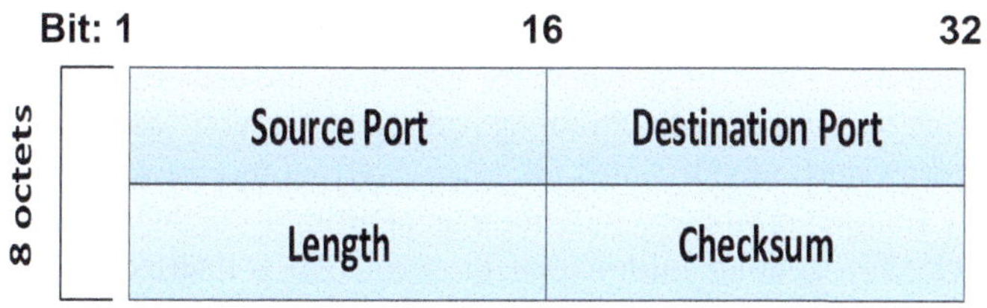

Fig. 4.16 UDP header structure

Table 4.3 List of popular TCP and UDP port IDs and corresponding applications, where many more exist

Port ID	Port application
20	File Transfer Protocol (FTP) data transfer
21	FTP control
23	Telnet
25	Simple Mail Transfer Protocol (SMTP)
35	Private printer server protocol
53	Domain Name System (DNS)
80	Hypertext Transfer Protocol (HTTP)
123	Network Time Protocol (NTP)
143	Internet Message Access Protocol (IMAP)

Table 4.3 provides a list of popular TCP and UDP port IDs and corresponding applications, where many more exist. The UDP ports help to connect the applications between the source and destination computers. The source and destination port numbers (16 bits) are the same as the port numbers used in TCP.

Length field (16 bits) contains the length of the entire UDP segment, which includes the UDP header and data.

Checksum field (16 bits) is used to detect bit errors in the UDP header. If an error is detected, the segment is discarded, and no error recovery action is taken. Use of the checksum field in UDP is optional. When it is not used, all bits are set to zero.

4.7 Internet Routing and Functions

Internet routing is supported by the Interior Gateway Protocol (IGP) and Exterior Gateway Protocol (EGP).

IGP is used by the routers to exchange routing information within (inside) an Autonomous System (AS). A typical AS is governed by an Internet Service Provider (ISP), or it could be a corporate network or campus network.

EGP is used by gateways to exchange routing information between (outside) ASs. EGP routing path selection is based on network policies, in which currently the Internet uses Border Gateway Protocol version 4 (BGP4), which was initially published in RFC 1654 in 1994, and has been upgraded to RFC 4271 in 2006 [2]. BGP4 has been used since 1994, where its operations are decided by the ISPs that are neighboring and share the EGP routing path.

IGP categories can be divided into Link-State (LS) routing protocols and Distance-Vector (DV) routing protocols. IGP routing algorithms include Open Shortest Path First (OSPF), Routing Information Protocol (RIP), Intermediate System to Intermediate System (IS-IS), and Enhanced Interior Gateway Routing Protocol (EIGRP).

BGP4 is the representative EGP, which has two types, which are the external BGP (eBGP) and internal BGP (iBGP). eBGP provides BGP routing used between ASs, and iBGP supports BGP routing within an AS.

Because BGP gateways commonly belong to different ISPs, the gateways may use different encryption and security schemes because the routers and gateways are administrated and managed by different ISPs. This is why BGP gateway security coordination is difficult. BGP routers need to exchange setup and updated information with each other, in which poor coordination can result in various types of security vulnerabilities. When different encryption and security schemes are used, it can be difficult to authenticate and protect against spoofed BGP messages and malware intrusions.

4.7.1 Open Shortest Path First (OSPF)

OSPF is the most widely used Internet IGP routing protocol. OSPF is used as the routing protocol for IPv4, IPv6, and CIDR addresses. OSPF is used by Internet gateways and routers worldwide. OSPF uses the Link State (LS) routing algorithm, which operates as described below [1, 2, 15, 16].

OSPF Operations

1. Routers collect LS information from other routers in the AS network.
2. Network Connection Map (Tree) is made including Cost values (i.e., the metric values in the routing tables) of links. The Network Connection Map will have tree roots from the Source node and branches (that make connection) to all other nodes in the AS network.
3. On the Network Connection Map, the LS routing algorithm is used to setup a loop-free Shortest Path Tree (SPT) routing path from the Source node to all Destination nodes in the network. SPT selects all the Minimum Cost routing paths (using Dijkstra's algorithm) from the Source node to each Destination node in the AS network.
4. Gateways/Routers will setup (and update) their Routing Tables (including the metric values) based on the SPT routing paths most recently computed.
5. When changes in the network are detected, steps 1–4 are repeated.

Some of the OSPF link cost influencing factors are listed below with units in the brackets (i.e., [units]).

- Distance of a router [seconds, meters, km, or hops]
- Round-Trip Time (RTT) [seconds]
- Number of hops (routers/switches) to reach destination [hops]
- Throughput [bits/second or packets/second]
- Utilization [%]
- Availability [%]
- Reliability [%].

OSPF operates using the following shortest path algorithm to create a loop-free Shortest Path Tree (SPT) routing path from the source node to all destination nodes in the AS network.

1. **M** is the set of nodes connected to the SPT. Initially **M** includes only the source node **S**.
2. Find the least cost route (by adding up link cost values on the routing path) from **S** to node **X** (node **X** is not in set **M** yet), using only nodes in **M** as intermediate nodes.
3. Find the least cost route (by adding up link cost values on the routing path) from **S** to node **X** (that is not in set **M** yet), using only nodes in **M** as intermediate nodes. This enables the SPT to grow, where in each step only one node can become a new member of set **M** and be added to the SPT.

   A. No loops are allowed, because loops can create network routing malfunctions.
   B. In case two (or more) link connections to a certain destination node have the same least cost route value, then only the node with the smallest IP address can connect to that destination node.

4. Once all nodes of the AS network are members of the set **M**, then the OSPF shortest path algorithm is stopped and the SPT is complete.

In the following, the OSPF LS operation step-by-step procedures (using Dijkstra shortest path algorithm) to set up a SPT are presented in the Figs. 4.17, 4.18, 4.19, 4.20, 4.21, 4.22, 4.23, 4.24, 4.25, 4.26, 4.27, 4.28, 4.29, 4.30, 4.31, 4.32, 4.33 and 4.34.

In Fig. 4.17, the OSPF LS procedure starts to build a SPT for source node **S**. When node **S** sends an IP packet, this SPT will be used to decide the route the IP packet will go through to reach the destination node (based on the destination IP address in the IP packet). At the beginning, only node **S** is a member of set **M**. A SPT will be built in the following steps using node **S** as the root node.

In Fig. 4.18, the link costs that are connected to the current SPT (based on set **M**) are compared, which include link cost 2 (from node **S** to node 1), link cost 5 (from node **S** to node 3), link cost 5 (from node **S** to node 4), and link cost 3 (from node **S** to node 2).

In Fig. 4.19, the minimum cost link 2 is selected, which connects node 1 to the SPT.

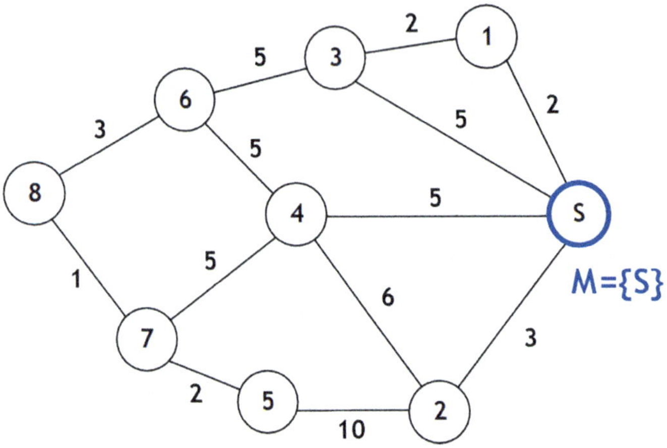

Fig. 4.17 OSPF LS procedures will build a SPT for source node **S**. As this will become a SPT for node **S** to use when routing IP packets, node **S** is the root node of this SPT. At the beginning only node **S** is a member of set **M**

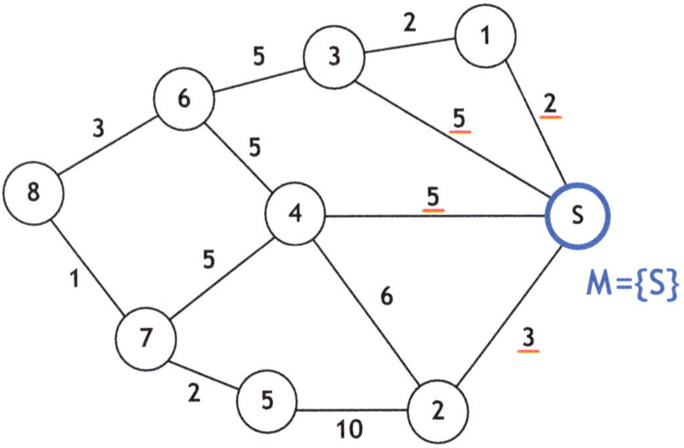

Fig. 4.18 OSPF LS procedures of comparing the link costs that are connected to the current SPT (based on set **M**) which include link cost 2 (from node **S** to node 1), link cost 5 (from node **S** to node 3), link cost 5 (from node **S** to node 4), and link cost 3 (from node **S** to node 2)

In Fig. 4.20, node 1 is included into set **M**, and then the algorithm compares the link costs to select the next minimum cost link. The link that connects node 1 to node 3 needs to have an added link cost from the source node **S**, which becomes 4 due to adding up the link costs of 2 (from node **S** to node 1) and 2 (from node 1 to node 3).

In Fig. 4.21, the minimum cost link 3 is selected, which connects node 2 to the SPT.

In Fig. 4.22, node 2 is included into set **M**, and then the algorithm compares the link values to select the next minimum cost link. The link that connects node 2 to

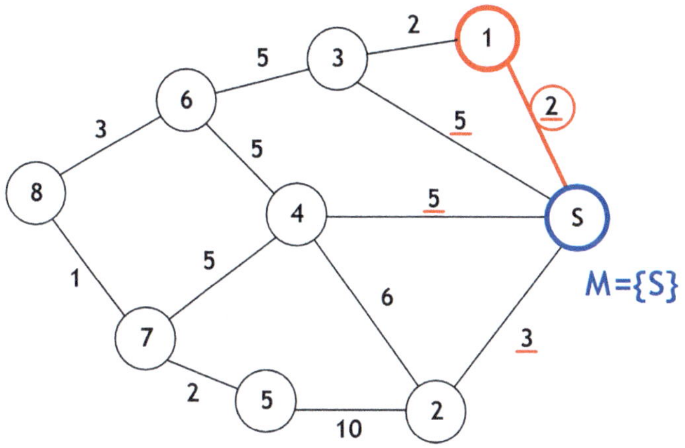

Fig. 4.19 OSPF LS procedures select the minimum cost link 2, which connects node 1 to the SPT

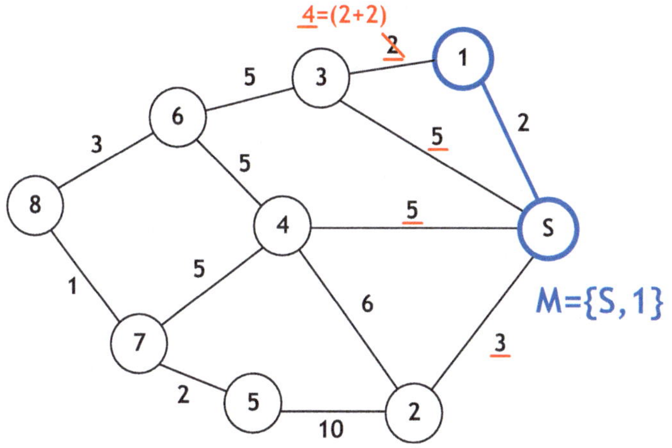

Fig. 4.20 OSPF LS procedures include node 1 into set **M**, and then compares the link costs to select the next minimum cost link. The link that connects node 1 to node 3 needs to have an added link cost from the source node **S**, which becomes 4 due to adding up the link costs of 2 (from node **S** to node 1) and 2 (from node 1 to node 3)

node 5 needs to have an added link cost from the source node **S**, which becomes 13 due to adding up the link costs of 3 (from node **S** to node 2) and 10 (from node 2 to node 5). In addition, the link that connects node 2 to node 4 needs to have an added link cost from the source node **S**, which becomes 9 due to adding up the link costs of 3 (node **S** to node 2) and 6 (from node 2 to node 4).

In Fig. 4.23, the minimum cost link 4 is selected, which connects node 1 to node 3.

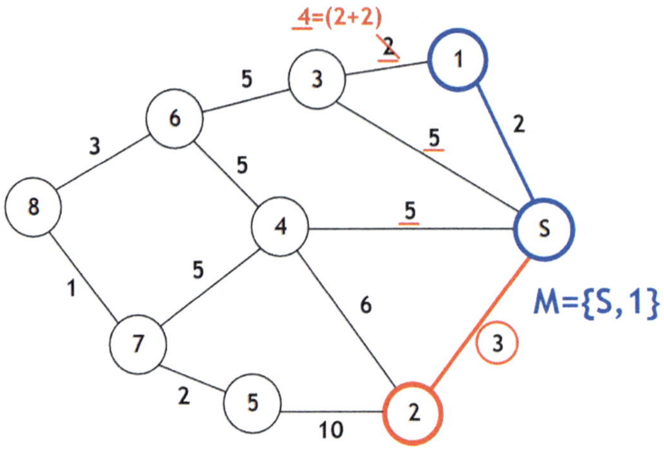

Fig. 4.21 OSPF LS procedures select the minimum cost link 3, which connects node 2 to the SPT

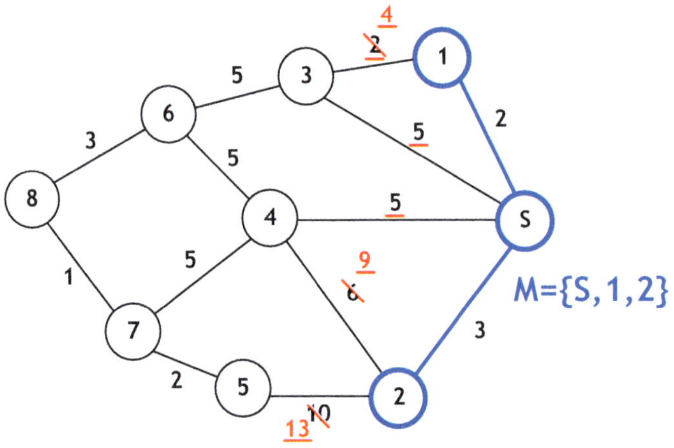

Fig. 4.22 OSPF LS procedures include node 2 into set **M**, and then compares the link values to select the next minimum cost link. The link that connects node 2 to node 5 needs to have an added link cost from the source node **S**, which becomes 13 due to adding up the link costs of 3 (from node **S** to node 2) and 10 (from node 2 to node 5). In addition, the link that connects node 2 to node 4 needs to have an added link cost from the source node **S**, which becomes 9 due to adding up the link costs of 3 (node **S** to node 2) and 6 (from node 2 to node 4)

In Fig. 4.24, node 3 is included into set **M**, and to avoid any loops in the SPT, the direct link between node **S** and node 3 is no longer selectable. This is to prevent a loop between node **S** and node 1 and node 3 from forming, where a routed IP packet may not be able to escape this loop if it is formed. Next the OSPF LS procedures will select the next minimum cost link. The link that connects node 3 to node 6 needs to have an added link cost from the source node **S**, which becomes 9 due to

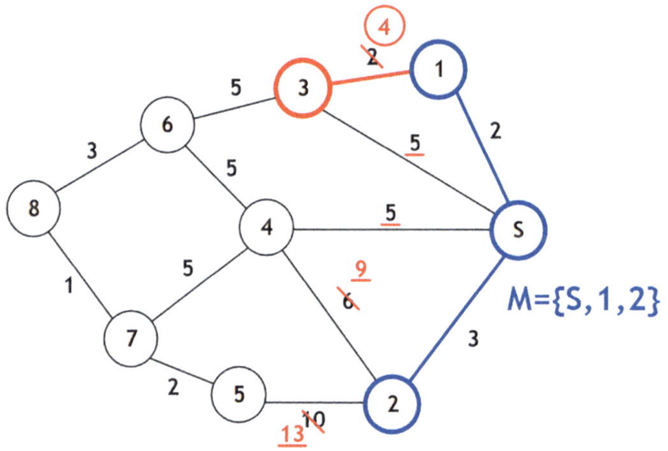

Fig. 4.23 OSPF LS procedures select the minimum cost link 4, which connects node 1 to node 3

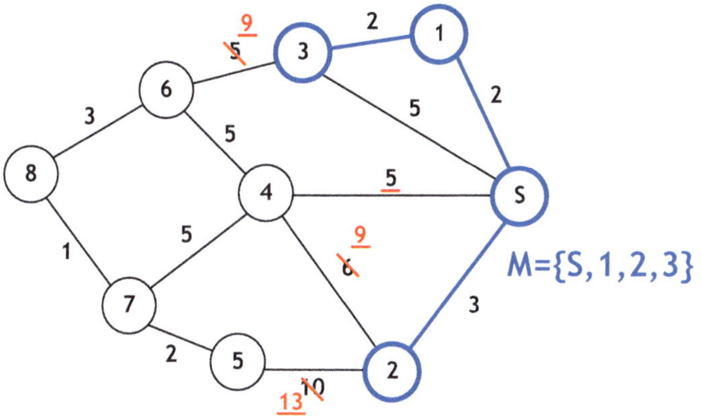

Fig. 4.24 OSPF LS procedures include node 3 into set **M**, and to avoid any loops in the SPT, the direct link between node **S** and node 3 is no longer selectable. This is to prevent a loop between node **S** and node 1 and node 3 from forming, where a routed IP packet may not be able to escape this loop if it is formed. Next the OSPF LS procedures will select the next minimum cost link. The link that connects node 3 to node 6 needs to have an added link cost from the source node **S**, which becomes 9 due to adding up the link costs of 2 (from node **S** to node 1) and 2 (from node 1 to node 3) and 5 (from node 3 to node 6)

adding up the link costs of 2 (from node **S** to node 1) and 2 (from node 1 to node 3) and 5 (from node 3 to node 6).

In Fig. 4.25, the minimum cost link 5 is selected, which connects node **S** to node 4.

In Fig. 4.26, node 4 is included into set **M**, and to avoid any loops in the SPT, the direct link between node 2 and node 4 is no longer selectable to prevent a loop from forming. Next the OSPF LS procedures will select the next minimum cost link. The

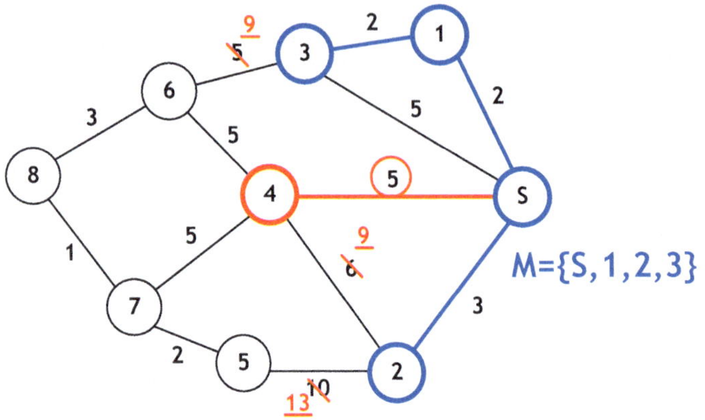

Fig. 4.25 OSPF LS procedures select the minimum cost link 5, which connects node **S** to node 4

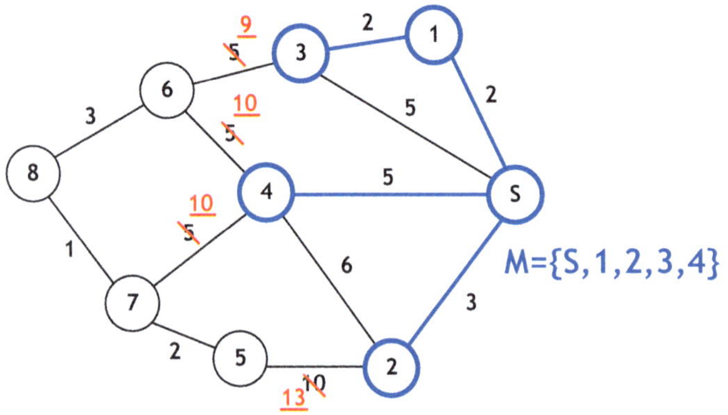

Fig. 4.26 OSPF LS procedures include node 4 into set **M**, and to avoid any loops in the SPT, the direct link between node 2 and node 4 is no longer selectable to prevent a loop from forming. Next the OSPF LS procedures will select the next minimum cost link. The link that connects node 4 to node 6 needs to have an added link cost from the source node **S**, which becomes 10 due to adding up the link costs of 5 (from node **S** to node 4) and 5 (from node 4 to node 6). In addition, the link that connects node 4 to node 7 needs to have an added link cost from the source node **S**, which becomes 10 due to adding up the link costs of 5 (from node **S** to node 4) and 5 (from node 4 to node 7)

link that connects node 4 to node 6 needs to have an added link cost from the source node **S**, which becomes 10 due to adding up the link costs of 5 (from node **S** to node 4) and 5 (from node 4 to node 6). In addition, the link that connects node 4 to node 7 needs to have an added link cost from the source node **S**, which becomes 10 due to adding up the link costs of 5 (from node **S** to node 4) and 5 (from node 4 to node 7).

In Fig. 4.27, the minimum cost link 9 is selected, which connects node 3 to node 6.

In Fig. 4.28, node 6 is included into set **M**, and to avoid any loops in the SPT, the direct link between node 4 and node 6 is no longer selectable to prevent a loop from

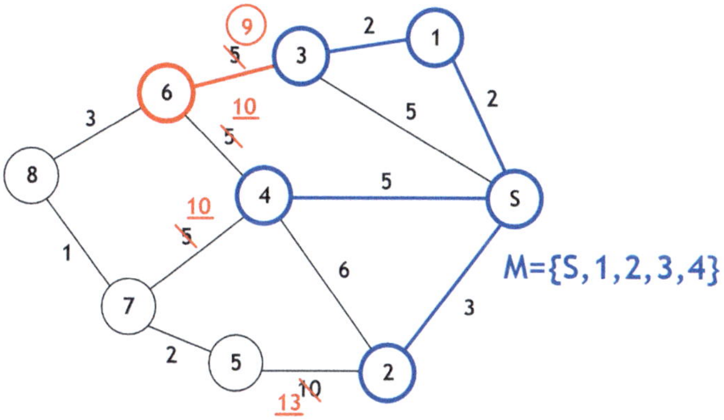

Fig. 4.27 OSPF LS procedures select the minimum cost link 9, which connects node 3 to node 6

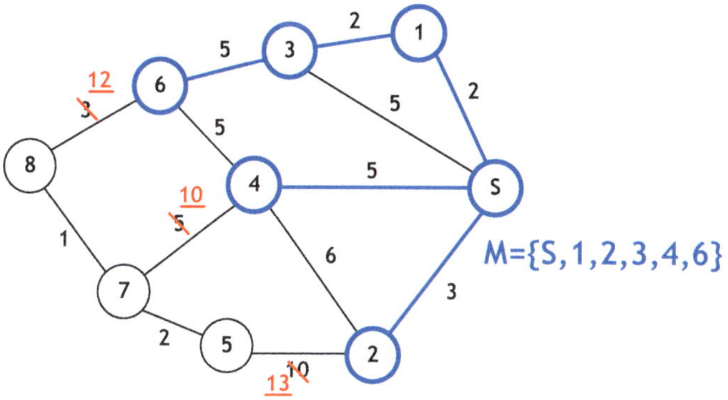

Fig. 4.28 OSPF LS procedures include node 6 into set **M**, and to avoid any loops in the SPT, the direct link between node 4 and node 6 is no longer selectable to prevent a loop from forming. Next the OSPF LS procedures will select the next minimum cost link. The link that connects node 6 to node 8 needs to have an added link cost from the source node **S**, which becomes 12 due to adding up the link costs of 2 (from node **S** to node 1) and 2 (from node 1 to node 3) and 5 (from node 3 to node 6) and 3 (from node 6 to node 8)

forming. Next the OSPF LS procedures will select the next minimum cost link. The link that connects node 6 to node 8 needs to have an added link cost from the source node **S**, which becomes 12 due to adding up the link costs of 2 (from node **S** to node 1) and 2 (from node 1 to node 3) and 5 (from node 3 to node 6) and 3 (from node 6 to node 8).

In Fig. 4.29, the minimum cost link 10 is selected, which connects node 4 to node 7.

In Fig. 4.30, node 7 is included into set **M**. Next the OSPF LS procedures will select the next minimum cost link. The link that connects node 7 to node 5 needs to have an added link cost from the source node **S**, which becomes 12 due to adding up the link costs of 5 (from node **S** to node 4) and 5 (from node 4 to node 7) and 2

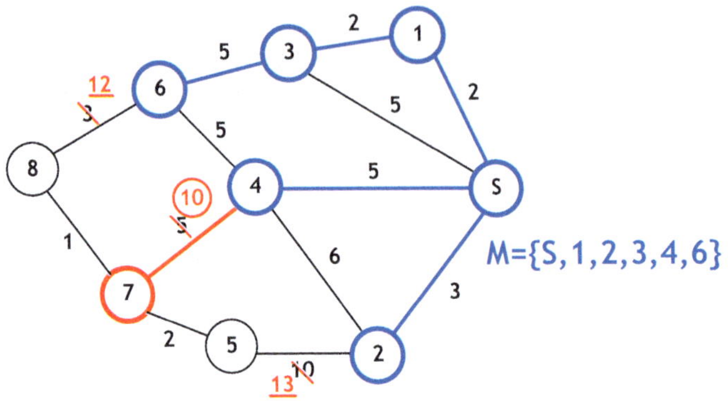

Fig. 4.29 OSPF LS procedures select the minimum cost link 10, which connects node 4 to node 7

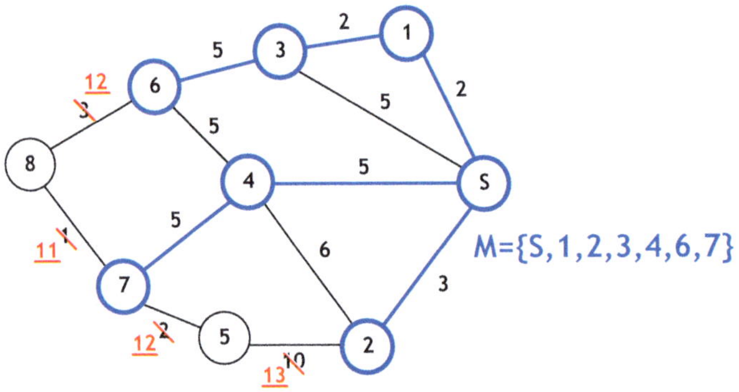

Fig. 4.30 OSPF LS procedures include node 7 into set **M**. Next the OSPF LS procedures will select the next minimum cost link. The link that connects node 7 to node 5 needs to have an added link cost from the source node **S**, which becomes 12 due to adding up the link costs of 5 (from node **S** to node 4) and 5 (from node 4 to node 7) and 2 (from node 7 to node 5). In addition, the link that connects node 7 to node 8 needs to have an added link cost from the source node **S**, which becomes 11 due to adding up the link costs of 5 (from node **S** to node 4) and 5 (from node 4 to node 7) and 1 (from node 7 to node 8)

(from node 7 to node 5). In addition, the link that connects node 7 to node 8 needs to have an added link cost from the source node **S**, which becomes 11 due to adding up the link costs of 5 (from node **S** to node 4) and 5 (from node 4 to node 7) and 1 (from node 7 to node 8).

In Fig. 4.31, the minimum cost link 11 is selected, which connects node 7 to node 8.

In Fig. 4.32, node 8 is included into set **M**. To avoid any loops in the SPT, the direct link between node 6 and node 8 is no longer selectable to prevent a loop from forming. Next the OSPF LS procedures will select the next minimum cost link. The only node that is not a member of set **M** is node 5, which can be connected to the

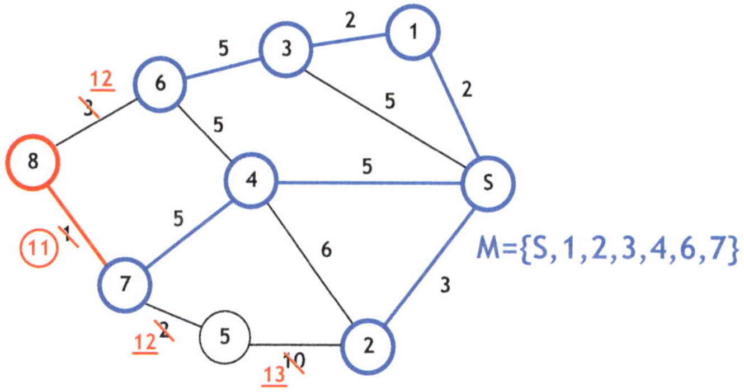

Fig. 4.31 OSPF LS procedures select the minimum cost link 11, which connects node 7 to node 8

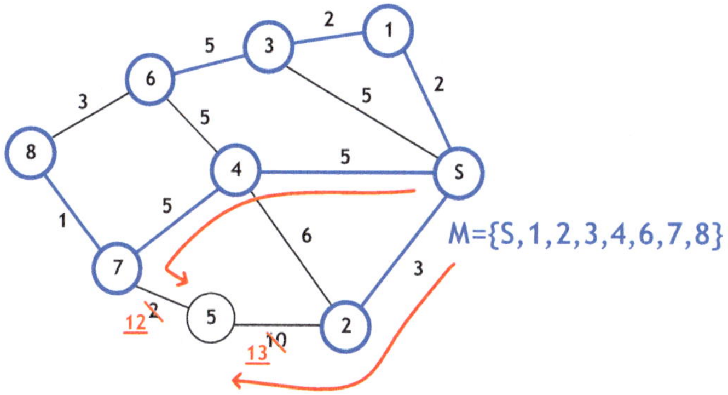

Fig. 4.32 OSPF LS procedures include node 8 into set **M**. To avoid any loops in the SPT, the direct link between node 6 and node 8 is no longer selectable to prevent a loop from forming. Next the OSPF LS procedures will select the next minimum cost link. The only node that is not a member of set **M** is node 5, which can be connected to the SPT from either node 2 or node 7

SPT from either node 2 or node 7. OSPF LS procedures will select the minimum cost link 12, which connects node 7 to node 5.

In Fig. 4.33, the minimum cost link 12 is selected, which connects node 7 to node 5. Now, all nodes are a member of set **M**, so the OSPF LS procedures end and the SPT for node **S** is completed.

In Fig. 4.34, the SPT for node **S** is presented. Node **S** will use this SPT when sending future IP packets. If a change in link cost is detected (i.e., a metric value change in a routing table is detected), then this SPT will need to be built again. The root node **S** will use this SPT to reach any node in the network as summarized in the routing table in Fig. 4.34.

Note that each node in the AS network needs to make its own SPT to use for IP packet routing when sending an IP packet. Each node will serve as a root node for its own SPT.

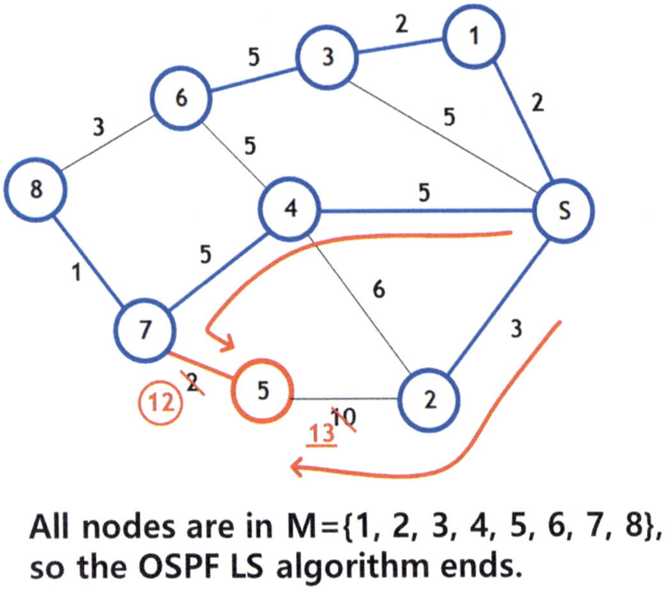

Fig. 4.33 OSPF LS procedures will select the minimum cost link 12, which connects node 7 to node 5. Now, all nodes are a member of set **M**, so the OSPF LS procedures end and the SPT for node **S** is completed

4.7.2 Additional OSPF Functions and Router Types

Multicast Open Shortest Path First (MOSPF) is an OSPF extension to support multicast routing. MOSPF enables routers to share Group Membership information in multicast routing path setup. Alternative multicasting schemes include a combination of OSPF with Protocol Independent Multicast (PIM).

OSPF subdivided networks are a division of an AS network based on administration and management requirements. OSPF subdivided networks are used to simplify administration, optimize traffic and resource utilization, enhance security, enable faster routing updates, as well as support network subdividing to divide subarea networks and the backbone network area.

OSPF Traffic Engineering (TE) is used to provide quality of service (QoS) routing control of IP packets from the ingress node to the egress node. TE enhances reliability by minimizing service outage from errors and failures. TE makes this possible by enabling measurement, characterization, and modeling for performance optimization of Internet traffic (RFC 2702). TE can be used for IP and non-IP networks (e.g., optical networks) (RFC 3630).

In addition, there are multiple OSPF router types, which are listed in the following.

- Internal Router (IR) is used when all routing interfaces belong to the same network area.

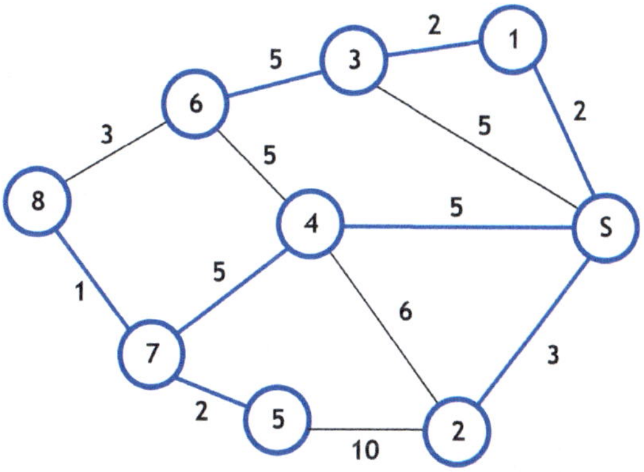

Root	Destination	SPT Route
S	1	S → 1
	2	S → 2
	3	S → 1 → 3
	4	S → 4
	5	S → 4 → 7 → 5
	6	S → 1 → 3 → 6
	7	S → 4 → 7
	8	S → 4 → 7 → 8

Fig. 4.34 The OSPF LS procedures have ended and the SPT for node **S** is presented. Node **S** will use this SPT when sending future IP packets. If a change in link cost is detected (i.e., if a metric value change in a routing table is detected), then this SPT will need to be built again. The root node **S** will use this SPT to reach any node in the network following the path in the routing table. Note that each node of the AS network needs to make its own SPT to use for IP packet routing when sending an IP packet. Each node will serve as a root node for its own SPT

- Area Border Router (ABR) is used to connect subarea networks to the backbone network.
- Backbone Router (BR) is used to connect to the backbone network.
- Autonomous System Boundary Router (ASBR) is used to connect between ASs using multiple routing protocols.

Routers can have multiple functionalities, so a router can be designed to serve multiple router roles. Especially gateways have multiple router roles, because they operate at the edge of an AS network interfacing other networks as well as serve routing within an AS network.

4.8 Summary

The TCP/IP model is the global standard for network operations and is the basis for all future network designs. One reason why it is beneficial to know the details of the IPv4, IPv6, TCP, and UDP protocols is because new applications and new devices will be developed in which new networking technologies will need to be created, which is our role. In addition, reliable and efficient routing from one end device to another end device will always be a challenge, due to factors like distance, various network types, different ASs, diverse security packages, different ISP policies, as well as numerous types of devices. In addition, all new network technologies have to be compatible with existing standards in order to enable access of existing information resources and also enable existing users to connect and use the new services and devices we create. In addition, we need to create new security technologies that can help to protect against new emerging cyber-attack methods.

References

1. J.-M. Chung, Introduction to TCP/IP. Coursera. https://www.coursera.org/learn/tcpip
2. W. Stallings, *Data and Computer Communications*, 10th edn. (Prentice Hall, 2014)
3. J. Postel, Internet Protocol DARPA Internet Program Protocol Specification. IETF RFC 791 (1981)
4. D. Clark, IP Datagram Reassembly Algorithms. IETF RFC 815 (1982)
5. Wikipedia, Internet Protocol. https://en.wikipedia.org/wiki/Internet_Protocol
6. Wikipedia, Internet Protocol version 4. https://en.wikipedia.org/wiki/Internet_Protocol_version_4
7. J. Mogul, S. Deering, Path MTU discovery. IETF RFC 1191 (1990)
8. S. Deering, R. Hinden, Internet Protocol, Version 6 (IPv6) Specification. IETF RFC 2460 (1998)
9. Wikipedia, Internet Stream Protocol. https://en.wikipedia.org/wiki/Internet_Stream_Protocol
10. Wikipedia, IPv6. https://en.wikipedia.org/wiki/IPv6
11. IPv6.com. http://ipv6.com
12. Cisco, Resolve IP Fragmentation, MTU, MSS, and PMTUD Issues with GRE and IPSEC (2015). http://www.cisco.com/c/en/us/support/docs/ip/generic-routing-encapsulation-gre/25885-pmtud-ipfrag.html. Accessed 1 June 2015
13. Microsoft technet, IPv6, IPv6 Technical Reference (2015). (https://technet.microsoft.com/en-us/library/cc738109(v=ws.10).aspx. Accessed 1 June 2015
14. Wikipedia, Border Gateway Protocol. https://en.wikipedia.org/wiki/Border_Gateway_Protocol
15. J. J. Metzner, *Reliable Data Communications* (Academic Press, 1998)
16. Wikipedia, Open Shortest Path First. https://en.wikipedia.org/wiki/Open_Shortest_Path_First

Chapter 5
Internet and Wireshark Project

In this chapter, a series of procedures are introduced, which describes how to check a computer's Internet connection configurations, conduct Internet ping and trace route measurements. Next a Wireshark Internet analysis project is presented to show how to check Internet Protocol (i.e., IPv4 and IPv6) packets that were transmitted and received by a computer. Next, this chapter explains the procedures of how one can check an IP packet's source and destination IP address as well as the Transmission Control Protocol (TCP) three-way handshake connection setup parameters, TCP window size, and transport security protocols that are being used. The chapter also describes how to generate file transfer Input and Output (I/O) graphs, analyze the data rate of the network, and conduct IP address based physical location (e.g., mailing address) identification experiments. Using the information of this chapter, the reader will be able to clearly understand the details of Chap. 3 (titled "Internet Access for my Personal Computer (PC) and Smartphone") and Chap. 4 (titled "TCP/IP Protocols & IP Routing"), which are actually implemented on one's computer.

5.1 Internet Connection Configuration Analysis

This chapter first introduces the method of how to check a computer's Internet connection. The procedures start with checking the network configuration of a computer. This is the most fundamental information that is required to understand how a computer can make a connection to the Internet.

A method of how to check the IP configuration of a computer is described in the following. The procedures in the following are based on a computer that uses **Microsoft Windows** as its **Operating System** (OS). Computers that use a different OS will have slightly different procedures, but the network and Internet information to check will all be the same.

J.-M. Chung, *Emerging Secure Networks, Blockchains and Smart Contract Technologies*, https://doi.org/10.1007/978-3-031-65866-2_5

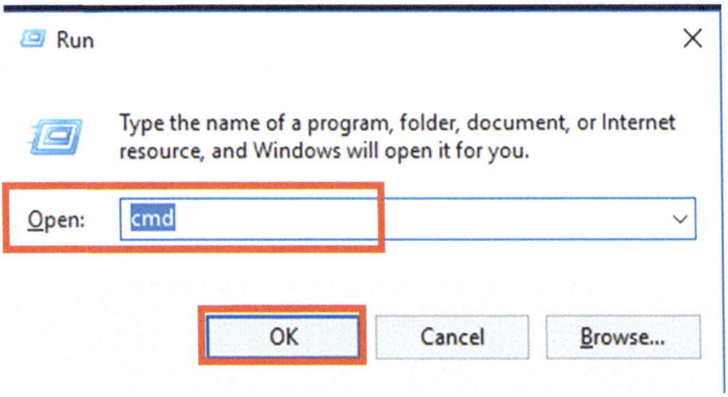

Fig. 5.1 How to open the command prompt using the run window

First, we need to open the **Command Prompt** window to check the IP configuration of the computer. There are a few ways to open the **Command Prompt** window, which are based on the following options.

If the computer uses Microsoft Windows and the keyboard has a **Windows key** (e.g., , ,), then press the **Windows key** and **R** key simultaneously to open the **Run** window, as shown in Fig. 5.1. Then in the **Run** window type in "**cmd**" and click **OK** to open the **Command Prompt** window.

But, if the keyboard does not have a **Windows key** on your keyboard, then the following procedures can be used, which are based on the version of Windows installed on the computer. In the following, examples are provided for a computer using Windows 7, Windows 10, and Windows 11.

For a computer using **Windows 7**, click the window button (e.g., , ,) in the left bottom corner to open the function menu, and then go to "**All programs**" followed by "**Accessories**" and then "**Command Prompt**" as shown in Fig. 5.2(a).

For a computer using **Windows 10**, click the window button (e.g., , ,) in the left bottom corner to open the function menu, and then go to "**All apps**" followed by "**Windows System**" and then "**Command Prompt**" as shown in Fig. 5.2(b).

For a computer using **Windows 11**, click the window button (e.g., , ,) in the left bottom corner to open the function menu, and then select "**All apps**" followed by "**Windows Tools**" and then "**Command Prompt**" as shown in Fig. 5.3.

Figure 5.4 shows an example of how the command prompt looks like.

Using the Command Prompt window, the IP configuration of the computer can be checked using the "IPconfig" command [1]. As presented in Fig. 5.5, the computer's IPv4 and IPv6 Address, Subnet Mask, Default Gateway, and more information will be provided. Because the computer used in this example makes connection to the Internet through its Ethernet Network Interface Card (NIC), the Internet connection information of the Ethernet adapter is presented.

As presented in Fig. 5.6, by using the "ipconfig /all" command, all IP configuration information for all network adapters of the computer will be presented.

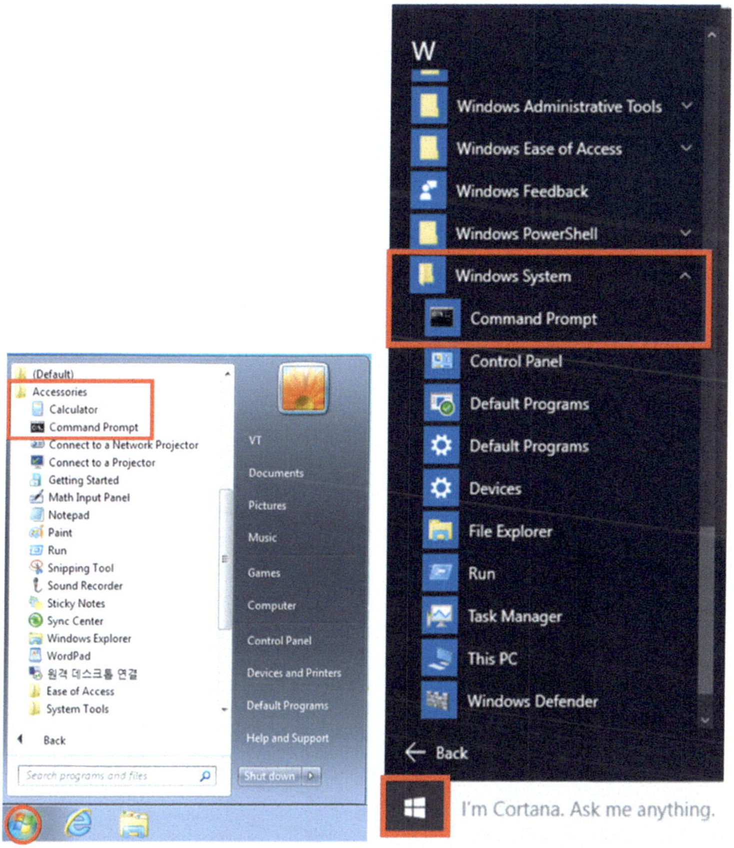

(a) Windows 7 (b) Windows 10

Fig. 5.2 How to open the command prompt in Windows 7 and Windows 10

Although there are many more, some of the useful IPconfig commands are described in the following.

- IPconfig /release
  - Release the computer's current IP information and obtain a new IP address from the DHCP server.
- IPconfig /renew
  - Renew the computer's IP address.
  - Can be used if the computer is set to automatic IP address setup mode (using DHCP).
- IPconfig /displaydns
  - Show the computer's current DNS Resolver Cache Logs.

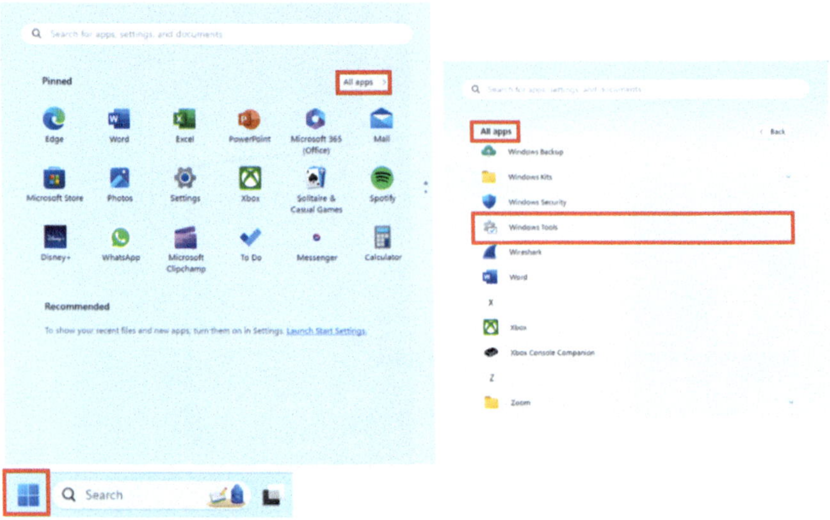

(a) Open the Windows menu and select "All apps." (b) Select "Windows Tools."

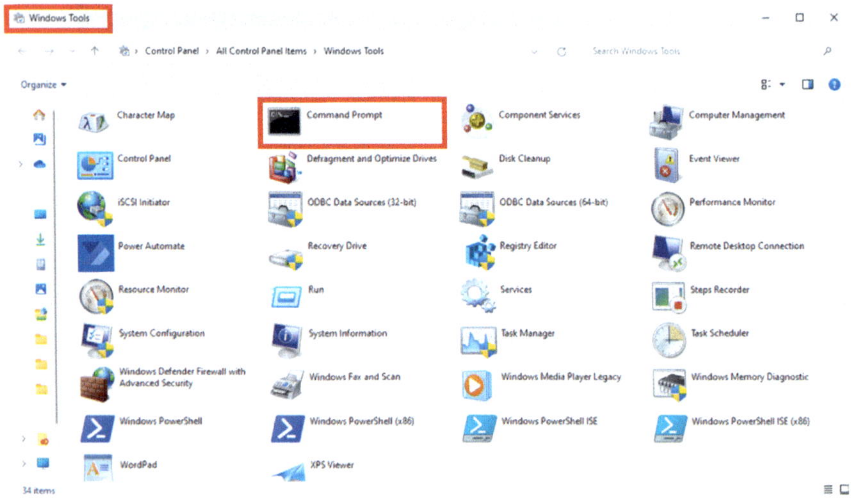

(c) Select "Command Prompt"

Fig. 5.3 How to open the command prompt in Windows 11

- IPconfig /flushdns
  – Flush or clear the computer's current DNS Resolver Cache Logs.
- IPconfig /registerdns
  – Update the DNS settings on the computer.

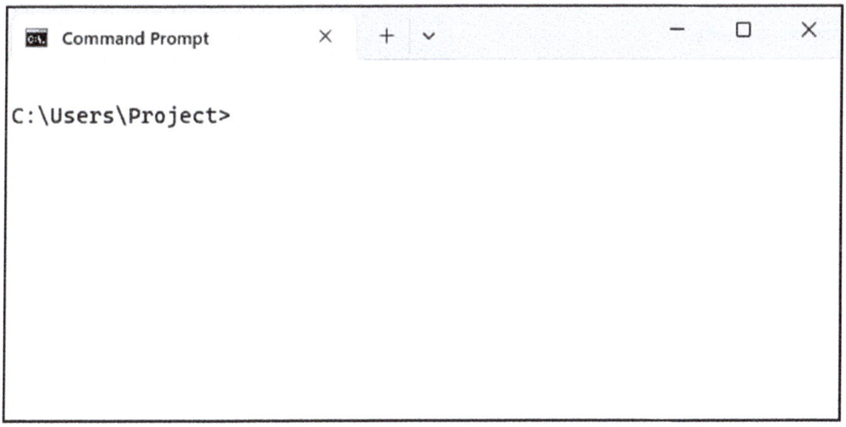

Fig. 5.4 Example of how the command prompt looks like

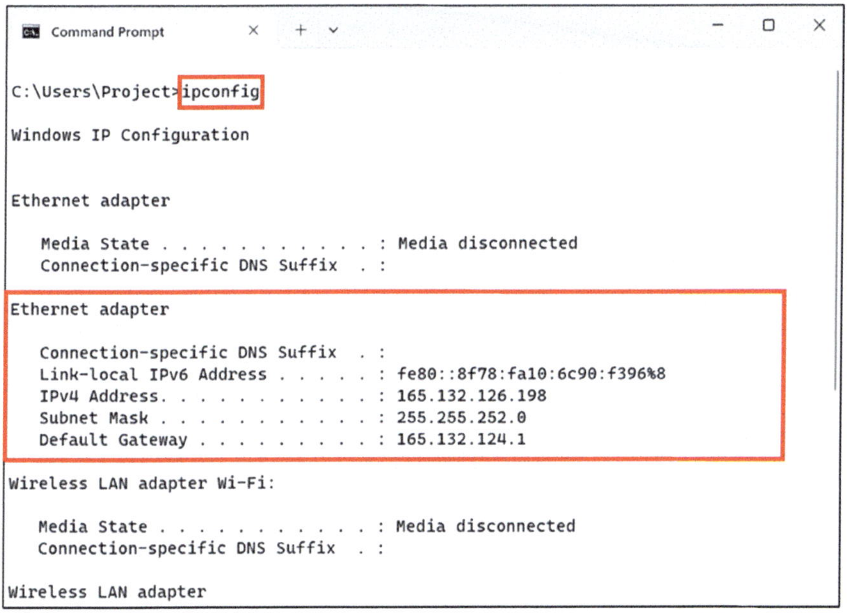

Fig. 5.5 Example of IP configuration of a computer

5.1.1 Ping Measurements

A very useful command is **Ping**, which is used in the 'ping (target address)' form in the Command Prompt [2]. Ping can be used to test the Round-Trip Time (RTT) from the source computer (which is the computer being used by the reader) to a

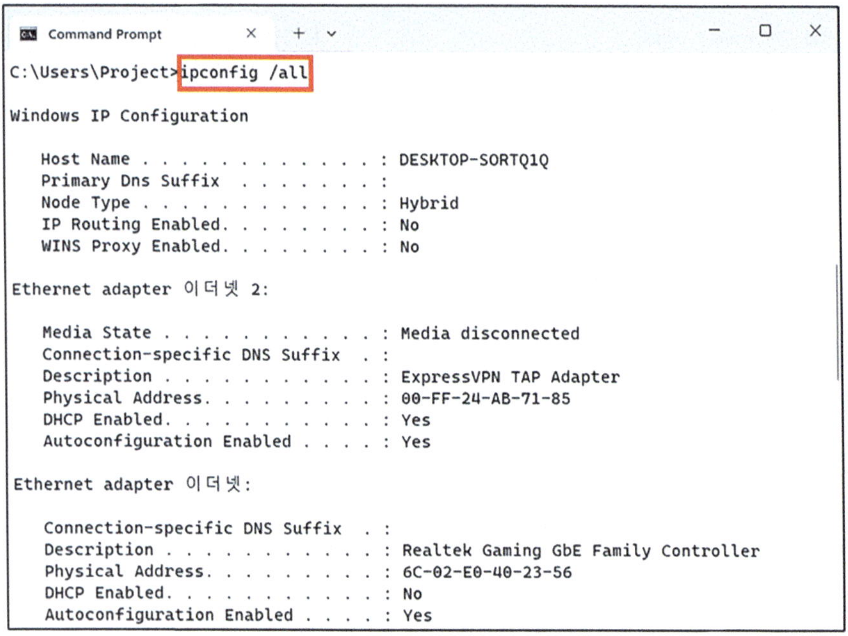

Fig. 5.6 Example of "ipconfig /all" used to display the configuration information of all network adapters of the computer

specified destination computer, server, etc. The 'target address' can be an IP address or a domain name.

Figure 5.7 presents an example of pinging the Google domain name using "ping www.google.com" to find the RTT from a Google website server. The local DNS server will see the website domain name "www.google.com" and will find the IPv4 or IPv6 address to use, which is the IPv4 address 142.251.42.196 in this example. As presented in Fig. 5.7, four ping messages were sent to the Google website server IPv4 address 142.251.42.196 where the measured RTT minimum value is 65 ms, the maximum RTT is 66 ms, and the average of the four RTTs is 65 ms.

In the example presented in Fig. 5.7, the Ping command is used to send four **ICMP Echo Request** messages to the Google website server. Four is the default number of ICMP Echo Request messages used by the Ping command, which can be changed using "ping -n" presented below. The Ping reply messages from the Google server are the **Echo Reply** messages, which has a default size of 32 octets (bytes). This can be changed to have a size from 32 octets to 65,527 octets using "ping -l *Size*" as presented below. This can be used to test the network's capability to transfer large IP packets. In addition, a Time to Live (TTL) of 49 was used in this example, where "ping -i *TTL*" can be used to set the TTL value (as presented below), where the TTL maximum is 255.

Although there are many more, some of the useful Ping commands are described in the following.

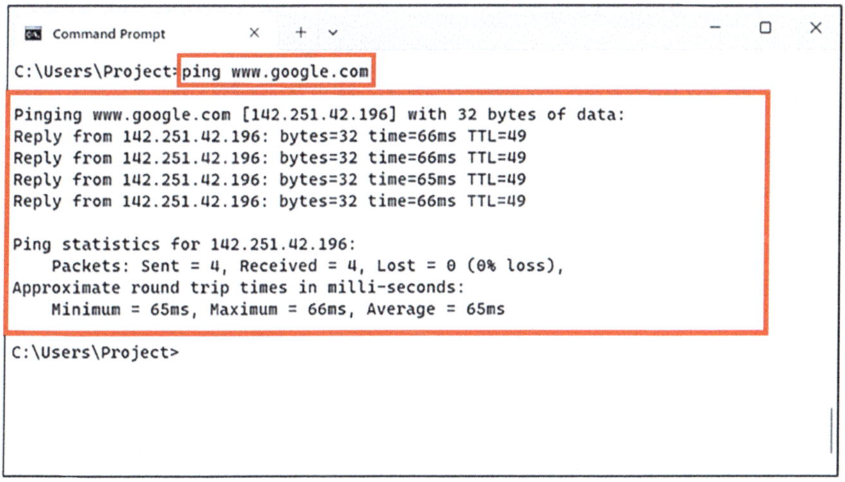

Fig. 5.7 Example of using Ping to a Google website server to find the RTT information

- ping -a
  - Resolve the hostname of an IP address target.
- ping -n
  - Sets the number of ICMP Echo Request messages to send.
  - Without this option, four requests will be sent.
- ping -l *Size*
  - Set the size of the echo reply packet from 32 octets to 65,527 octets (bytes).
  - Without this option, a 32 octet echo reply will be sent.
- ping -i *TTL*
  - Used to set the Time to Live (TTL) value.
  - TTL maximum is 255.
- ping -w *Timeout*
  - Specifying a timeout value when executing the Ping command.
  - Adjusts the amount of time (in ms) that Ping waits for each reply.
  - Without this option, the default timeout value is 4,000 ms.
- ping -p
  - Used to ping a Hyper -V Network Virtualization provider address.

5.1.2 Trace Route Measurements

Another very useful command is the **Tracert**, which is used in the "tracert (target address)" form in the Command Prompt [3]. Figure 5.8 shows an example of how Tracert is used to find the IP routing path information of a Google website server.

Tracert is short for "Trace Route" which displays the **route** and measures the **transit** delays from the source to the destination. The target address used in Tracert can be an **IP address** or a **domain name**. If a domain name is used, the local DNS server will change the domain name into an IPv4 or IPv6 address.

The Tracert used in Fig. 5.8 shows the results of three route measurements with the Google website server (www.google.com) in which the first two hops were completed in less than 1 ms through local gateways/routers.

The basic Tracert information is described in the following.

- Hop #
  - Number of hops along the route.
- RTT
  - RTT for a Tracert packet to reach the route and return to the source.
  - Traceroute checks the RTT three times for each hop.
- Name/IP Address
  - Domain name or IP address of the gateway/router.

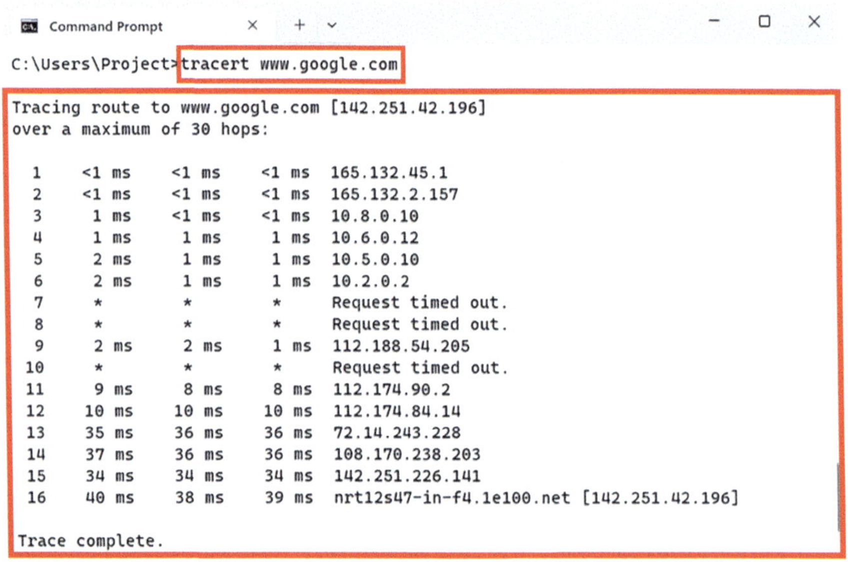

Fig. 5.8 Example of using Tracert to a Google website server to find the IP routing path information

Although there are many more, two useful Tracert commands are described in the following.

- tracert -h *MaxHops*
  - Specify the maximum number of hops in the search for the target.
- tracert -w *Timeout*
  - Specify the time to allow each reply before timeout (in milliseconds).

5.2 Wireshark Internet Analysis Project

In this section, details of the Wireshark Internet project are introduced [4]. This includes a method to conduct Internet packet analysis using Wireshark, where the project goals are listed in the following. First, using Wireshark, real Internet packets exchanged along with network operations will be observed. Second, the Internet network will be analyzed using Wireshark, which will include use of Hypertext Transfer Protocol Secure (HTTPS) transfer. Third, TCP, TLS, UDP, DNS operations will be analyzed and the location of the IP address will be checked.

The Internet packet analysis procedures using Wireshark will be conducted in the following order.

1. Capture packets of HTTPS protocol using Wireshark when accessing a website of the reader's choice.
2. Identify TCP and TLSv1.3 handshake.
3. Find if any retransmission packets were sent.
4. Analyze the average data rate and identify the total elapsed time used in downloading.

In order to conduct this project, first Wireshark needs to be installed on your computer. So, first we need to go to the website https://www.wireshark.org/download.html [4] and download the Wireshark install file according to your computer. The type of OS used on your computer needs to be known and used in the selection. Please note that newer version of Wireshark may be available, but the basic procedures will be similar to the procedures presented below.

Figure 5.9 shows the Wireshark install options. In this example, we will select the file to download (Wireshark-win64-2.2.5.exe in this example) to start the installation process and click the "Next" button in Fig. 5.10 to start the Wireshark setup process.

Figure 5.11 shows the Wireshark setup license agreement, where the "I Agree" is selected to accept the License Agreement to use Wireshark.

Figure 5.12 shows the Wireshark setup component selection, where we will choose the components to install (e.g., I selected all components) and click the "Next" button.

Figure 5.13 shows the Wireshark setup selection of additional tasks. Here we can select the options to **Create Shortcuts** for Wireshark and select the **File Extensions** you want to connect to Wireshark and then click the "Next" button.

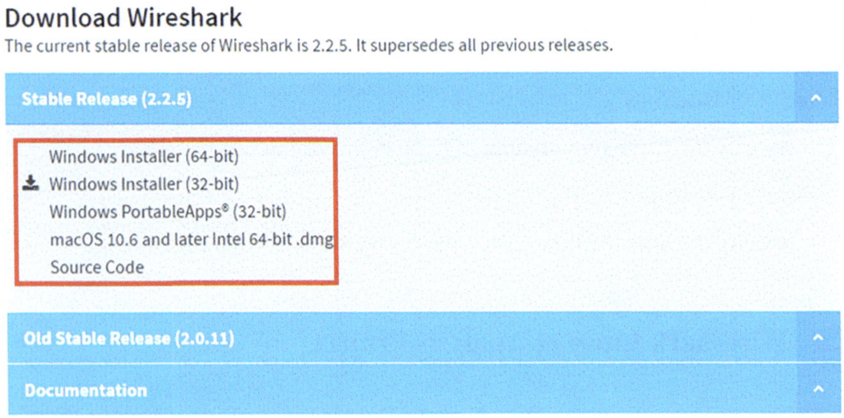

Fig. 5.9 Wireshark install options

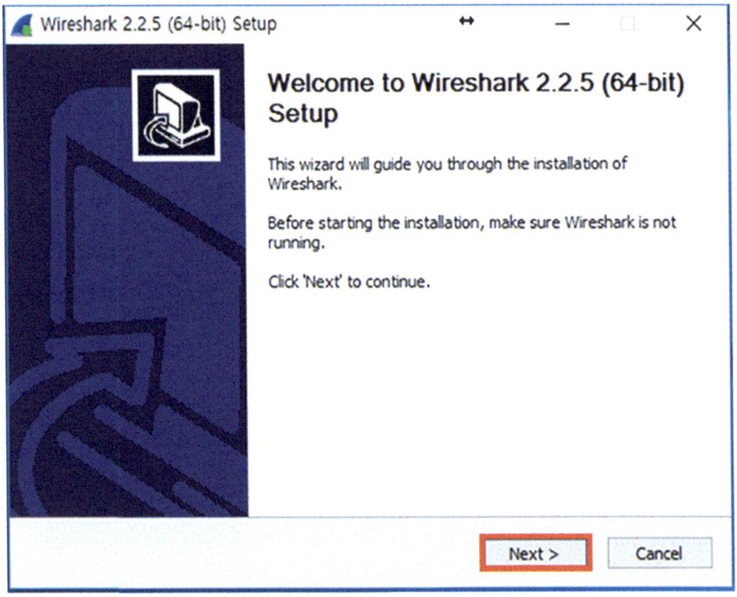

Fig. 5.10 Wireshark setup window

Figure 5.14 shows the Wireshark setup install location selection. Here we can choose the installation location (i.e., **Destination Folder**) for Wireshark and then click the "Next" button.

Figure 5.15 shows the Wireshark setup install of WinPcap. Here we need to select the **Install WinPcap** option, which will help us capture live network packet data, and then click the "Next" button.

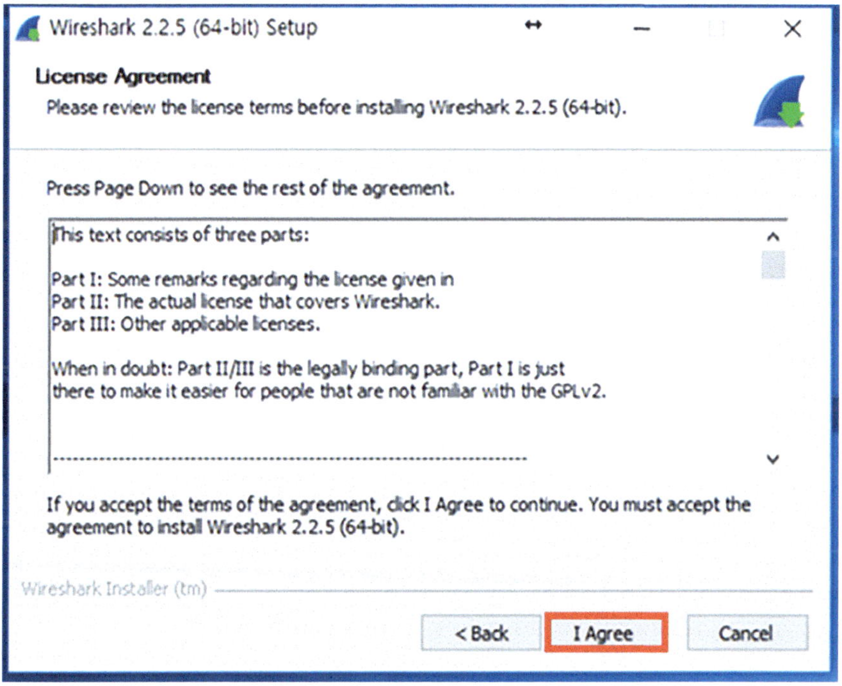

Fig. 5.11 Wireshark setup license agreement

Figure 5.16 presents the Wireshark installation complete window. We need to wait until the installation is completed and then click the "Next" button.

Figure 5.17 presents the Wireshark setup completion reboot window. Rebooting your computer will complete the installation process.

5.2.1 Source and Destination IP Addresses

In this section, we will use the Wireshark that we installed to conduct Internet packet analysis of HTTPS transfer. We will need to use the procedures in Sect. 5.1 "Internet Connection Configuration Analysis" to open the Command Prompt window. As a reminder, if you have a **Windows key** (e.g., , ,) on your keyboard, then press the **Windows key** and **R** key simultaneously to open the **Run** window and then type in "**cmd**" and click **OK**. Then in the Command Prompt window type "ipconfig" to find your computer's IP address. Figure 5.18 shows an example of a computer's source IP address (165.132.126.198) found using ipconfig.

Figure 5.19 shows how to find the destination IP address using Tracert, where we need to type the Tracert command "tracert (target address)" to find a website's IP

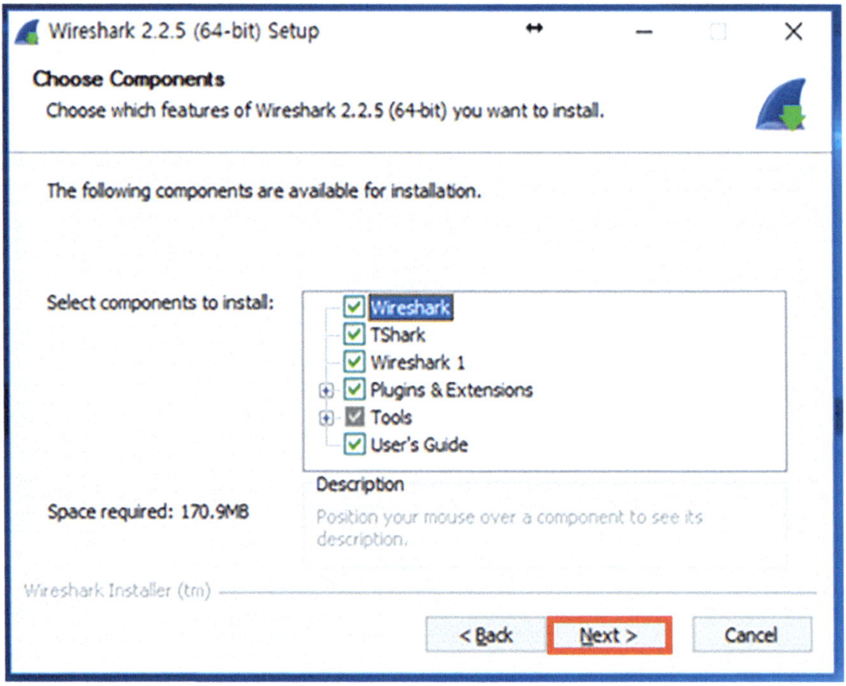

Fig. 5.12 Wireshark setup component selection

address. Here we will use the example of the USA Today website (www.usatoday.com) which is a representative newspaper of the USA.

Although "www.usatoday.com" was used in this example, any website can be used in your experiment. Next, run Wireshark and apply a filter using your IP address "ip.add == (address)" which is a filter (the "ip.add ==" part) used to focus only on packets that use the IP address written in the "(address)" area. Figure 5.20 shows an example of Wireshark using a source IP address filter.

In addition, we can apply another filter just to focus on your IP address and the website's IP address you choose by using the filer "ip.addr == (address 1) && ip.addr == (address 2)" in Wireshark. Figure 5.21 shows an example of Wireshark using a source IP address and destination address filter.

5.2.2 TCP Three-Way Handshake, Window Size and Security

By accessing a website of your choice, using Wireshark, you can identify the TCP handshake between your computer and the server being setup. Figure 5.22 presents an example of a TCP three-way handshake used to setup a TCP session. More details on a TCP three-way handshake can be found in Chap. 4 (Sect. 4.5 TCP) of this book.

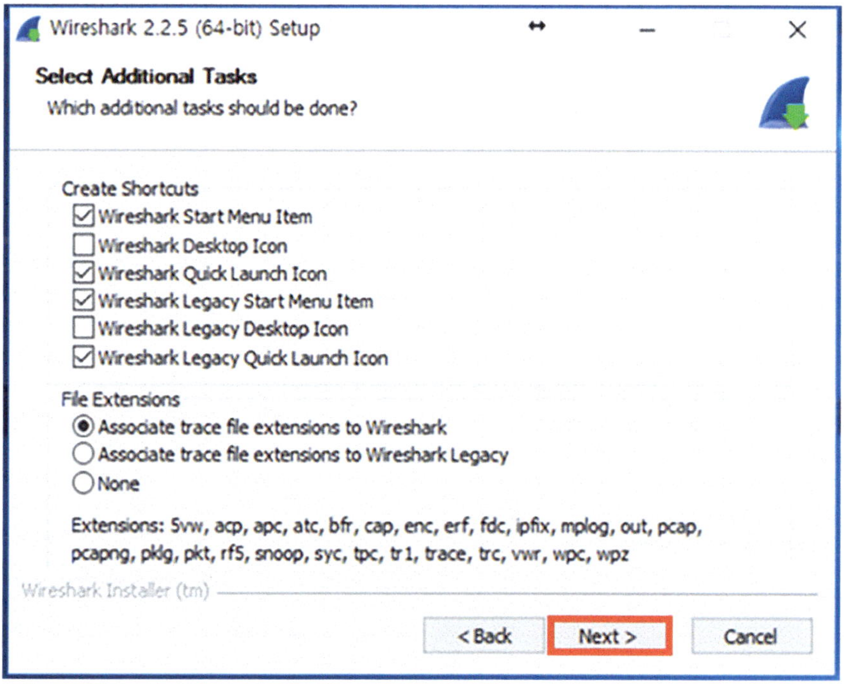

Fig. 5.13 Wireshark setup selection of additional tasks

We can apply the filter "ip.addr == (address 1) && ip.addr == (address 2) && tls" in Wireshark to identify the TLSv1.3 handshake between your computer and the server. Transport Layer Security (TLS) is a transport-layer (i.e., layer 4 in the TCP/IP 5 layer model) cryptographic protocol, where we can see that TLS version 1.3 (v1.3) is being used in Fig. 5.23. The TLSv1.3 standard was published by the Internet Engineering Task Force (IETF) in the RFC 8446 document in 2018, and has been in use since its publication.

Figure 5.23 presents the TLSv1.3 handshake between client (165.132.126.198) and destination (146.75.50.62). Figure 5.24 presents the TCP handshake information between the client (165.132.126.198) and USA Today website (146.75.50.62).

Figure 5.25 presents the TCP window size changes, when the window size increases, TCP data segments can be transported to the destination at a higher rate. On the other hand, when the window size decreases, TCP data segments can be transported to the destination at a slower rate. In Fig. 5.25, the TCP window size increases significantly enabling TCP data segment transportation at a much higher rate.

Figure 5.26 presents the TCP retransmission information. Retransmission occurs when the TCP data segments are not received at the destination, resulting in a segment delivery failure. In this case, the destination server/computer's TCP system will send duplicated ACKs to the source computer that is sending the TCP data segments. When

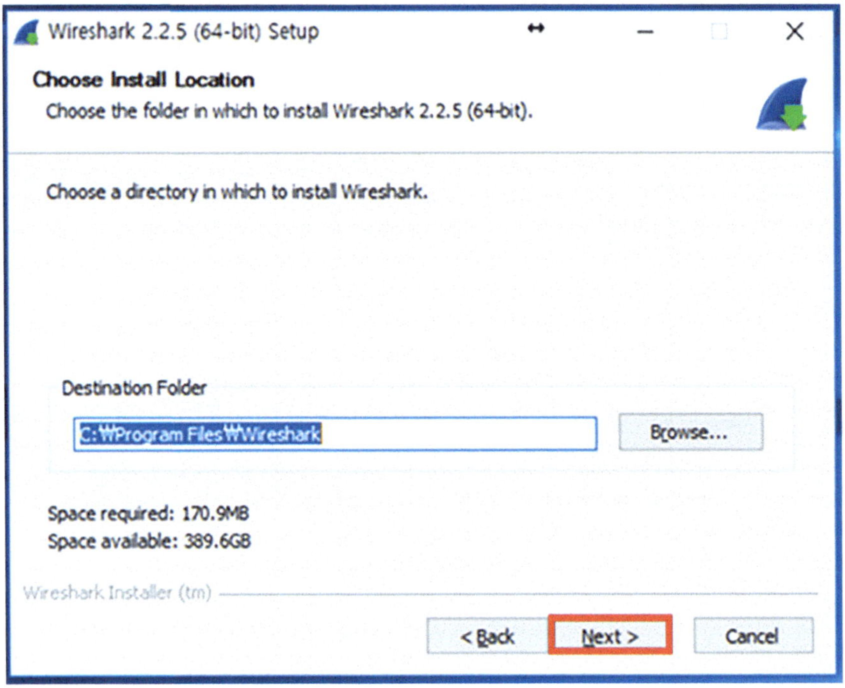

Fig. 5.14 Wireshark setup install location selection

three ACKs with the same TCP sequence number are sent back to the source computer (that is sending the TCP data segments), then the source computer will send that data segment to the destination again, which is a **TCP Retransmission**. The three ACKs with the same TCP sequence number indicate that three segments of data that were sent after this missing TCP sequence number segment have been already received at the destination, and the missing segment is needed.

Much more details of TCP are explained in Chap. 4 of this book.

5.2.3 File Transfer I/O Graph and Data Rate Analysis

Figure 5.27 presents the Wireshark I/O graph selection in Wireshark, where the data rate graph can be used by selecting **Statistics** from the top menu and then select the "**I/O graph**" option.

In the **I/O Graph**, you can enter your IP address and the Server's IP address in the "**Display filter tab**" and change the "**Y Axis**" tab to "**Bytes**" as shown below in Fig. 5.28.

Figure 5.29 presents the Wireshark I/O graph example with data transfer rate and total elapsed download time information marked on the graph.

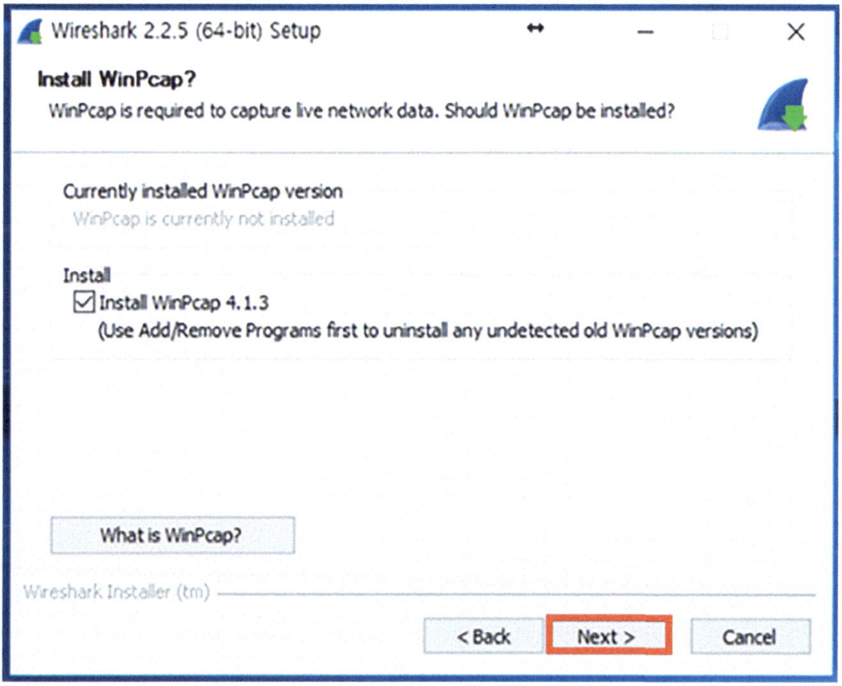

Fig. 5.15 Wireshark setup install of WinPcap

If you use a Google Chrome web browser, sometimes the Google drive uses a different transport protocol such as Quick UDP Internet Connections (QUIC). Figure 5.30 presents an example of Wireshark used on a Google Chrome web browser that applies QUIC instead of TCP for transport session control between the Client (165.132.126.158) and the Google Drive (172.217.25.225).

5.2.4 IP Address Based Physical Location Identification

In this section, we will use the **InfoSniper** application to find the website's location using its IP address [5]. First, we need to go to the website http://www.infosniper.net and enter the IP address we want to check and press the **Check** Button.

Figure 5.31 presents an example of the InfoSniper website used to find the location of the IP address. Figure 5.32 shows the InfoSniper results when used to find the location of the Google server using its IP address. Information like Provider, Hostname, Timezone, Country, Latitude, Longitude, etc. can be provided by the InfoSniper. From the location of the destination we identified using the InfoSniper application, and using the RTT information obtained from the Ping command, we can check if

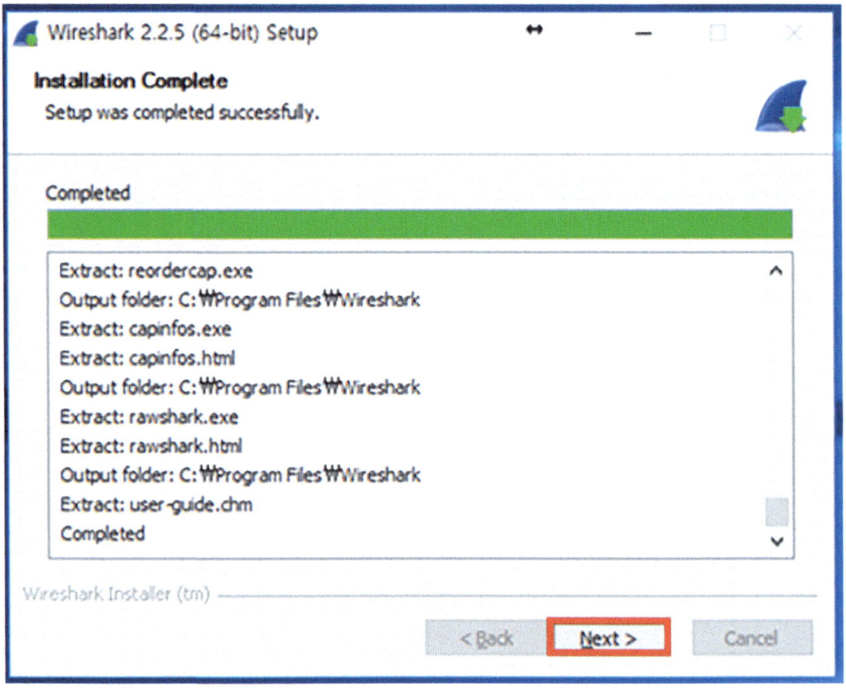

Fig. 5.16 Wireshark setup installation complete

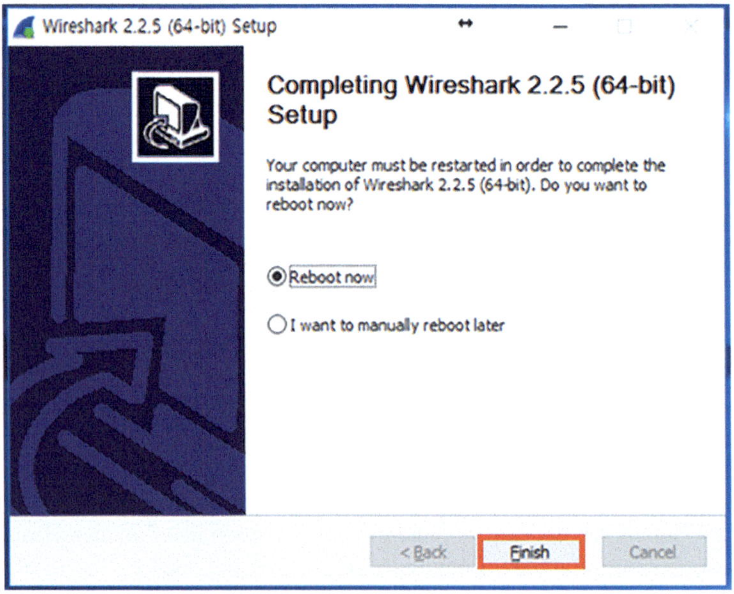

Fig. 5.17 Wireshark setup completion reboot

```
Command Prompt

C:\Users\Project>ipconfig

Windows IP Configuration


Ethernet adapter

   Media State . . . . . . . . . . . : Media disconnected
   Connection-specific DNS Suffix  . :

Ethernet adapter

   Connection-specific DNS Suffix  . :
   Link-local IPv6 Address . . . . . : fe80::8f78:fa10:6c90:f396%8
   IPv4 Address. . . . . . . . . . . : 165.132.126.198
   Subnet Mask . . . . . . . . . . . : 255.255.252.0
   Default Gateway . . . . . . . . . : 165.132.124.1
```

Fig. 5.18 Find the computer's source IP address using IPconfig

```
Command Prompt

C:\Users\Project>Tracert www.usatoday.com

Tracing route to domains.gannett.map.fastly.net [146.75.50.62]
over a maximum of 30 hops:

  1     2 ms     3 ms     3 ms  165.132.45.1
  2     1 ms    <1 ms    <1 ms  165.132.2.157
  3    <1 ms    <1 ms    <1 ms  10.8.0.10
  4     1 ms     1 ms    <1 ms  10.6.0.12
  5     1 ms     1 ms     1 ms  10.4.0.10
  6     2 ms     1 ms     1 ms  10.2.0.3
  7     2 ms     2 ms     1 ms  210.98.171.201
  8     2 ms     2 ms     2 ms  1.213.1.173
  9     3 ms     2 ms     2 ms  1.213.2.173
 10     8 ms    15 ms    17 ms  1.213.113.173
 11     3 ms     2 ms     2 ms  1.208.175.10
 12     4 ms     3 ms     3 ms  61-111-61-62.kidc.net [61.111.61.62]
 13     3 ms     3 ms     3 ms  61-111-34-102.kidc.net [61.111.34.102]
 14     *       20 ms     5 ms  211.234.118.46
 15     4 ms     3 ms     3 ms  203.246.169.175
 16     3 ms     3 ms     3 ms  121.78.30.146
 17     4 ms     3 ms     3 ms  146.75.50.62

Trace complete.
```

Fig. 5.19 Find the destination IP address using Tracert

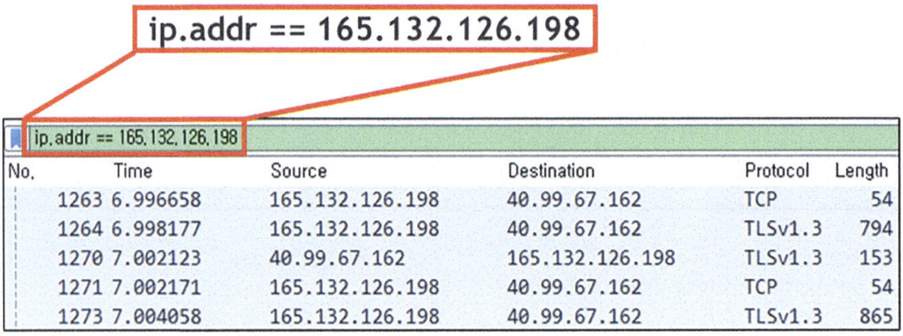

No.	Time	Source	Destination	Protocol	Length
1263	6.996658	165.132.126.198	40.99.67.162	TCP	54
1264	6.998177	165.132.126.198	40.99.67.162	TLSv1.3	794
1270	7.002123	40.99.67.162	165.132.126.198	TLSv1.3	153
1271	7.002171	165.132.126.198	40.99.67.162	TCP	54
1273	7.004058	165.132.126.198	40.99.67.162	TLSv1.3	865

Fig. 5.20 Example of Wireshark using a source IP address filter

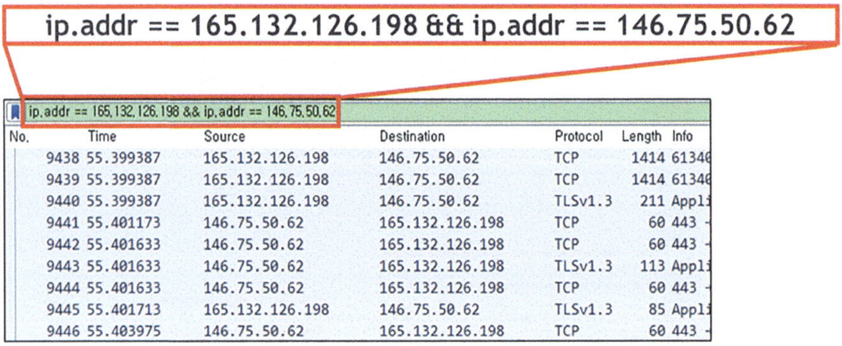

No.	Time	Source	Destination	Protocol	Length	Info
9438	55.399387	165.132.126.198	146.75.50.62	TCP	1414	6134
9439	55.399387	165.132.126.198	146.75.50.62	TCP	1414	6134
9440	55.399387	165.132.126.198	146.75.50.62	TLSv1.3	211	Appli
9441	55.401173	146.75.50.62	165.132.126.198	TCP	60	443 -
9442	55.401633	146.75.50.62	165.132.126.198	TCP	60	443 -
9443	55.401633	146.75.50.62	165.132.126.198	TLSv1.3	113	Appli
9444	55.401633	146.75.50.62	165.132.126.198	TCP	60	443 -
9445	55.401713	165.132.126.198	146.75.50.62	TLSv1.3	85	Appli
9446	55.403975	146.75.50.62	165.132.126.198	TCP	60	443 -

Fig. 5.21 Example of Wireshark using a source IP address and destination address filter

Fig. 5.22 TCP three-way handshake used to setup the TCP session

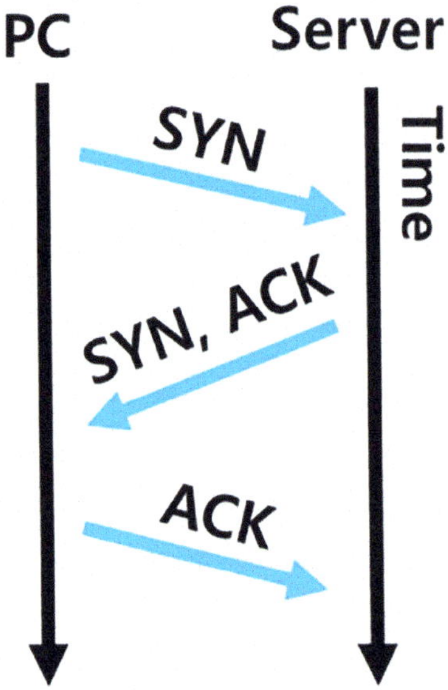

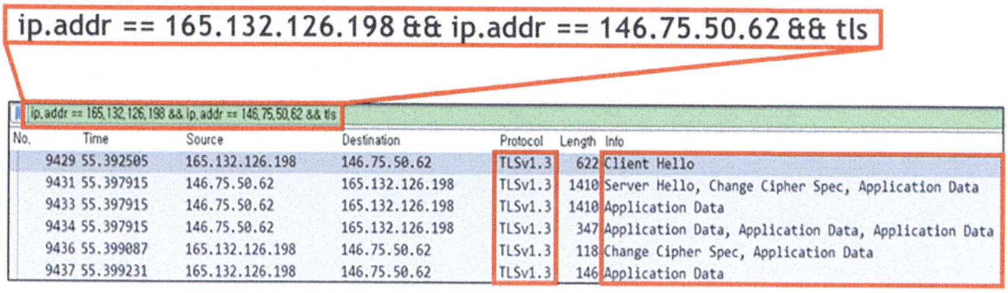

Fig. 5.23 TLSv1.3 handshake between client and destination

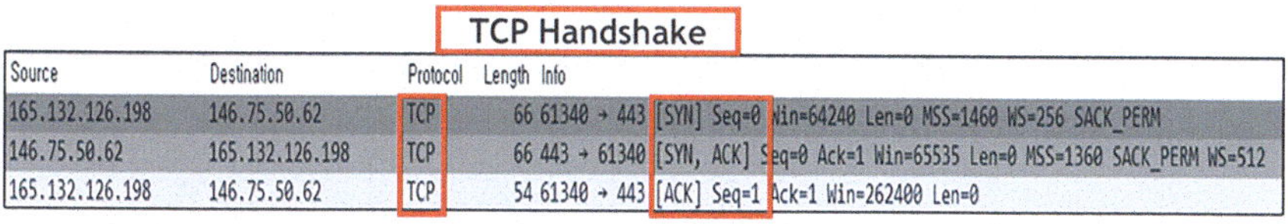

Fig. 5.24 TCP handshake information

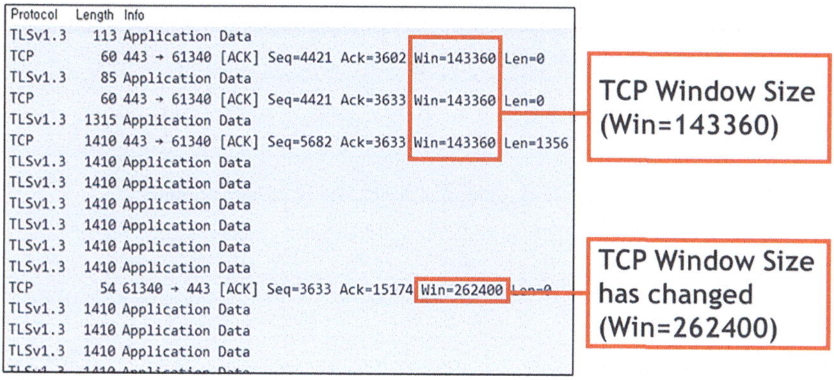

Fig. 5.25 TCP window size changes

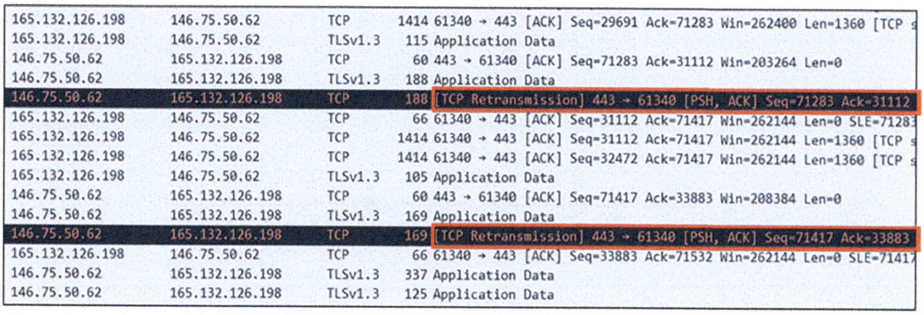

Fig. 5.26 TCP retransmission information

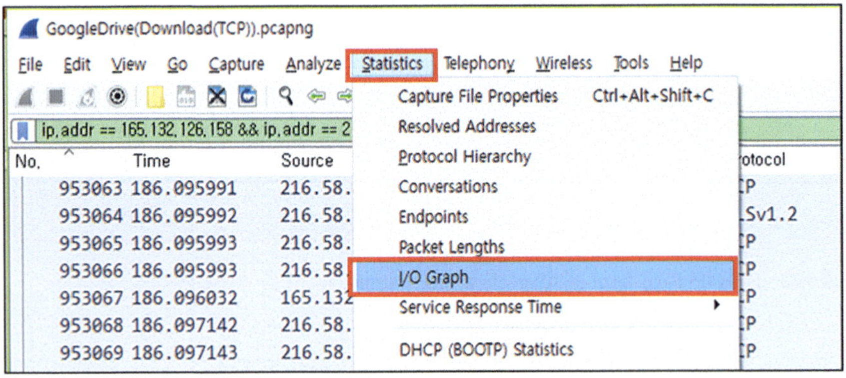

Fig. 5.27 Wireshark I/O graph selection

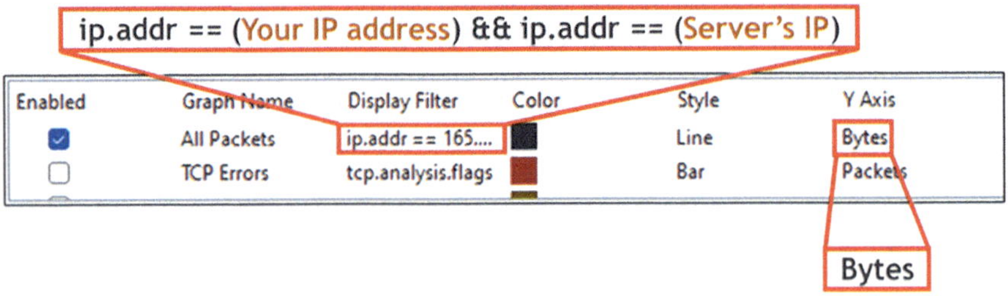

Fig. 5.28 Wireshark I/O graph setting source and destination IP address selection

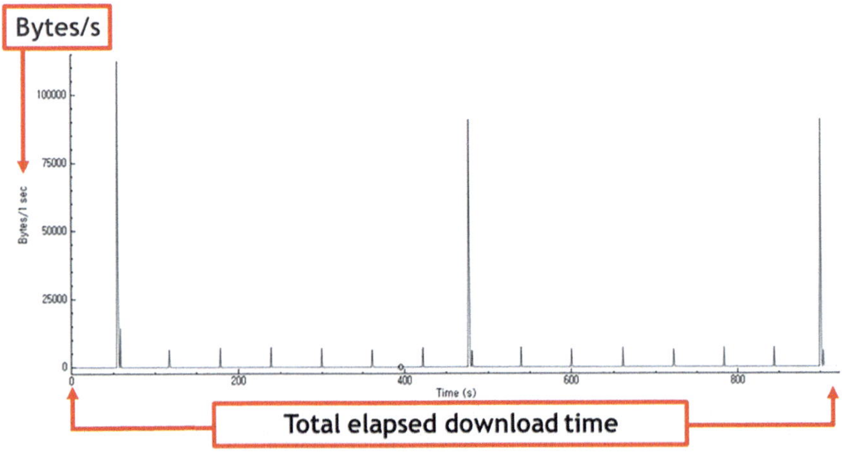

Fig. 5.29 Wireshark I/O graph example with data transfer rate and total elapsed download time information

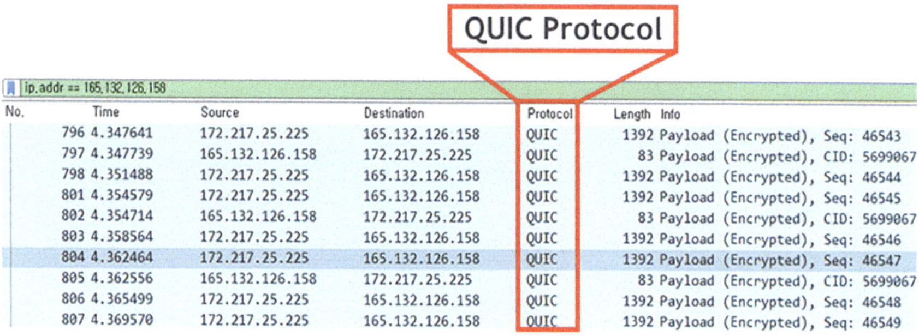

No.	Time	Source	Destination	Protocol	Length	Info
796	4.347641	172.217.25.225	165.132.126.158	QUIC	1392	Payload (Encrypted), Seq: 46543
797	4.347739	165.132.126.158	172.217.25.225	QUIC	83	Payload (Encrypted), CID: 5699067
798	4.351488	172.217.25.225	165.132.126.158	QUIC	1392	Payload (Encrypted), Seq: 46544
801	4.354579	172.217.25.225	165.132.126.158	QUIC	1392	Payload (Encrypted), Seq: 46545
802	4.354714	165.132.126.158	172.217.25.225	QUIC	83	Payload (Encrypted), CID: 5699067
803	4.358564	172.217.25.225	165.132.126.158	QUIC	1392	Payload (Encrypted), Seq: 46546
804	4.362464	172.217.25.225	165.132.126.158	QUIC	1392	Payload (Encrypted), Seq: 46547
805	4.362556	165.132.126.158	172.217.25.225	QUIC	83	Payload (Encrypted), CID: 5699067
806	4.365499	172.217.25.225	165.132.126.158	QUIC	1392	Payload (Encrypted), Seq: 46548
807	4.369570	172.217.25.225	165.132.126.158	QUIC	1392	Payload (Encrypted), Seq: 46549

Fig. 5.30 Wireshark used on a Google Chrome web browser that applies QUIC instead of TCP for transport session control

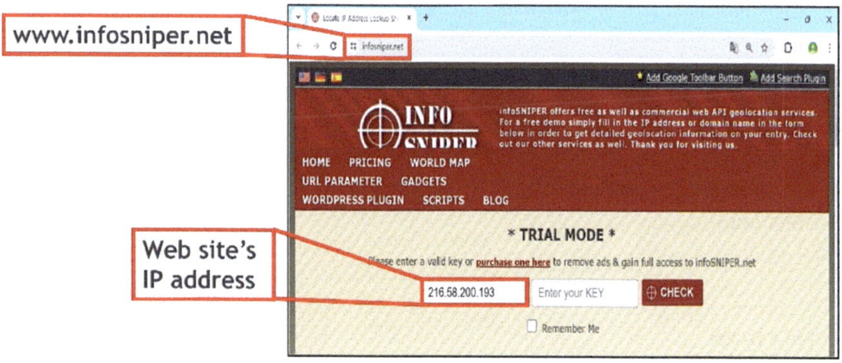

Fig. 5.31 InfoSniper website used to find location of the IP address

the distance to the destination is proportional to the RTT duration. In addition, we can check if the number of hops to reach the destination using the Tracert command is also proportional to the distance of the destination (e.g., check for possible routing path detour). From these observations, we can see if the distance is the main cause of time delay (i.e., signal propagation time) with the destination server, or if the server or network is slow (i.e., due to congestion or poor network conditions, etc.).

5.3 Summary

In this chapter, details of how to check a computer's Internet connection configurations as well as conduct Internet ping and trace route measurements are provided, followed by a Wireshark Internet analysis project. The project goals accomplished by this project are listed in the following. First, we checked real Internet packets exchanged and observe network operations using Wireshark. Second, we observed

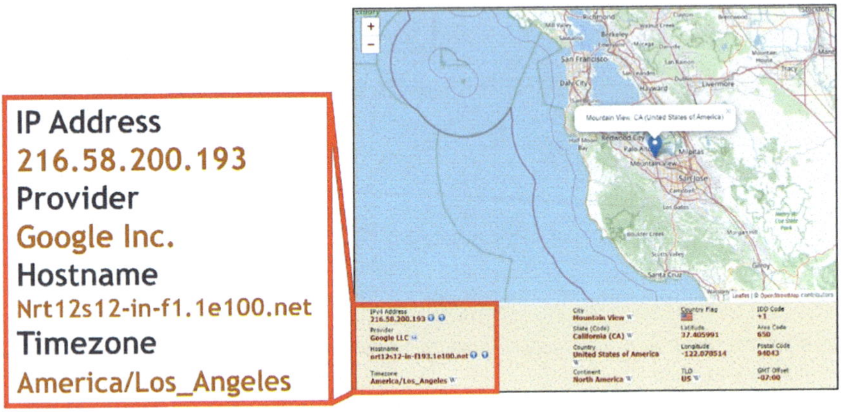

Fig. 5.32 InfoSniper website used to find the location of the Google server from its IP address

TCP, TLS, UDP, and DNS operations conducted in an active network connected to the Internet. In addition, we checked the location of a website server using its IP address and analyzed the (possible) delay factors. This chapter provided an experimental project that can be used to confirm the information described in Chaps. 3 and 4 of this book.

References

1. Microsoft Windows IT Center, Docs, Administration, ipconfig. https://docs.microsoft.com/en-us/windows-server/administration/windows-commands/ipconfig. Accessed 4 May 2017
2. Microsoft Windows IT Center, Docs, Administration, ping. https://docs.microsoft.com/en-us/windows-server/administration/windows-commands/ping. Accessed 4 May 2017
3. Microsoft Windows IT Center, Docs, Administration, tracert. https://docs.microsoft.com/en-us/windows-server/administration/windows-commands/tracert. Accessed 4 May 2017
4. Wireshark. https://www.wireshark.org/download.html
5. Infosniper. http://www.infosniper.net

Chapter 6
Hash Technology

This chapter introduces hash technologies that are commonly used in databases, blockchains, Bitcoin, cryptocurrencies, and smart contracts. First, Sect. 6.1 introduces hash table operations. In Sect. 6.2, hash table open addressing technologies are described, which include linear probing, plus 3 rehash, quadratic probing, and double hashing. In Sect. 6.3, hash table closed addressing technology is introduced. Next, examples of each of these hash table operations are described in Sect. 6.4. In Sect. 6.5, hash table load factor and dynamic data structures are introduced, which are used in determining when a hash table or database needs to be expanded. Section 6.6 introduces the Secure Hash Algorithm (SHA) family protocols, which are used in all blockchains, Bitcoin, cryptocurrencies, Ethereum, and smart contracts. SHA based preimage resistance and second-preimage resistance as well as preimage attacks and second-preimage attacks are described followed by details on SHA preimage attack defense methods and double hashing. Considering that Bitcoin uses the double hashed SHA-256 (SHA256D) mechanism of the SHA-2 family protocols, and Ethereum uses the SHA-3 family protocols, evidently the content of this chapter is very important.

A 'Hash Function' is a function that can map an arbitrary size data file into a fixed size hash value. Hash functions are used in hash maps and hash tables. Hash functions are used in locating a data record based on a search key (a.k.a., headword) in a database or memory system. A Key is used as the input to the hash function, which outputs a hash value that is used as the index position of where the data file (Value) corresponding to the Key value will be stored in the database. This is why hash functions, hash maps, and hash tables are based on (Key, Value) pairs [1–3].

Hashing is needed because without using a hash function, finding a data file in a database will require a sequential search process, which will take too much time and processing energy. This is because a sequential search process requires a search time that depends on the database size and the position of the data file in the database. Especially, as the database becomes larger (which is very common in the Big Data and Artificial Intelligence (AI) era of today's cloud computing based industries), using a hash function to assign an index position for a data file (Value) to be stored

J.-M. Chung, *Emerging Secure Networks, Blockchains and Smart Contract Technologies*, https://doi.org/10.1007/978-3-031-65866-2_6

in a database is known to be much more efficient compared to using a sequential search down through a large database.

Finding a data file using the file's Index number generated from a hash value in a database can be conducted quickly. Hash value index search speeds are fast because the process is independent of the database size and the location of the data file within the database. Because hash functions enable easy index number generation using the data file's own characteristics, files are distributed randomly throughout the database, so certain directories in the database do not get overloaded easily. This is why all modern large databases use hash tables, which use hash functions to conduct very fast index assignment, file retrieval, and confirmation of a data file's location in the database regardless of the data size. Hash tables are used in database indexing, caching, program compilation, metadata, authentication, error checking, and all memory devices.

Hash functions use the data file itself to create an index number for itself to be located within the database, which allows immediate database access of the data.

The hash function **H(·)** is a function that generates a fixed (bit) length output hash value regardless of the input (Key or file) size. Hash functions use the modulo 'mod' operator to generate a hash value with fix bit length to be used as the index of the file to store in the database.

For example, Key can be converted using various codes (e.g., ASCII: American Standard Code for Information Interchange [4]) and folding can be applied. The folding operation divides the file into equal parts and then adds the parts together. For example, the Key 0,123,456,789, which is a ten digit number, is divided into a pair of two digits and the values are folded every two digits and added up, resulting in $(01 + 23 + 45 + 67 + 89) = 225$, which can be used by the hash function to create an index value.

In hash tables and hash functions, 'hash security' is based on the level of difficulty that the original input file could be recreated from its output hash value (i.e., ideally we want it to be impossible to recreate the original file from its hash value). Hash security and the hashing speed are important factors in deciding what hash function (i.e., algorithm) to use. Many algorithms can be used to hash a file. Some of the hash function types that have been used include CRC32, MD2, MD5, Tiger, Alder32, RipeMD128, etc. Especially, the Secure Hash Algorithm (SHA) group standards like SHA-1, SHA-2, and SHA-3 are used very frequently.

A 'Hash Map' is formed based on a collection of (Key, Value) pairs. An object-oriented database can be formed using a hash map, where the 'Key' represents the instance of a class and the 'Value(s)' represent the property(s) of the class. The Key is used as input to the hash function to generate the hash value, which is used as the index value. The index value is used to determine where the file (Value) is to be placed in a database. The Key value is also used to find the file (Value) in the database later. It is desired for the index values to be all unique in a hash map, but the Values (files) can be any value, where some may overlap.

Hash map and hash table are used in the same terms in many documents. There are a few differences between the two. One difference is that a hash map is called 'thread-unsafe' because if a Key is accessed at the same time by two or more operations (e.g., read and/or save operations) then a conflict in operation may occur. A hash table can be seen as a 'thread-safe' enhanced version of a hash map, as it allows multiple Key operations in a stable and secure manner based on synchronization of executable operations.

The term 'thread' is a simple sequence of operations executed by a scheduler. A thread is commonly executed on a single processing core through the operating system (OS) that is controlled by the central processing unit (CPU) of a computer. Computers that have multiple processing cores can process multiple threads at the same time, which makes the computer more versatile and enhances the performance to enable multi-tasking functionality.

6.1 Hash Table Operations

It is desired for a hash function to generate a unique output hash value for each input file. However, hash tables have an index size limit. Therefore, as the number of input files increase, due to the limitation in the database index size, some of the input files will result in the same output hash value. Therefore, the input files will need to have an additional indexing rule applied before they can be assigned to a database index location. There are two types of addressing schemes that are used to solve the overlapping index positions in a database. One is the 'Open Addressing' scheme and the other is the 'Closed Addressing' scheme. The Open Addressing scheme includes algorithms like Linear Probing, Plus 3 Rehash, Quadratic Probing, and Double Hashing. The Closed Addressing scheme is represented by the Chaining algorithm. A description of each algorithm is provided in the following [1–3].

6.2 Hash Table Open Addressing Types

The representative hash table open addressing types include linear probing, plus 3 rehash, quadratic probing, and double hashing. A brief description of these techniques are provided below and examples of hash table open addressing schemes are provided in the following section.

6.2.1 *Linear Probing*

If the index value generated from the hash function is a value that overlaps with a previously generated hash value, then an index 'collision' has occurred, which needs to be solved. In linear probing, the next nearest index location that is empty is used.

Linear probing can create 'primary clustering,' which is a phenomenon that generated hash values are placed in indexes that are all located in clusters near each other, creating over populated areas in the database. It is preferable for the database to be equally utilized in a random scattered fashion, in which linear probing does not perform well in this manner.

6.2.2 Plus 3 Rehash

If the index value generated from the hash function is already occupied (collision), then the 3rd nearest empty index location is used.

Plus 3 Rehash will help alleviate 'primary clustering' somewhat because a little more spacing between collided index positions will help to spread out a little more since two additional empty index positions will be skipped before the (Key, Value) is placed in the database's next empty location.

6.2.3 Quadratic Probing

If the index value generated from the hash function is already occupied (collision), then the number of failed attempts (number of collisions) **X** is used, where a $\mathbf{X}^2$ number of index slots down will be checked to see if that index location is empty. If it is empty, then it will be used to store the (Key, Value). If that slot is not empty then **X** will increase to $(\mathbf{X} + 1)$, and $(\mathbf{X} + 1)^2$ number of index slots down will be checked to see if that index location is empty. If it is empty, then it will be used to store the (Key, Value). This is repeated until an empty index position is found where the (Key, Value) can be stored, where every time a collision occurs, the **X** value is increased by 1.

Quadratic Probing will result in an increasing $\mathbf{X}^2$ pattern of 1, 4, 9, 16, 25, 36, 49, 64, … in finding the next index position to use. This will help to alleviate 'primary clustering' because there will be more spacing between index positions.

6.2.4 Double Hashing

Double hashing uses two hash functions $\mathbf{H}_1(\cdot)$ and $\mathbf{H}_2(\cdot)$. If the index value generated from the first hash function $\mathbf{H}_1(\cdot)$ is already occupied (collision), then the $\mathbf{H}_2(\cdot)$ is used to compute another hash value, which will be the number of slots down that will be checked to see if that index location is empty. If it is empty, then it will be used to store the (Key, Value).

The index hash values generated from the two hash functions $\mathbf{H}_1(\cdot)$ and $\mathbf{H}_2(\cdot)$ are independent and individually have a (pseudo) random nature. The original index position determined from $\mathbf{H}_1(\cdot)$, and the number of slots from this position to the next index position obtained from $\mathbf{H_2}(\cdot)$, creates a well distributed index spread-out throughout the database. Therefore, primary clustering can be avoided, and the database index positions will most likely to be equally utilized in a random scattered fashion. The downside of this method is that two hash functions $\mathbf{H_1}(\cdot)$ and $\mathbf{H_2}(\cdot)$ have to be computed, which is more complex.

More details on double hashing is provided in the following section on SHA-2 hashing algorithm's preimage attack defense methods.

6.3 Hash Table Closed Addressing Types

Chaining is a representative technology of closed addressing. Chaining will store the data file (Value) in the index position based on the hash value even if the index position has already been assigned to another (Key, Value) pair. If an index position collision occurs, the chaining process will allocate an additional memory space for the additional (Key, Value) pair to be stored in a chained position to the original index position in the database.

Chaining will simplify the process of needing to change the index position if an index position collision occurs. However, chaining adds on a new dimension of complexity to the memory allocation process, as the memory needs to be able to quickly allocate the chained data files in the index position assignment process and find the index position quickly in search processes.

Because closed addressing schemes like chaining can expand the database index and storage dimension based on the number of collisions in indexes that occur, 'closed addressing' is also called 'open hashing' as the hashing table has to have an open structure, such that it can be expanded based on index assignments that collide.

6.4 Examples of Hash Table Techniques

To assist the understanding of this description, examples are provided below and illustrated using Figs. 6.2, 6.3, 6.4, 6.5, 6.6, 6.7, 6.8, 6.9, 6.10, and 6.11. There are ten names listed in the left-end column of Fig. 6.1. Each name is three letters in this example, where each person's name is used as the Key to store a datafile on that person based on the hash value based index position [1–3].

First, each Key (i.e., each name) is converted into a binary value using the ASCII code of each letter [4], as presented in Fig. 6.1.

Second, the ASCII code value of each Key (name) is added up for the three letters, which is the 'ASCII Sum' value in the table.

Third, by applying a simple modulo $\mathbf{N} = 13$ process (mod13) as the hash function, the right-end column presents the hash value indexes of the Keys (names).

Name	First Letter	ASCII code	Second Letter	ASCII code	Third Letter	ASCII code	ASCII Sum	mod13
Mia	**M**	77	**i**	105	**a**	97	279	6
Tim	**T**	84	**i**	105	**m**	109	298	12
Bea	**B**	66	**e**	101	**a**	97	264	4
Zoe	**Z**	90	**o**	111	**e**	101	302	**3**
Sue	**S**	83	**u**	117	**e**	101	301	**2**
Len	**L**	76	**e**	101	**n**	110	287	1
Moe	**M**	77	**o**	111	**e**	101	289	**3***
Lou	**L**	76	**o**	111	**u**	117	304	5
Rae	**R**	82	**a**	97	**e**	101	280	7
Noc	**N**	78	**o**	111	**c**	99	288	**2***

Fig. 6.1 Keys (names) converted into indexes using ASCII code values and the hash function mod13. Indexes (hash values) that experience a collision (i.e., Moe and Noc) have an asterisk (*) marking

0	**1**	**2**	**3**	**4**	**5**	**6**	**7**	**8**	**9**	**10**	**11**	**12**
	Len	Sue	Zoe	Bea		Mia						Tim

(a)

0	**1**	**2**	**3**	**4**	**5**	**6**	**7**	**8**	**9**	**10**	**11**	**12**
	Len	Sue	Zoe	Bea	**Moe**	Mia						Tim

(b)

Fig. 6.2 Open addressing with linear probing example **a** before assignment of Moe, and **b** although Moe is originally assigned to index position 3, due to an index collision in the hash table, Moe is linearly shifted up in to the next empty index position 5

0	**1**	**2**	**3**	**4**	**5**	**6**	**7**	**8**	**9**	**10**	**11**	**12**
	Len	Sue	Zoe	Bea	Moe	Mia	**Lou**					Tim

Fig. 6.3 Open addressing with linear probing example, where Lou is assigned to an index position after an index collision in the hash table

0	**1**	**2**	**3**	**4**	**5**	**6**	**7**	**8**	**9**	**10**	**11**	**12**
	Len	Sue	Zoe	Bea	Moe	Mia	Lou	**Rae**				Tim

Fig. 6.4 Open addressing with linear probing example, where Rae is assigned to an index position after an index collision in the hash table

0	1	2	3	4	5	6	7	8	9	10	11	12
	Len	Sue	Zoe	Bea	Moe	Mia	Lou	Rae	**Noc**			Tim

Fig. 6.5 Open addressing with linear probing example, where Noc is assigned to an index position after an index collision in the hash table

0	1	2	3	4	5	6	7	8	9	10	11	12
	Len	Sue	Zoe	Bea		Mia			**Moe**			Tim

Fig. 6.6 Open addressing with plus 3 rehash example, where Moe is assigned to an index position after an index collision in the hash table

0	1	2	3	4	5	6	7	8	9	10	11	12
	Len	Sue	Zoe	Bea	Lou	Mia	Rae	**Noc**	Moe			Tim

Fig. 6.7 Open addressing with plus 3 rehash example, where Noc is assigned to an index position after an index collision in the hash table

0	1	2	3	4	5	6	7	8	9	10	11	12
	Len	Sue	Zoe	Bea		Mia	**Moe**					Tim

Fig. 6.8 Open addressing with quadratic probing example, where Moe is assigned to an index position after an index collision in the hash table

0	1	2	3	4	5	6	7	8	9	10	11	12
	Len	Sue	Zoe	Bea	Lou	Mia	Moe	**Rae**				Tim

Fig. 6.9 Open addressing with quadratic probing example, where Rae is assigned to an index position after an index collision in the hash table

There are two collisions in the indexes, as the index 3 is same for the Keys (names) Zoe and Moe, and index 2 is the same for the Keys (names) Sue and Noc. Therefore, the hash table collision resolution schemes need to be used.

Since **N** = 13, this means that the database size is 13, the names (Keys) will be assigned to the database index positions from 0 to 12. This is because the hash function applied is mod13 which can only result in remainder values after dividing with **N** = 13, which are the values from 0 to 12. The Keys (names) are assigned to the database index positions from the top to bottom, so first Mia will be assigned to the hash table followed by Tim, Bea, Zoe, Sue, Len, Moe, Lou, Rae, and Noc.

So, the index assignments in the hash table proceed with first Mia being assigned to index position 6, followed by Tim to index position 12, Bea to index position 4, Zoe to index position 3, Sue to index position 2, and Len is assigned to index position 1. So far there were no collisions in the index positions of the hash table, as presented in Fig. 6.2(a). But, when Moe is assigned to index position 3, a collision occurs with

0	1	2	3	4	5	6	7	8	9	10	11	12
	Len	Sue	Zoe	Bea	Lou	Mia	Moe	Rae			**Noc**	Tim

Fig. 6.10 Open addressing with quadratic probing example, where Noc is assigned to an index position after an index collision in the hash table

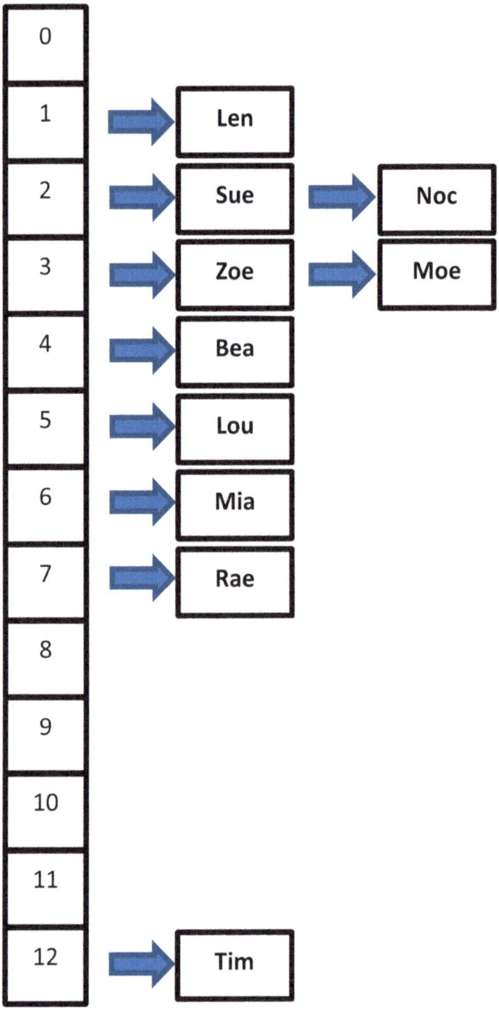

Fig. 6.11 Example of closed addressing with chaining applied

Zoe, and hash table collision resolution procedures have to be conducted, which are described in the following.

6.4.1 Open Addressing with Linear Probing Example

Figures 6.2 through 6.5 show the procedures for the open addressing with linear probing example based on the Keys (names) and mod13 hash value indexes listed in Fig. 6.1.

In the hash table, Mia is assigned to index position 6, Tim to index position 12, Bea to index position 4, Zoe to index position 3, Sue to index position 2, and Len is assigned to index position 1, as presented in Fig. 6.2(a).

Because Moe is also assigned to index position 3, a collision occurs with Zoe who is already in index position 3 in the hash table. Based on the linear probing procedures, Moe will check the next higher index position 4, but it is already occupied by Bea. So, based on the linear probing procedures, Moe will check the next higher index position 5, which is empty. So, Moe will use this index position, as presented in Fig. 6.2(b).

Based on Fig. 6.1, Lou is the next Key (name) to be assigned to index position 5, which was not a colliding index position. But due to the linear probing of Moe which already went into index position 5, now Lou's index position 5 collides with Moe. Based on the linear probing procedures, Lou will check the next higher index position 6, but it is already occupied by Mia. So, based on the linear probing procedures, Lou will check the next higher index position 7, which is empty. So, Lou will use this index position, as presented in Fig. 6.3.

Based on Fig. 6.1, Rae is to be assigned to index position 7, but due to a collision with Lou, Rae will check the next higher index position 8, which is empty. Based on the linear probing procedures, Rae will use this index position, as presented in Fig. 6.4.

Based on Fig. 6.1, Noc is to be assigned to index position 2, which collides with Sue. Due to the linear probing procedures, Noc will check the next higher index positions 3, 4, 5, 6, 7, and 8 which are all already preoccupied. So, based on the linear probing procedures, Noc will check the next higher index position 9, which is empty. So, Noc will use this index position, as presented in Fig. 6.5. Since all the Keys (names) in Fig. 6.1 are assigned to the hash table, the procedures end.

6.4.2 Open Addressing with Plus 3 Rehash Example

Figures 6.6 and 6.7 show the procedures for the open addressing with the plus 3 rehash example based on the Keys (names) and mod13 hash value indexes listed in Fig. 6.1.

In the hash table, Mia is assigned to index position 6, Tim to index position 12, Bea to index position 4, Zoe to index position 3, Sue to index position 2, and Len is assigned to index position 1.

Because Moe is also assigned to index position 3, a collision occurs with Zoe who is already in index position 3 in the hash table.

Based on the plus 3 rehash procedures, Moe will check the next +3 higher index position $(3 + 3) = 6$, but it is already occupied by Mia. So, based on the plus 3 rehash procedures, Moe will check the next +3 higher index position $(3 + 3 + 3) = 9$, which is empty. So, Moe will use this index position, as presented in Fig. 6.6.

Afterwards, based on Fig. 6.1, Lou is assigned to index position 5 and Rae is assigned to index position 7 without any collision. Next, because Noc has an index of 2, a collision occurs with Sue who is already in index position 2 in the hash table.

Based on the plus 3 rehash procedures, Noc will check the next +3 higher index position $(2 + 3) = 5$, but it is already occupied by Lou. So, based on the plus 3 rehash procedures, Noc will check the next +3 higher index position $(2 + 3 + 3) = 8$, which is empty. So, Noc will use this index position, as presented in Fig. 6.7. Since all the Keys (names) in Fig. 6.1 are assigned to the hash table, the procedures end.

6.4.3 Open Addressing with Quadratic Probing Example

Figures 6.8 through 6.10 show the procedures for the open addressing with quadratic probing example based on the Keys (names) and mod13 hash value indexes listed in Fig. 6.1.

In the hash table, Mia is assigned to index position 6, Tim to index position 12, Bea to index position 4, Zoe to index position 3, Sue to index position 2, and Len is assigned to index position 1.

Because Moe is also assigned to index position 3, a collision occurs with Zoe who is already in index position 3 in the hash table.

Based on the quadratic probing procedures, Moe will check the next $+(1)^2 = +1$ higher index position $(3 + 1) = 4$, but it is already occupied by Bea. So, based on the quadratic probing procedures, Moe will check the next higher $+(2)^2 = +4$ index position $(3 + 4) = 7$, which is empty. So, Moe will use this index position, as presented in Fig. 6.8.

Afterwards, based on Fig. 6.1, Lou is assigned to index position 5, but Rae has a collision in index position 7 because Moe already went into this index position.

Based on the quadratic probing procedures, Rae will check the next $+(1)^2 = +1$ higher index position $(7 + 1) = 8$, which is empty. So, Rae will use this index position, as presented in Fig. 6.9.

Because Noc has the index 2, a collision occurs with Sue who is already in index position 2 in the hash table.

Based on the quadratic probing procedures, Noc will check the next $+(1)^2 = +1$ higher index position $(2 + 1) = 3$, but it is already occupied by Zoe. So, based on the quadratic probing procedures, Noc will check the next higher $+(2)^2 = +4$ index position $(2 + 4) = 6$, but it is already occupied by Mia. So, based on the quadratic probing procedures, Noc will check the next higher $+(3)^2 = +9$ index position $(2 + 9) = 11$, which is empty. So, Noc will use this index position, as presented in Fig. 6.10. Since all the Keys (names) in Fig. 6.1 are assigned to the hash table, the procedures end.

6.4.4 Closed Addressing with Chaining Example

In Fig. 6.11, an example of closed addressing with chaining is presented for the Keys (names) and mod13 hash value indexes listed in Fig. 6.1. Because Zoe and Moe have the same index 3, and Zoe was assigned to the hash table before Moe, Moe is chained to the index position of Zoe. Likewise, because Sue and Noc have the same index 2, and Sue was assigned to the hash table before Noc, Noc is chained to the index position of Sue.

Therefore, if the file for Moe is searched by someone in the database, the Key 'Moe' will be used for the search process. Then the ASCII code sum of Moe will be derived and the hash function 'mod13' will be applied to compute the hash value of 3. The index position 3 in the database will be searched. Since Zoe will be found in this index position, next a search down the chained list will be conducted, where the file for Moe will be found.

6.5 Load Factor and Dynamic Data Structure

Considering the massive amount of data that is being generated and stored, there is a definite need to monitor the status of the hash table and conduct an expansion when needed. For this purpose, load factor monitoring and dynamic data structures are needed.

The 'Load Factor' provides information of how much the hash table is filled up. The Load Factor is obtained from dividing the 'Total number of items to store' by the 'Size of the Hash Table (**N**)' as presented below. The Load Factor will result in a number between 0 and 1.

$$\text{Load Factor} = \frac{\text{Total number of items to store}}{\text{Size of the Hash Table } (\mathbf{N})}$$

If the hash table is empty then the load factor will equal 0, and if the hash table is full, the load factor will equal 1. To avoid/reduce index collisions, the size (**N**) of the hash table needs to be much larger than the total number of items to store.

If the hash table has a dynamic data structure, then the size of the hash table (**N**) can be enlarged/expanded when the Load Factor reaches a preset threshold level.

6.6 Secure Hash Algorithm (SHA)

Among various hash functions that exist, currently the Secure Hash Algorithm 2 (SHA-2) and Secure Hash Algorithm 3 (SHA-3) family of standards are the most frequently used. For this reason, more details on the Secure Hash Algorithm (SHA) family of hash functions will be described in the following [5].

In 2001, SHA-2 was published by the National Institute of Standards and Technology (NIST). SHA-2 was designed by the United States (US) National Security Agency (NSA). The initial values (IVs) of SHA-2 are generated using the US Federal Information Processing Standards (FIPS) PUB 180-4 federal standards [6]. SHA certification is based on the FIPS PUB 180-4, CRYPTREC, and NESSIE.

SHA have a series of family standards, which are the SHA-0, SHA-1, SHA-2, SHA-3, and more will come in the future as technology evolves.

SHA-2 algorithms are patented in US patent 6,829,355, but the US has released the patent under a royalty-free license, which makes it much easier to use. SHA-2 uses the Merkle-Damgard construction scheme, which uses the Davies-Meyer (one-way) compression function based on a specialized block cipher.

SHA-2 is a family that includes 6 hash functions that will result in hash values (digests) lengths of 224, 256, 384, or 512 bits. Based on these digests, the SHA-2 family of 6 hash functions are SHA-224, SHA-256, SHA-384, SHA-512, SHA-512/224, and SHA-512/256.

- SHA-256 is a hash functions computed with 32 bit words.
- SHA-224 is a truncated version of SHA-256 with different IVs.
- SHA-512 is a hash function computed with 64 bit words.
- SHA-384, SHA-512/224, and SHA-512/256 are truncated versions of SHA-512 with different IVs.

Please note that SHA-256 is also written as SHA256 in various documents, and both will be used in this book. This applies to the other SHA protocols as well. SHA-3 was released on 2015 August 5, where SHA-3 does not use the MD5 base architecture like SHA-1 and SHA-2 does. SHA-3 was used in the Ethash PoW of Ethereum, but is now used in Ethereum's Proof-of-Stake (PoS) operations, where more details on Ethereum are provided in Chap. 10 [7].

6.6.1 Preimage Resistance and Preimage Attacks

The security level against attacks on a hash table are measured based on the **Preimage Resistance** and the **Second-Preimage Resistance**. Hash functions need to be able to resist these types of attacks on its preimage [8].

Preimage resistance is the feature of being computationally infeasible to find the original input file that resulted in the hash value output. Preimage resistance is very important because when looking at the hash value, the original file needs to be unrecoverable to protect the content of the original file.

For example, assuming that the hash function is **H**, **X** is the input file, and **Y** is the hash value output, then the relation becomes **H**(**X**) = **Y**. Because the hash value **Y** is used as the index in the hash table, it is known to others. The desired **preimage resistance** is that although **Y** is known, it is impossible to regenerate the original input file **X** from **Y**, so the content of **X** is kept secure.

Preimage Attack is an attack that tries to find the original file that will result in this specific hash value, based on a specific hash function.

6.6.2 Second-Preimage Resistance and Second-Preimage Attacks

Second-preimage resistance is the feature of being computationally infeasible to find any second input which has the same hash value output as that of a specified input.

Second preimage resistance is the feature of being computationally infeasible to find an alternative input file (a file different from the original input file) that will result in the same hash value output as the hash value output of the original input file. Second preimage resistance is very important in defending against **second preimage attacks**, because when looking at the hash value, the original file could be forged with a file that generates the same hash value output. This is especially important in Blockchains where the hash value outputs are used to secure the content of the new Merkle tree root (of TXIDs) and the previously chained block in the newly chaining block to ensure the data of that block is unforgeable.

For example, assuming that the hash function is **H**, **X** is the original input file, and **Y** is the hash value output, then the relation becomes $\mathbf{H}(\mathbf{X}) = \mathbf{Y}$. The second preimage attacker will try to generate a forged input file **Z** that results in the same hash value output **Y**, such that the relation becomes $\mathbf{H}(\mathbf{Z}) = \mathbf{Y}$. Then the second preimage attackers will try to use the forged **Z** file to attack the hash table or blockchain. Therefore, second-preimage resistance is to make it nearly impossible to find a file **Z** that results in the same hash value output **Y** as the file **X**, in which **Z** is different to **X** (i.e., $\mathbf{X} \neq \mathbf{Z}$).

6.6.3 Preimage Attack Defense Methods

Defense against 'brute-force attacks' and 'collision attacks' need to be considered, which are descried in the following.

Brute-force attacks are commonly used to conduct preimage attacks. A brute-force attack is when an attacker tries to conduct a preimage attack or a secondary preimage attack by using repeated searching until a forged file that generates the same hash value as the original file is found. For an **n** bit hash, this attack has a time complexity proportional to $2^{\mathbf{n}}$. Currently, a sufficiently large **n** value is needed to be secure, in which a **n** value of 128 bits or larger is recommended (i.e., $128 \leq \mathbf{n}$). However, as technology evolves (e.g., considering the future influence of quantum computing), a much larger **n** bit number will be needed in the near future.

To prevent **hash value collisions** and **preimage attacks**, **double hashing** is a popular technology to use. As mentioned above, in open addressing schemes for hash tables, double hashing is also a very popular technology to use to prevent index collisions. Double hashing is also used in Blockchains. Especially, the Bitcoin system uses double hashed SHA256 (i.e., SHA256D) to avoid preimage attacks and double preimage attacks.

6.6.4 Double Hashing Method

Double hashing is a very effective technique to prevent hash collisions in hash tables as it approximates uniform open address hashing characteristics. Double hashing also provides enhanced security against preimage attacks and double preimage attacks. Details of the double hashing algorithm are provided below.

- **T**: Hash table **T** uses the double hashing function **H(i,k)** (which is based on two hash functions $\mathbf{H_1(k)}$ and $\mathbf{H_2(k)}$) for indexing data files.
- **n**: Number of elements stored in the hash table **T**. The value of **n** is updated as more data files are included in the hash table.
- α: Load factor of **T** that is obtained from $\alpha = \mathbf{n}/|\mathbf{T}|$, where $|\mathbf{T}|$ is the norm (size) of **T**.
- **H(i,k)**: Double hashing function for the $(\mathbf{i} + 1)$th index location and the key **k**, which uses two specially selected hash functions $\mathbf{H_1(k)}$ and $\mathbf{H_2(k)}$ based on the following function.

$$\mathbf{H(i,k)} = (\mathbf{H_1(k)} + \mathbf{iH_2(k)})\mathrm{mod}|\mathbf{T}|$$

Hash functions $\mathbf{H_1(k)}$ and $\mathbf{H_2(k)}$ are randomly, uniformly, and independently selected from a library of universal hashing functions **H***. The hash functions in **H*** are specially chosen hash functions that map keys from universe **U** to **m** bins, in which any two keys of the universe have a collision probability less than or equal to 1/**m**. There are some additional properties in double hashing, where only the core properties have been described.

6.7 Summary

Hash technology is commonly used in databases, blockchains, Bitcoin, cryptocurrencies, and smart contracts. So, the content of this chapter serves as a foundation for the following Chap. 8 'Blockchains' and Chap. 9 'Bitcoin' as well as Chap. 10 'Smart Contracts and Ethereum'.

The hash table and hash function operations introduced in Sects. 6.1, 6.2, and 6.3 are individually demonstrated using examples in Sect. 6.4. The hash table open addressing technologies introduced include linear probing, plus 3 rehash, quadratic probing, and double hashing. In addition, a closed addressing example is also provided in Sect. 6.4.

In order to be able to determine when a hash table or database needs to be expanded, Sect. 6.5 provides details on hash table load factor and dynamic data structures.

In addition, details of the Secure Hash Algorithm (SHA) are introduced in Sect. 6.6, which is very important because SHA-2 and SHA-3 family protocols are used in Bitcoin and Ethereum as well as all other blockchains, cryptocurrencies, and smart contracts.

References

1. Wikipedia, Hash table. https://en.wikipedia.org/wiki/Hash_table
2. Wikipedia, Hash function. https://en.wikipedia.org/wiki/Hash_function
3. Computer Science, Hash Tables and Hash Functions. https://www.youtube.com/watch?v=KyUTuwz_b7Q
4. ASCII Codes Table. https://ascii.cl/
5. Wikipedia, Secure Hash Algorithms. https://en.wikipedia.org/wiki/Secure_Hash_Algorithms
6. Wikipedia, SHA-2. https://en.wikipedia.org/wiki/SHA-2
7. Wikipedia, SHA-3. https://en.wikipedia.org/wiki/SHA-3
8. Wikipedia, Preimage attack. https://en.wikipedia.org/wiki/Preimage_attack

Chapter 7
Cryptography Technology

This chapter provides details of cryptography technologies, with a focus on blockchains, Bitcoin, cryptocurrencies, Ethereum, and smart contract systems. In Sect. 7.1, a brief history of asymmetric cryptography is provided. Section 7.2 introduces public key cryptography types, including ElGamal cryptography, Rabin cryptography, and elliptic-curve cryptography. In addition, a brief introduction of authentication protocols, point-to-point authentication protocols (PPP), authentication, authorization, accounting (AAA) protocols, and centralized network authentication are provided. Section 7.3 provides a description of Rivest–Shamir–Adleman (RSA) public key and private key cryptography, including details of cryptography procedures, example of RSA encryption and decryption, and an example case of Man-In-The-Middle (MITM) attacks. Since MITM attacks are frequently used in cryptocurrency and smart contract intrusions, it is important to understand its mechanism and how it can use asymmetric cryptography in its advantage to obtain important user information, including login IDs and passwords. As asymmetric cryptography is used in blockchains, Bitcoin, cryptocurrencies, Ethereum, and smart contract systems, the content of this chapter serves as a foundation for the following Chap. 8 'Blockchains' and Chap. 9 'Bitcoin' as well as Chap. 10 'Smart Contracts and Ethereum.'

Cryptography technology can be divided into symmetric-key and asymmetric-key algorithms. Symmetric-key cryptography will use one secret key to both encrypt and decrypt a file or a fixed-size part of a file, which is called a 'block.' Therefore, the symmetric-key cryptography process is sometimes called 'block ciphering.' Therefore, when symmetric-key cryptography is used, the encrypting party and the decrypting party have the same key, which can be vulnerable to security breaches and cyber-attacks.

On the other hand, in asymmetric-key cryptography, two keys (i.e., a 'public key' and 'private key') are generated, where the public key is used only for encryption of a file and the private key is only used for decryption of that encrypted file.

In some cases of symmetric-key cryptography, a small transformation may be applied to one of the keys making the encryption key and decryption key a little

J.-M. Chung, *Emerging Secure Networks, Blockchains and Smart Contract Technologies*, https://doi.org/10.1007/978-3-031-65866-2_7

different, but in general, the encryption key and decryption key are the same in symmetric-key cryptography [1].

Symmetric-keys are in general smaller than asymmetric-keys, so the symmetric-key encryption and decryption processes are faster and result in smaller encrypted file sizes, resulting in a higher efficiency in processing, networking, and storage.

However, asymmetric-key cryptography is easier to use and less vulnerable to security breaches and cyber-attacks. For these reasons, in symmetric-key encryption setup processes, the secret key of the symmetric-key algorithm is commonly delivered using asymmetric-key cryptography [2].

Some of the representative symmetric-key cryptography protocols include Twofish, Serpent, Advanced Encryption Standard (AES), Camellia, Salsa20, ChaCha20, Blowfish, CAST-128 (a.k.a. CAST5 where 'CAST' comes from Carlisle Adams and Stafford Tavares), Kuznyechik, RC4 (where 'RC' comes from Rivest Cipher), Data Encryption Standard (DES), 3DES, Skipjack, Safer, International Data Encryption Algorithm (IDEA), etc.

In this book, due to popular usage, more focus will be on asymmetric-key cryptography technologies, as presented in the following sections.

7.1 Asymmetric Cryptography History

Asymmetric cryptography is a technology that uses a 'public key' and 'private key,' in which the public key is sent out (or placed on a public/private website) so it can be used to encrypt a file, where the encrypted file is only decryptable using the private key. For this reason, the private key is also called the 'secure key' as it must be kept a secret and protected safely. Because the public key and private key have different roles and have different key values, this technology is called asymmetric cryptography. Asymmetric cryptography is used widely to protect important data file transfers over the Internet as well as satellite and mobile communication networks.

In 1969, James H. Ellis, who worked for the British intelligence agency Government Communications Headquarters (GCHQ), began to work on a 'non-secret encryption' technology, which is the foundation of the asymmetric cryptography. An English mathematician Clifford Cocks was assigned to work on this GCHQ project, which used prime factorization based on number theory to develop an encryption algorithm. In 1973, a very powerful asymmetric cryptography algorithm was developed by Cocks. However, due to having very little computing capability at the time, GCHQ could not find a way to use this newly developed algorithm and kept this algorithm as classified information. Later, the encryption algorithm developed by Cocks was introduced to the National Security Agency (NSA) of the USA by GCHQ, but due to having very limited computing capability at the time, the militaries of the United Kingdom (UK) and the USA could not use Cocks' cryptography algorithm. During this period, an independent effort to develop cryptography technology was in progress in the USA.

In 1976, the concept of asymmetric cryptography using public key encryption technology was introduced by Whitfield Diffie and Martin Hellman. In 1977, the knapsack public key encryption algorithm was introduced by Ralph Merkle and Martin Hellman.

The knapsack public key encryption algorithm is related to the 'knapsack problem' which is a popular combinatorial optimization problem in mathematics. The knapsack problem is based on a fixed number of items that each have weights and values, where the objective is to find a way to put the items in the knapsack such that the sum of the item values is maximized but the weight does not exceed the knapsack's limit. In other words, the goal of the knapsack problem is to use the knapsack to take the largest amount of item value but prevent the knapsack from ripping apart due to excessive weight.

In 1978, the RSA encryption algorithm was introduced by Ron Rivest, Adi Shamir, and Leonard Adleman, in which RSA is an acronym of the three inventors last names. In addition, many other encryption algorithms were proposed [3].

The asymmetric cryptography algorithm developed by Cocks in 1973 shows many aspects of equivalence to these algorithms, however due to being kept as classified information, unfortunately Cocks' algorithm was unknown at that time. Eventually, Cock's asymmetric cryptography algorithm was declassified in 1997.

7.2 Public Key Cryptography Types

Asymmetric cryptography based public key techniques are now widely used by various secure file transfer systems, signature algorithms, and authentication protocols. Public key cryptography algorithm types include ElGamal cryptography, Rabin cryptography, elliptic-curve cryptography (ECC), and RSA cryptography. Currently, RSA is the most popular asymmetric encryption algorithm [2].

7.2.1 ElGamal Cryptography

In 1984, Taher ElGamal used the Diffie–Hellman key exchange scheme to create the ElGamal cryptography algorithm. ElGamal cryptograph was implemented in the GNU Privacy Guard software (GnuPG) and is used in Pretty Good Privacy (PGP) encryption. Currently, PGP encryption is used to secure email transactions, which are commonly assisted with PGP digital signature verification technology. Digital Signature Algorithm (DSA) is a widely used extension of the ElGamal signature algorithm. The ElGamal cryptography algorithm was developed using high levels of complexity in computing discrete logarithms based on a cyclic group G defined in an abstract algebra computational space.

7.2.2 *Rabin Cryptography*

In 1979, Michael O. Rabin used the difficulty of integer factorization to develop the Rabin cryptography algorithm. Each output of the Rabin algorithm produces an encrypted code that can be created from four possible input options. This is why the Rabin cryptography algorithm may require additional decryption processing so the true original file can be identified.

7.2.3 *Elliptic-Curve Cryptography*

In 1985, Neal Koblitz and Victor S. Miller independently developed elliptic curve-based cryptography algorithms, which are now the basis for elliptic-curve cryptography technology. Elliptic curves used in elliptic curve cryptography algorithms are commonly defined over a finite Galois field, in an abelian group, which has an identity element that is a distinguished point located at infinity (∞). An elliptic plane curve equation example is $y^2 = x^3 + ax + b$, where the Galois field structure is based on a divisor group formed from the underlying algebraic group's variety. Elliptic curve cryptography commonly uses shorter encryption keys compared to other types of encryption algorithms. Starting from 2004, elliptic curve cryptography algorithms were popularly used. For example, some of the author's research on elliptic curve cryptography applied to ZigBee based Internet of Things (IoT) wireless sensor networks can be found in [4] and [5].

7.2.4 *Authentication Protocols*

Authentication protocols are used by the network to confirm that a connecting user is a valid user. In addition, authentication protocols are used by the user to confirm that the network is a valid network that can be safely connected to [6].

There are three authentication protocol types, which include Point-to-Point (PPP) authentication protocols, Authentication, Authorization, Accounting (AAA) protocols, and Centralized network authentication protocol. More information on these authentication protocols is provided in the following.

7.2.4.1 PPP Authentication Protocols

PPP authentication protocols include the Password Authentication Protocol (PAP), Challenge-Handshake Authentication Protocol (CHAP), and Extensible Authentication Protocol (EAP).

Among PPP protocols, PAP is one of the earliest protocols. CHAP builds up on PAP functionality, where it has been frequently used by servers when sending a random bit string (e.g., 128 octets) to a client device. PAP can be repeatedly used and can be initialized and executed anytime during an ongoing session. Client devices can use a password and a string based on a Message-Digest version 5 (MD5) hash function. More details on hash functions are provided in Chaps. 6 and 9.

EAP has been extensively used as it is part of the IEEE 802.1x authentication framework, where an updated version is standardized in RFC 5247. EAP has been used in IEEE 802.3 for Ethernet networks, IEEE 802.11 for Wi-Fi networks, IEEE 802.16 for WiMAX and WiBro networks, etc.

7.2.4.2 Authentication, Authorization, Accounting (AAA) Protocols

AAA protocols include Terminal Access Controller Access-Control System (TACACS), Extended TACACS (XTACACS), TACACS Plus (TACACS+), Remote Authentication Dial-In User Service (RADIUS), and DIAMETER.

Authentication protocols are used in the initial connection stages of a communication session, to assist user and network authentication and approval procedures. AAA protocols are also used to support the authorization and accounting functions. Authorization is conducted after the (user or network) authentication process, where the process considers the accounting issues of the user. For example, even if a user is confirmed to be a valid user of the network through the authentication process, the network administrator and management system need to confirm what level of priority, security, service data rate, data usage amount, and access network type can be served to this user.

For example, for smartphones, the mobile communication network type (i.e., 2G, 3G, 4G/LTE, 5G, 6G) needs to be selected based on the user's enrolled mobile service plan, where the network type may change during communication due to the amount of accumulated roaming data that has been used.

TACACS is the original AAA protocol, which uses IP based authentication without encryption of usernames and passwords. XTACACS is an extended version of TACACS, where authorization and accounting functions were added. TACACS+ is a proprietary of Cisco. TACACS+ separates the AAA components, and provides TCP based transport with full packet encryption. RADIUS is a popular protocol that has been commonly used by Internet Service Providers (ISPs). In RADIUS, the UDP protocol and Network Access Server (NAS) were used to support transport operations, and username-password combination-based credentials were frequently used. DIAMETER is an enhanced version of RADIUS that supports higher throughput services and enhanced security features. DIAMETER uses TCP or Stream Control Transmission Protocol (SCTP) and Transport Layer Security (TLS) to conduct AAA operations.

7.2.4.3 Centralized Network Authentication

A representative example of a centralized network authentication protocol is Kerberos. Kerberos is programmed in C language and can provide mutual authentication of the user and the server using an encrypted message called a 'ticket.' Kerberos enables both the client and server to securely authenticate the identity of each other over a non-secure network, thereby providing high levels of protection against replay attacks and information eavesdropping. Kerberos uses symmetric-key cryptography and a trusted third party in its authentication process, but may also use asymmetric-key cryptography during the authentication process when needed [7].

More details on cryptography with an example of public-key cryptography will be provided in the following section.

7.3 RSA Public Key and Private Key Cryptography

The RSA encryption algorithm was first introduced in 1978 by Ron Rivest, Adi Shamir, and Leonard Adleman. RSA is an asymmetric-key cryptography technology where the encryption algorithm is based on using a 'public key' and 'private key.'

The RSA encryption algorithm has the following components. The 'original file' is the file before any encryption is applied, which is also called 'plain text.' The 'original file' is encrypted using the public key, where the 'encrypted file' is also called 'cipher text.' The 'encrypted file' is decrypted using the private key, where the decrypted file is also called 'decipher text.' If the asymmetric-key cryptography process is conducted without any errors, the decrypted file will be identical to the original file.

7.3.1 RSA Cryptography Procedures

The RSA cryptography procedures are described in the following.

The RSA procedures start with selecting the modulo operation (hashing) base value N, which is a common part of both the public key and the private key. Then the public key's $\boldsymbol{E}$ value and the private key's $\boldsymbol{D}$ value are derived, using the following procedures. Selecting a large N will enable encryption of larger data segments, which will be demonstrated in the example presented in Sect. 7.3.2.

[Step 1] Derive N from $N = \boldsymbol{P} \times \boldsymbol{Q}$.

Select two large prime numbers $\boldsymbol{P}$ and $\boldsymbol{Q}$. Then N is obtained from $N = \boldsymbol{P} \times \boldsymbol{Q}$. A prime number is a natural number (positive integer number) that is greater than 1 and cannot be divided by any other natural number except for 1 and itself.

[Step 2] Derive $\boldsymbol{L}$ from $\boldsymbol{L} = \text{lcm}(\boldsymbol{P} - 1, \boldsymbol{Q} - 1)$.

Compute $\boldsymbol{L}$ from $\boldsymbol{L} = \text{lcm}(\boldsymbol{P} - 1, \boldsymbol{Q} - 1)$, where 'lcm' stands for 'least common multiple.' Thus, $\boldsymbol{L}$ is the least common multiple of the numbers $\boldsymbol{P} - 1$ and $\boldsymbol{Q} - 1$. $\boldsymbol{L}$

will be used to derive the public and private keys. However, $\boldsymbol{L}$ is not used as part of a key or in the encryption or decryption process.

[Step 3] Derive the prime number $\boldsymbol{E}$ that satisfies $1 = \gcd(\boldsymbol{E}, \boldsymbol{L})$ and $1 < \boldsymbol{E} < \boldsymbol{L}$.

$\boldsymbol{E}$ is a prime number that has to be in the range of $1 < \boldsymbol{E} < \boldsymbol{L}$ while satisfying $1 = \gcd(\boldsymbol{E}, \boldsymbol{L})$, where 'gcd' is the 'greatest common divisor.' This means that $\boldsymbol{E}$ is a prime number chosen such that the greatest common divisor of $\boldsymbol{E}$ and $\boldsymbol{L}$ equals 1. The $1 = \gcd(\boldsymbol{E}, \boldsymbol{L})$ condition guarantees the existence of $\boldsymbol{D}$, which is derived in the next stage and is a part of the private key.

[Step 4] Derive the prime number $\boldsymbol{D}$ that satisfies $1 = (\boldsymbol{E} \times \boldsymbol{D})\text{mod}\boldsymbol{L}$ and $1 < \boldsymbol{D} < \boldsymbol{L}$.

$\boldsymbol{D}$ is a prime number that has to be in the range of $1 < \boldsymbol{D} < \boldsymbol{L}$. The condition $1 = (\boldsymbol{E} \times \boldsymbol{D})\text{mod}\boldsymbol{L}$ guarantees that the decrypted file will be the same as the original file. The 'mod$\boldsymbol{L}$' is the 'modulo $\boldsymbol{L}$' operation. In other words, $\boldsymbol{D}$ is selected to satisfy $1 = (\boldsymbol{E} \times \boldsymbol{D})\text{mod}\boldsymbol{L}$, which means that $\boldsymbol{D}$ is a prime number that when multiplied with $\boldsymbol{E}$ and then divided by $\boldsymbol{L}$ the remainder will become 1.

In this step, the modulo 'mod' operator is used. For example, in $\boldsymbol{Y} = \boldsymbol{X}\text{mod}\boldsymbol{L}$ the 'mod' operator will divide the $\boldsymbol{X}$ value by $\boldsymbol{L}$, and the remainder will be the resulting $\boldsymbol{Y}$ value. In digital encryption operations, $\boldsymbol{Y}$, $\boldsymbol{X}$, and $\boldsymbol{L}$ are all binary values, in which commonly $\boldsymbol{X}$ has a longer bit length (i.e., it is a larger binary number) than $\boldsymbol{Y}$ or $\boldsymbol{L}$, and $\boldsymbol{Y}$ is smaller than $\boldsymbol{L}$ in binary value because it is the remainder value after the division by $\boldsymbol{L}$.

[Step 5] The Public Key is $(\boldsymbol{E}, \boldsymbol{N})$ and Private Key is $(\boldsymbol{D}, \boldsymbol{N})$.

Using the $\boldsymbol{E}$, $\boldsymbol{D}$, and $\boldsymbol{N}$ values already derived, the key pairs are set to Public Key $= (\boldsymbol{E}, \boldsymbol{N})$ and Private Key $= (\boldsymbol{D}, \boldsymbol{N})$.

[Step 6] The Public Key $(\boldsymbol{E}, \boldsymbol{N})$ is placed on a website for public usage or sent to someone that wants to send you an encrypted file.

[Step 7] The Public Key $(\boldsymbol{E}, \boldsymbol{N})$ is used to encrypt a file 'Original' using the following formula Encrypted $= (\text{Original})^{\boldsymbol{E}}\text{mod}\boldsymbol{N}$.

It is worthy to note that all binary data files can be seen as a large binary number. Therefore, binary number addition, subtraction, multiplication, and division can be applied.

Using the Public Key $(\boldsymbol{E}, \boldsymbol{N})$, the 'Original' file is a binary number that is multiplied to itself $\boldsymbol{E}$ times (i.e., the $(\text{Original})^{\boldsymbol{E}}$ operation) and the resulting binary number is divided by $\boldsymbol{N}$ (i.e., the mod$\boldsymbol{N}$ operation) where the remainder becomes the Encrypted file.

[Step 8] The Private Key $(\boldsymbol{D}, \boldsymbol{N})$ is used to decrypt the Encrypted file to obtain the 'Original' file using Decrypted $= (\text{Encrypted})^{\boldsymbol{D}}\text{mod}\boldsymbol{N}$.

Using the process of Decrypted $= (\text{Encrypted})^{\boldsymbol{D}}\text{mod}\boldsymbol{N}$, if the 'Decrypted' file is the same as the 'Original' file, then the RSA encryption and decryption process was a success.

A summary of the RSA encryption procedures is presented in Fig. 7.1.

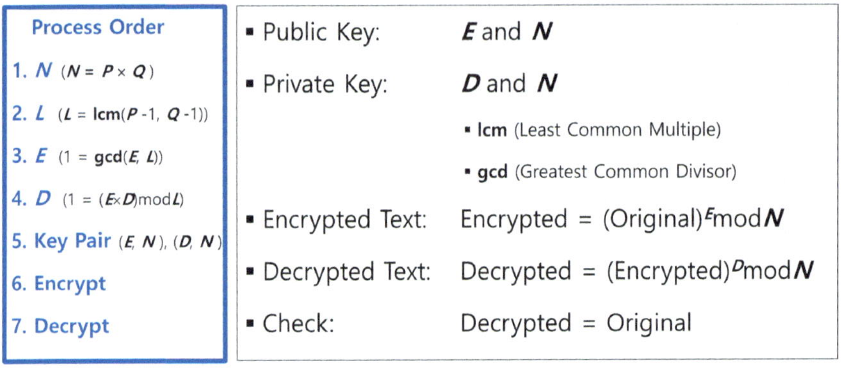

Fig. 7.1 Summary of RSA encryption algorithm procedures

7.3.2 Example of RSA Encryption and Decryption

This process of RSA encryption and decryption is supposed to be conducted with very long binary numbers, normally using very big binary numbers, that have long bit sequences. But to show an example of the RSA encryption and decryption process with long binary numbers makes it very hard to understand. So, an example of RSA encryption and decryption using small numbers is presented.

[Step 1] Derive N from $\boldsymbol{N} = \boldsymbol{P} \times \boldsymbol{Q}$.

First, we need to select 2 large prime numbers $\boldsymbol{P}$ and $\boldsymbol{Q}$. If $\boldsymbol{P}$ and $\boldsymbol{Q}$ are large, then N will be large, which will enable encryption of a large data segment. As mentioned above, a prime number is a natural number (positive integer number) that is greater than 1 that cannot be divided by any other natural number except for 1 and itself. Prime numbers below 500 are 2, 3, 5, 7, 11, 13, 17, 19, 23, 29, 31, 37, 41, 43, 47, 53, 59, 61, 67, 71, 73, 79, 83, 89, 97, 101, 103, 107, 109, 113, 127, 131, 137, 139, 149, 151, 157, 163, 167, 173, 179, 181, 191, 193, 197, 199, 211, 223, 227, 229, 233, 239, 241, 251, 257, 263, 269, 271, 277, 281, 283, 293, 307, 311, 313, 317, 331, 337, 347, 349, 353, 359, 367, 373, 379, 383, 389, 397, 401, 409, 419, 421, 431, 433, 439, 443, 449, 457, 461, 463, 467, 479, 487, 491, and 499.

Although we should select 2 large prime numbers, to make this example easy to understand, we will select two small prime numbers $\boldsymbol{P} = 17$ and $\boldsymbol{Q} = 19$ among the list of prime numbers above. Then N becomes $\boldsymbol{N} = \boldsymbol{P} \times \boldsymbol{Q} = 17 \times 19 = 323$.

[Step 2] Derive $\boldsymbol{L}$ from $\boldsymbol{L} = \text{lcm}(\boldsymbol{P} - 1, \boldsymbol{Q} - 1)$.

Then we can derive $\boldsymbol{L}$ from $\boldsymbol{L} = \text{lcm}(\boldsymbol{P} - 1, \boldsymbol{Q} - 1) = \text{lcm}(17 - 1, 19 - 1) = \text{lcm}(16, 18) = 144$. The ‘lcm’ stands for ‘least common multiple.’ In lcm(16, 18), for the two numbers 16 and 18, the smallest (least) number that is dividable by both 16 and 18 is the number 144.

[Step 3] Derive the prime number $\boldsymbol{E}$ that satisfies $1 = \text{gcd}(\boldsymbol{E}, \boldsymbol{L})$ and $1 < \boldsymbol{E} < \boldsymbol{L}$.

$\boldsymbol{E}$ has to be a prime number in the range of $1 < \boldsymbol{E} < 144$ (since $\boldsymbol{L} = 144$) that satisfies $\text{gcd}(\boldsymbol{E}, \boldsymbol{L}) = \text{gcd}(\boldsymbol{E}, 144) = 1$. In $\text{gcd}(\boldsymbol{E}, \boldsymbol{L})$, ‘gcd’ is the ‘greatest common divisor.’ This means that $\boldsymbol{E}$ is a prime number chosen such that the greatest common divisor of E and L equals 1. To make this example easy, we will select (a small value) $E = 5$.

For $E = 5$, gcd(E, 144) = gcd(5, 144) = 1 means that the greatest (largest) common divisor of the two numbers 5 and 144 is 1. This means that the largest natural number that can divide both 5 and 144 is 1.

[Step 4] Derive the prime number D that satisfies $1 = (E \times D)\text{mod}L$ and $1 < D < L$.

To obtain D, we need to select a prime number D in the range of $1 < D < 144$ (since $L = 144$) that satisfies $(E \times D)\text{mod}L = (5 \times D)\text{mod}144 = 1$. Candidate values of D include 2, 3, 5, 7, 11, 13, 17, 19, 23, 29, 31, 37, 41, 43, 47, 53, 59, 61, 67, 71, 73, 79, 83, 89, 97, 101, 103, 107, 109, 113, 127, 131, 137, and 139. To make this example easy, we will select $D = 29$, as it satisfies $(E \times D)\text{mod}L = (5 \times 29)\text{mod}144 = (145)\text{mod}144 = 1$.

[Step 5] The Public Key is (E, N) and the Private Key is (D, N).

The Public Key and Private Key pairs are obtained.

Public Key = (E, N) = (5, 323)
Private Key = (D, N) = (29, 323)

[Step 6] The Public Key (E, N) is placed on a website for public usage or sent to someone that wants to send you an encrypted file securely.

[Step 7] The Public Key (E, N) is used to encrypt the file 'Original' using the following formula Encrypted = $(\text{Original})^E\text{mod}N$.

To use the Public Key and Private Key properly, the 'Original' files' binary number has to be smaller than N. Considering that $N = 323$, in this example we will use 'Original' = 123, since 123 is smaller than $N = 323$. Then from

Encrypted = $(\text{Original})^E\text{mod}N = 123^5\text{mod}323 = 225$

the 'Encrypted' file becomes 225. In the $123^5\text{mod}323$ computation, the 123^5 operation means that the 'Original' file 123 is multiplied to itself 5 times. The mod323 means that the resulting number is divided by 323, where the remainder number becomes the 'Encrypted' file.

This 'Encrypted' file (i.e., 225) is sent over the network to the destination where the secret private key will be used to decipher the 'Encrypted' file.

[Step 8] The Private Key (D, N) is used to decrypt the Encrypted file to obtain the 'Original' file using Decrypted = $(\text{Encrypted})^D\text{mod}N$.

Using the Private Key (D, N) = (29, 323), we can obtain the 'Decrypted' file.

Decrypted = $(\text{Encrypted})^D\text{mod}N = 225^{29}\text{mod}323 = 123$

In this example, the 'Decrypted' file equals the 'Original' file 123. So, this is a successful encryption and decryption process.

In real networking systems, the communicating users will not be able to know what the 'Original' file was, so there is no way to make a conclusion that the 'Decrypted' file equals the 'Original' file. Therefore, additional procedures and messages may be used to make additional confirmation of a successful secure communication process over the network.

7.3.3 RSA Procedures and Man-in-The-Middle (MITM) Attacks

In Fig. 7.2, an example of how Alice and Bob communicate using RSA encrypted messages is described. The communication begins with Alice needing to send Bob a confidential file. So, Alice sends Bob a request for Bob's public key. Bob generates a RSA public key and a private key and sends the public key (***E***, ***N***) to Alice. Alice encrypts the original file using the public key and sends the encrypted file to Bob. Bob uses the private key (***D***, ***N***) to decrypt the encrypted file and obtains the original file sent from Alice. Due to the encryption process, others on the network may be able to obtain the encrypted file that Alice has sent to Bob, however, they will not be able to read the file as it can only be decrypted using the private key secretly kept by Bob.

Figure 7.3 shows a Man-in-the-Middle (MITM) attack, and how it can be used to create a serious security breach for both Alice and Bob. Alice and Bob are attempting the same communication procedures as in Fig. 7.2, so Alice can send an encrypted file securely to Bob using Bob's RSA public key and a private key. However, a MITM is planning to hijack the communication sequence and gain access into both Alice and Bob's computers.

The communication begins with Alice needing to send Bob a confidential file. So, Alice sends Bob a request for Bob's public key. The MITM sees this request by

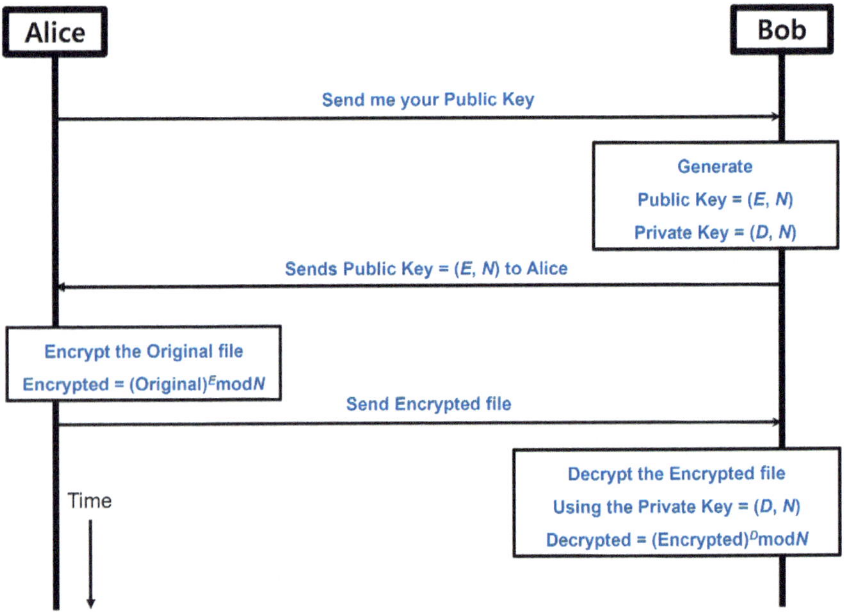

Fig. 7.2 Example of communicating using RSA encrypted messages

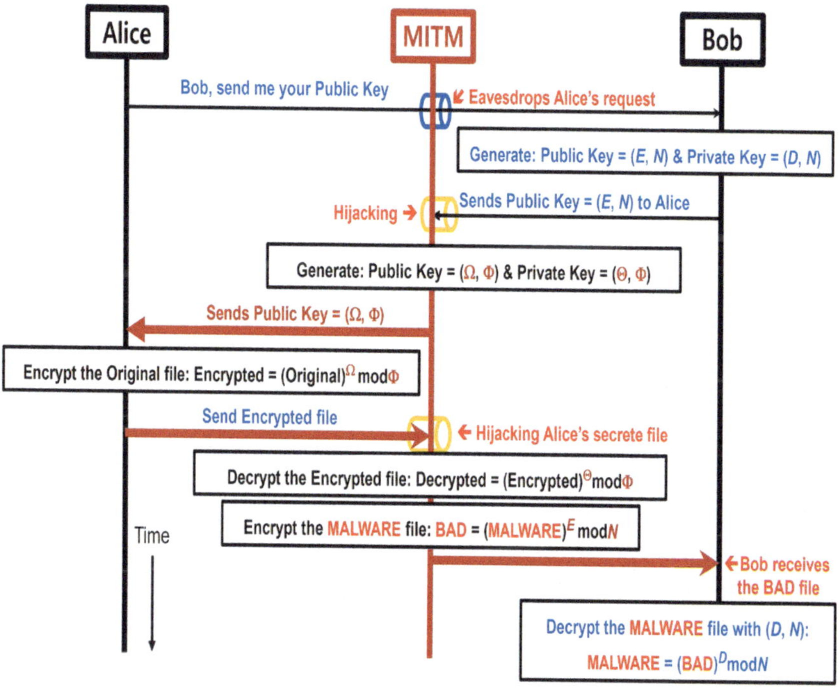

Fig. 7.3 Example of MITM hijacking RSA encrypted messages and spreading malware

eavesdropping on Alice's request. Bob generates an RSA public key and a private key and sends the public key $(\boldsymbol{E}, \boldsymbol{N})$ to Alice.

However, the MITM hijacks the public key $(\boldsymbol{E}, \boldsymbol{N})$ sent to Alice and generates its own public key $(\boldsymbol{\Omega}, \boldsymbol{\Phi})$ and private key $(\boldsymbol{\Theta}, \boldsymbol{\Phi})$ and sends the public key to Alice as if it is sent from Bob. Because Alice does not know what type of a public key will arrive from Bob, Alice accepts the MITM's public key and uses it to encrypt its original file and sends it to Bob. The MITM node hijacks Alice's encrypted file and uses its private key $(\boldsymbol{\Theta}, \boldsymbol{\Phi})$ to decode the original file which was only intended for Bob to receive and read.

Then the MITM encrypts a malware infected file using Bob's public key $(\boldsymbol{E}, \boldsymbol{N})$ and send this file to Bob as if it was sent from Alice. As Bob was waiting to receive a file from Alice, and also because the file is encrypted with Bob's public key $(\boldsymbol{E}, \boldsymbol{N})$, the firewall on Bob's computer will not be able to check for malware in this encrypted file. Therefore, when Bob decrypts this file using its private key, it will become contaminated with the malware sent from the MITM.

As a result, the MITM has stolen the Original file from Alice and has infected Bob with malware. By repeating this procedure in a network, a MITM can steal many important files and also infect many computers with malware, eventually resulting in the entire network being compromised by the MITM.

7.4 Summary

This chapter provides the basic information for cryptography technologies that are used in blockchains, Bitcoin, and smart contracts. So, there is more focus on asymmetric cryptography technology, compared to symmetric cryptography. Especially RSA public key and private key cryptography is a major interest. This chapter includes details of cryptography procedures, example of RSA encryption and decryption, and an example case of Man-In-The-Middle (MITM) attacks. MITM attacks have been used in various cryptocurrency and smart contracts intrusions, which is why it is important to understand the mechanism used in MITM attacks, which are described in this chapter. The details of this chapter are used in Chap. 8 'Blockchains,' Chap. 9 'Bitcoin,' and Chap. 10 'Smart Contracts and Ethereum.'

References

1. Wikipedia, Symmetric-key algorithm. https://en.wikipedia.org/wiki/Symmetric-key_algorithm
2. Wikipedia, Public-key cryptography. https://en.wikipedia.org/wiki/Public-key_cryptography
3. Wikipedia, RSA (cryptosystem). https://en.wikipedia.org/wiki/RSA_(cryptosystem)
4. H. Kim, J.-M. Chung, C. H. Kim, Secured communication protocol for internetworking ZigBee cluster networks. Comput. Commun. **32**(13–14) (2009)
5. H. Kim, C. H. Kim, J.-M. Chung, A novel elliptical curve ID cryptography protocol for multi-hop ZigBee sensor networks. Wirel. Commun. Mobile Comput. **12**(2) (2010)
6. Wikipedia, Authentication protocol. https://en.wikipedia.org/wiki/Authentication_protocol
7. Wikipedia, Kerberos (protocol). https://en.wikipedia.org/wiki/Kerberos_(protocol)

Chapter 8
Blockchains

This chapter focuses on the details of blockchains. First, the characteristics of blockchains are introduced in Sect. 8.1. In Sect. 8.2, the environmental influence of cryptocurrencies is described, which is based on a report from the Columbia Climate School. This section emphasizes on the needs for much more efficient blockchain technologies. In Sect. 8.3, important blockchain attacks are introduced, which include attacks on Nxt, Decentralized Autonomous Organization (DAO), and Binance. For the case of DAO, issues regarding compensation, delistings, as well as the U.S. Securities and Exchange Commission (SEC) rulings and Simple Agreement for Future Tokens (SAFT) investment contract information are provided. In Sect. 8.4, a comparison of Bitcoin and Ethereum is provided. In Sect. 8.5, details of the Byzantine Fault Tolerance (BFT) consensus mechanism are introduced, including the BFT operation and examples of the Byzantine Generals problem. The details of this chapter are abundantly used in Chap. 9 'Bitcoin' and Chap. 10 'Smart Contracts and Ethereum.'

8.1 Blockchain Characteristics

Blockchains are blocks of data that are securely chained to each other on a periodic (or on-demand) time basis. When a new blockchain is created, the first block is called the 'Genesis Block.' The Genesis Block contains a list of transactions that produces a unique hash value. The following chained (added) block includes a hash value of the former block. Therefore, the sequence of secure hash values in the blocks are what create the secure chaining effect that forms a blockchain, which starts from the Genesis Block.

Because a newly chained block will generate and include a new hash value (based on new transactions and the former block's hash value), the record of new transactions and the former block's hash value cannot be changed without other users knowing it,

J.-M. Chung, *Emerging Secure Networks, Blockchains and Smart Contract Technologies*, https://doi.org/10.1007/978-3-031-65866-2_8

which helps to keep the data in the blockchain safe. In addition, using the blockchain's hash function, data in a new block can be checked, and since the block generation procedures are all publicly known, anyone can recheck if everything in the blockchain is correct. This feature of the blockchain helps to keep the data in the blockchain reliable and secure.

Hash functions are used in blockchains to make them secure and immutable while keeping the data in the blocks transparent. The term 'immutable' means that data in the blockchain cannot be erased or modified, which is essential in keeping the data secure and protected against data forging.

For example, in Bitcoin, a blockchain hash value named the 'Transaction ID' (TXID) is used to confirm that the ledger data (financial record) is valid and was not tampered with or forged. More details on hash values can be found in Chap. 6 of this book.

The term 'transparent' refers to the characteristic that data in the blockchain can easily be accessed and read by anyone. The combination of data being both secure as well as being immutable and transparent allows many new applications to be built on blockchains with trust, which were not possible when using traditional databases, because data in traditional databases are much different.

Many blockchains are also built to be operated in a distributed manner, which means that many users share the responsibility to include all recent transaction records in each block and also check and reconfirm the block of data to be chained to the blockchain.

A Bitcoin reward is given to the first miner to publicly report a block header that can generate a hash value (using SHA256D) that is equal or less than the nBits target threshold, while including all of the verified Bitcoin transactions in their block of ledger data.

More details of Bitcoin's blockchain are presented in Chap. 9. In addition, the block to be chained to the blockchain must be submitted within a limited time period (approximately every 10 min in Bitcoin) to be considered as a candidate in the competition of being selected and chained to the blockchain (and receive a block reward). Table 8.1 summarizes some of the characteristics of blockchains.

Purpose wise, blockchains were made to be expandable, immutable, and secure to store blocks of data periodically to support a specific distributed application, such as, cryptocurrency, Non-Fungible Token (NFT), smart contract, etc.

Management wise, blockchains use distributed management, where new data is included in a new block that is chained (added) to the blockchain if a consensus from a majority of its users can be obtained.

From an openness point of view, blockchains provide protected open data with unlimited public access (transparent) using hash functions (Chap. 6) and security keys (Chap. 7).

Programming for blockchains is platform dependent, however, C++, Python, R, and Java scripts are commonly used. Bitcoin is the original blockchain cryptocurrency, which is programmed using C++ with functions added using Python and Java scripts. In addition, Ethereum is the original smart contract blockchain platform, that

Table 8.1 Blockchain characteristics

Category	Blockchain characteristics
Purpose	Expandable, immutable, and secure data system used to store blocks of data periodically to support a distributed application (e.g., cryptocurrency, smart contract, etc.)
Management	Distributed management based on consensus of a majority of the blockchain users
Openness	Open (protected) data with unlimited public access (transparent)
Programming	C++, Python, R, and Java scripts are commonly used. Bitcoin (the original blockchain cryptocurrency) is programmed using C++, Python, and Java scripts. Ethereum (the original blockchain smart contract) is programmed using Solidity, which is a high-level language created for Ethereum
Data	Structured data. New data records are collected and periodically included in a new block that is chained to the blockchain
Pros	Excellent in security, reliability, immutability, transparency, and expandability
Cons	Less effective in data analysis. Low efficiency of data record keeping due to growing file size. Used for application specific data saving based on consensus management. Consensus mechanisms cannot be completely secure

is programmed using Solidity, which is a high-level language created for Ethereum, which was implemented using C++.

Data wise, blockchains use structured data. In blockchains, new data records are collected and periodically included in a new block that is chained to the blockchain (that includes all past data records). The period of a new block chaining and the consensus procedures are based on the blockchain platform and application. In this book, details of the consensus process are provided in Sect. 8.5 (i.e., Byzantine Fault Tolerance (BFT)) and new block formation rules are described in Chap. 9 for Bitcoin cryptocurrency and Chap. 10 for Ethereum Smart Contracts.

Next, we look at some of the pros and cons of blockchains. Blockchains are excellent in security, reliability, immutability, transparency, and expandability. Blockchains are optimal for cryptocurrencies and smart contracts. However, blockchains are less efficient for data analysis. In addition, memory management wise, blockchains have low efficiency of data record keeping due their growing file size and Merkle Tree algorithm used in chaining data records (using a sequence of hash values). Blockchains are used for application specific data record keeping and consensus management. For example, the Bitcoin blockchain is only focused on keeping a record of all BTC exchanges. In addition, blockchains rely on consensus mechanisms, in which tampering is possible (considering large mining pools) that poses a risk in trustworthiness and security.

8.2 Environment Influence of Cryptocurrencies

It is worthy to note that there is a significant environmental downside to cryptocurrencies that needs to be considered. Based on a report from the Columbia Climate School on 2022 May 4th, there are about 19,000 cryptocurrencies that are in use, where the approximate summed value comes to a total of $1.75 trillion USD. This summed value approximately equals Italy's Gross Domestic Product (GDP), which is the 8th largest in the world. Among these cryptocurrencies, Bitcoin is the original and has the largest usage and overall value. Mining for Bitcoin will annually consume an approximate 150 Terawatt-hours (i.e., 150×10^{12} Watt-hours) of electricity. This consumption size of electricity approximately equals the annual electricity consumption of Argentina, which has a population of approximately 45 million people. The 150 Terawatt-hours of electricity consumed for Bitcoin mining results in a carbon dioxide (CO_2) emission of approximately 65 Megatons (i.e., 65×10^6 tons). This size of CO_2 emission approximately equals the amount emitted by the country Greece, which has an approximate population of 10.38 million people [1].

In order to solve this issue while making the blockchain process more reliable, stable, secure, and efficient, Ethereum has switched from using Proof-of-Work (PoW) based Ethash mining to Proof-of-Stake (PoS) in September of 2022. In addition, other smart contract and cryptocurrency companies have also changed to PoS or similar schemes to become more energy efficient. More details on Ethereum PoS are provided in Chap. 10 of this book. Such transitions to technologies with higher efficiency are strongly recommended, which is one of the objectives of this book.

8.3 Blockchain Attacks

There have been many cyber-attacks on blockchains in the past, where only a few are described in this section based on their characteristics.

8.3.1 NXT

In 2014, a ransomware cyber-attack was made on the Nxt community. NXT is the cryptocurrency of the Nxt community, where at the time, NXT was one of the most popular cryptocurrencies in the market. 50 million NXT was robbed through a cryptocurrency exchange, where a hard fork was requested such that the blockchain records could be rolled back and fixed (to compensate for the stolen NXTs), but the hard fork was rejected and some funds were recovered through paying the ransom and (undisclosed) negotiations (more details on hard fork are provided in Chaps. 9 and 10 of this book).

8.3.2 *DAO*

In 2016 April, through the effort of a few members of Ethereum, the Decentralized Autonomous Organization (DAO) was announced. The Genesis DAO was established on the Ethereum blockchain using its smart contract platform, where the software was programmed to be open source and was released on GitHub. The DAO objective was to help venture companies or initiatives receive venture capital funding more easily without a centralized authority so costs could be reduced, and the investors would have control over the company proportional to the amount of funding provided to the company or initiative. The DAO was focused on supporting crypto and decentralized space projects.

When a project was created, participants could make an investment of ETHs by sending it to the smart contract wallet address of the project, which was exchanged (based on a 1–100 scale) into DAO tokens. ETH is Ethereum's cryptocurrency, which is commonly pronounced as 'Ether.'

The DAO received great attention and popularity, which resulted in an investment of 12.7 million ETHs, which was worth approximately $150 million USD at that time. Later, the DAO broke a record and became the largest crowdfund of all time by achieving a total value that exceeded $250 million USD at that time (in which the approximate ETH trading value was $20/ETH).

The DAO smart contract model was formed such that any person could make a project proposal to the DAO community and receive funding through the DAO platform. In addition, investors could vote on a project plan, and in return, the profit would be shared among the investors.

8.3.2.1 DAO Hacker Attack

On 2016 June 17, the DAO received a cyber-attack in which 3.6 million ETHs were stolen, which was approximately worth $70 million USD at that time. The hacker self-elected to stop after exploiting 3.6 million ETHs. A loophole in the program code of DAO was found and exploited, where the hacker quickly made multiple recursive calls to receive ETH returns before the DAO platform's smart contract updated its balance. This is a typical case of the 'double spending problem' that blockchains were designed to prevent. It was claimed that this hackable loophole was found in the DAO application that was built on Ethereum, and not in Ethereum itself.

8.3.2.2 Compensation and Delisting

Due to this event, a holding period of 28 days was applied to the accounts of the DAO for protection. In addition, Ethereum hard forked its blockchain and refunded the stolen funds by returning it to the account owners using an exchange rate of 1 ETH to 100 DAO tokens, which was the same rate applied at the Initial Coin

Offering (ICO) of the DAO. This was criticized by many Ethereum users because it was believed that the tenets of blockchain technology were violated by the hard fork executed by Ethereum (more details on hard forks in Ethereum are provided in Chap. 10).

Following this unfortunate event, DAO tokens were delisted from cryptocurrency exchange platforms. On 2016 September 5, DAO tokens were delisted from Poloniex and in 2016 December, DAO tokens were delisted from Kraken.

8.3.2.3 SEC Rulings and SAFT Investment Contracts

On 2017 July 25, the U.S. Securities and Exchange Commission (SEC) made a ruling that the tokens offered and sold by the DAO were considered as securities and were subject to the federal securities laws. As a result of the SEC securities laws, the issuers of a distributed ledger or blockchain securities should register all offers and report sales unless it can be proven that a valid exemption applies to the securities.

The distributed nature and anonymous user privileges obtained from avoid using a centralized bank like entity were seriously challenged by these SEC rulings. This influenced not only all users of the DAO, but also all users of various cryptocurrencies and smart contracts that used cryptocurrencies.

The SEC used the 'Howey Test' (among several criteria) to evaluate cryptocurrency venture companies, which is based on the 1946 U.S. Supreme Court's ruling of the SEC versus W. J. Howey Co. case, which has been the guideline used to determine if a transaction should be considered a security or not. Based on this, over the years, a set of rules were made to help confirm 'security' qualifications, which are called the 'Howey Test.'

To pass the SEC's Howey Test, a Simple Agreement for Future Tokens (SAFT) investment contract is used by cryptocurrency developers to establish an investment contract with its accredited investors. SAFT investment contracts help investors to confirm that their cryptocurrency investment complies with SEC securities regulations. This helps investors to invest and helps cryptocurrency developers to legally collect investments.

The SAFT investment contract used for cryptocurrencies was modeled after the Simple Agreement for Future Equity (SAFE) investment contract used by startup company investors who would make an early investment for a security equity return later [2].

In relation to SAFT, the tokens issued by the investors need to have a legitimate 'utilitarian value' to the cryptocurrency user when using the blockchain platform. The term 'utilitarian value' is a value that the user receives when conducting a task related activity or a rational consumption act. Utilitarian value is related to the user's value of living necessities from a consumer perspective [3].

The SAFT method would work as follows.

1. For a new cryptocurrency (or smart contract), a SAFT securities investment contract is established with qualified investors to receive early money investments in return for cryptocurrency returns in the future.
2. After the securities contract is established, the contracted investors invest funds into a wallet on the blockchain platform.
3. The cryptocurrency (or smart contract) platform company will register the SAFT securities investment contract to the SEC.
4. After the company has been launched and its blockchain service network is activated, then the cryptocurrency tokens will be issued to the investors (i.e., Initial Coin Offerings (ICOs)) based on the SAFT securities investment contract. Since most SAFT securities investment contracts are smart contracts, this process will be conducted automatically.
5. The cryptocurrency tokens issued to the investors will be used for blockchain and network usage fees as well as other service utility fees on the cryptocurrency service platform. This aspect will satisfy the utilitarian value requirement of the sold cryptocurrency tokens, properly corresponding to the SEC regulatory requirements.
6. Because only accredited investors are allowed to establish a SAFT securities investment contract with the cryptocurrency (or smart contact) company, unfortunately retail investors cannot make investments in the early stages of the company. Retail investors will need to wait until the service platform is open to the public to make investments. This has the effect of protecting retail investors from loss of investment in case the cryptocurrency company fails and goes bankrupt, which does happen to start-up companies.

It needs to be noted that the SEC has not fully confirmed that no other regulations to cryptocurrency (or smart contract) companies as well as its users and investors will be applied in the future. In addition, the SEC regulations only apply within the USA, and other countries may have significantly different policies and regulations that apply to cryptocurrencies or smart contracts.

For example, 8 countries have already banned use of cryptocurrencies, which include P.R. China, Nepal, Afghanistan, Bangladesh, Morocco, Algeria, Egypt, and Bolivia [4].

8.3.3 Binance

Currently, based on daily trading cryptocurrency volume, Binance Holdings Ltd. (Binance) is one of the largest cryptocurrency exchange companies in the world. In 2017, Binance was founded with its headquarters located in P.R. China. However, on 2021 September 24, P.R. China's ban on cryptocurrency trading and mining became effective [5].

Due to these government regulations to be imposed, Binance had its headquarters moved out of China before the ban was officially enforced [6].

8.3.3.1 Binance Robbing

On 2019 May 8, it was reported that Binance received an attack that resulted in over 7,000 Bitcoins being stolen, which was worth over $40 million USD at that time [7].

The hackers obtained user account login information and used this information to withdraw Bitcoins quickly from Binance's 'hot wallet.' Binance's 'hot wallet' is an account accessible through the Internet, which held approximately 2% of Binance's Bitcoin holdings (Binance's 'cold wallet' is used for offline cryptocurrency savings). Hot wallet user account login information are two-factor authentication codes, in which the hackers had collected in advance and used to withdraw the Bitcoins when its value was approximately 9% higher compared to the previous week.

This attack was made on relatively independent accounts using procedures that passed the security check points, such that it was not easily detectable. In addition, the attack was conducted at a time when the gain of the robbery would be maximized.

Because this cyber-attack was conducted using valid user login information obtained through various hacking methods in advance (e.g., MITM, phishing, data exploitation, sniffing, etc. explained in Chaps. 1 and 7), Binanace and external legal authorities were not sure if further attacks would be made, or if the hackers would attempt to control the price of cryptocurrencies and had to keep alert.

Binance quickly entered damage management procedures, where all cryptocurrency withdraws were blocked and deposits were suspended, while trades were permitted. Binance also announced that the company would fully reimburse the robbed accounts.

8.4 Bitcoin and Ethereum Comparison

The origin of blockchain is based on the whitepaper titled "Bitcoin: A Peer-to-Peer Electronic Cash System" that was released under the author's name Satoshi Nakamoto in October of 2008 [8]. Based on this paper, the initial version of Bitcoin was released in January of 2009. Although many people tried, the true identity of the author Satoshi Nakamoto has not been confirmed yet, and is therefore considered to be a pseudonym.

In addition, in this whitepaper, the term "blockchain" is not mentioned, but was derived from the readers based on the description of technologies used to chain blocks of ledger data using hash technology and asymmetric encryption (i.e., public key and private key), where more details are provided in Chap. 7 'Cryptography Technology.'

In addition, Ethereum is the first company to use blockchain technology to provide smart contract applications to be created and enabled access to the general public. On 2015 July 30, Vitalik Buterin and Gavin Wood released Ethereum [9].

Ethereum expands on Bitcoin's blockchain technology to enable various types of decentralized applications to be implemented on its smart contract platform.

Considering that Bitcoin is the origin of all cryptocurrencies and Ethereum is the origin of all blockchain smart contracts, a comparison of Bitcoin and Ethereum is provided below in Table 8.2 [9, 10].

Functionality wise, Bitcoin (BTC) is to be used for digital money, which is the first cryptocurrency. Ethereum is the first blockchain based smart contract platform, which also supports digital money functionalities using ETH (Ether) for transactional fees and Gas for computation fees.

As noted in Tables 8.1 and 8.2, Bitcoin is programmed using C++, Python, and Java scripts. Ethereum is programmed using Solidity, which was specifically developed for Ethereum smart contracts and its blockchain platform support purposes.

Computability wise Bitcoin is considered to be Turing Incomplete (but there is debate that Bitcoin is Turing complete) and Ethereum is Turing complete.

A computing system is considered to be 'Turing complete' if it can be programmed to simulate any Turing machine. The meaning of 'being able to simulate any Turing machine' means that the system is capable of reproducing the computation results, data-manipulation results, decision results, and control procedures of another machine if it is given a sufficient amount of time and computing resources, and is supported by a qualified programmer. Such equivalent simulation capabilities of two computing systems can be confirmed based on a 'Turing equivalence' test. Turing equivalence is achieved between two computers if either computer can simulate the

Table 8.2 Bitcoin and Ethereum comparison

Characteristics	Bitcoin	Ethereum
Functionality	Digital money, cryptocurrency	Smart contracts, digital money, blockchain services
Cryptocurrency token	BTC (Bitcoin)	ETH (Ether) for transactional fees, gas for computation fees
Programming language	C++ (Python and Java script added)	Solidity
Computability	Turing incomplete (debatable)	Turing complete
Hashing algorithm	SHA256D (double hashed SHA256) based PoW	Ethash PoW was used but was changed to PoS in 2022 Sept. which uses SHA-3 based Keccak256
Coin release method	Early mining based Proof-of-Work (PoW)	Initial Coin Offerings (ICO) of $18 M in Bitcoins and PoS rewards
Creators	Satoshi Nakamoto (unknown)	Vitalik Buterin, Gavin Wood, Joseph Lubin
Release date	2009 January	2015 July
Average block time	10 min	12–15 s for Ethash and 12 s for PoS

operation of the other computer and produce the same results [11]. Turing equivalence is considered as a basic qualification of modern '**digital twin**' modeling which use various artificial intelligent (AI) and simulation technologies, where more details on Digital Twin technology are provided in Chap. 2.

The hashing algorithm used in Bitcoin is the SHA256D (Double Hashed SHA256) while Ethereum used Ethash for mining support but changed to Proof-of-Stake (PoS) in September of 2022 for improved computational efficiency, reliability, security, and energy efficiency.

The coin release method of Bitcoin is based on an early mining competition to be the first to derive a block header hash value that satisfies the nBits target threshold. Ethereum used Initial Coin Offerings (ICOs) based on $18 million USD in Bitcoins (BTC). This is why fluctuations in the value of BTC will influence the value of ETH, which can be observed by comparing Fig. 9.2 (Bitcoin BTC annual value changes) and Fig. 10.1 (Ethereum ETH annual value changes) of this book. Ethereum used to provide Ethash PoW rewards, but now provides PoS rewards.

Bitcoin was proposed by Satoshi Nakamoto, who is unknown, and was initially released in 2009 January. Ethereum was initially released in 2015 July by co-creators Vitalik Buterin, Gavin Wood, and Joseph Lubin.

Bitcoin has an average block chaining period of approximately 10 min. Ethereum Ethash PoW had a new block chaining period of 12–15 s. However, since Ethereum switched to PoS in September of 2022, a block is generated every 12 s, one block for each slot, where 32 slots form an epoch.

More details on Bitcoin can be found in Chap. 9 and more details on Ethereum are provided in Chap. 10.

8.5 Byzantine Fault Tolerance (BFT)

Blockchain based systems like Bitcoin, cryptocurrencies, Ethereum, and smart contracts need to make reliable decisions, but because they are based on distributed systems, the process can be difficult due to disagreements, communications errors, or time delays among participating nodes. In addition, when some of the nodes are compromised and conduct malicious behavior, these nodes can make it very difficult to make a correct decision or arrive at a proper consensus. This applies to both when nodes maybe located in a vicinity, but are independent, or when some or all nodes are remotely located. For such cases, Byzantine Fault Tolerant (BFT) algorithms are commonly used in blockchains and decentralized networks to enhance the security and reliability level.

8.5.1 BFT Operation

Distributed systems like Bitcoin, Ethereum, various blockchain services, big data systems, edge computing systems, and sharded databases periodically have to make decisions on the validness of information received from distributed (which are commonly remote) systems. Information that is determined true and valid will be saved in the blockchain or database and will be used in making future decisions. Therefore, it is very important to validate the information that is received from external distributed sources. Information collected from distributed and remote systems may have parts that do not match the information from other sources, in which an information selection process needs to be conducted [12].

BFT helps to derive the level of fault that a system can tolerate and still make a proper decision or arrive at a correct consensus. In the BFT model, it is assumed that parts of the system can be faulty (due to an error or malicious activity) while all other parts of the system act honest and proper. It is desired for all the parts of the system that act honest and proper to make the same correct decision on the actions to take on behalf of the overall system (disregarding the faulty parts of the system).

Because distributed systems may not have a central authority entity to conduct information validation and selection, a fast and reliable determination mechanism is needed.

Byzantine fault is a very common computer system problem in which components of the system may (unintentionally or intentionally) submit erroneous or incorrect information (or may not reply at all) and the computer system has to make a decision based on the collected data. This is a common issue that all distributed systems like Bitcoin, Ethereum, various blockchain services, and sharded databases experience.

Similar terminology to the Byzantine fault problem includes the Byzantine generals problem, Byzantine agreement problem, and Byzantine failure. The Byzantine fault problem is also related to error avalanche, interactive consistency, and source congruency problems.

Byzantine fault techniques have been used to enhance the reliability in decision making when using information received from distributed (and commonly remote) systems. This is why Bitcoin, cryptocurrencies, Ethereum, smart contracts, autonomous vehicles (i.e., cars, trucks, buses, drones, Unmanned Aerial Vehicles (UAVs)) as well as Boeing 777 and 787 flight control systems use BFT algorithms in making control decisions [13].

For example, an airplane can have multiple independent computers assisting the autopilot controller that is piloting the aircraft based on a BFT consensus mechanism to maximize the reliability, efficiency, and safety. Commonly, such systems might use 4, 7, 10, 13, ..., or ($3\mathbf{M} + 1$) number of independent computers when using BFT to make autopilot control decisions. The reason for this is because when you have ($3\mathbf{M} + 1$) independent multiple autopilot computers (where **M** is a positive integer number) then up to **M** autopilot computers can 'malfunction' and still the remaining ($2\mathbf{M} + 1$) computers will be able to provide correct information to the autopilot controller, which will be able to make a correct autopilot control decision based on the BFT consensus mechanism.

For example, if you use 4 independent computers with BFT to make autopilot control decisions, then up to 1 computer can malfunction and provide erroneous information (i.e., $\mathbf{M} \leq 1$) and still the remaining $(2\mathbf{M} + 1) = 3$ autopilot computers will be able to make a correct autopilot control decision based on the BFT consensus mechanism.

If you were to use 5 (or 6) independent computers with BFT to make autopilot control decisions, then still up to 1 autopilot computer can malfunction (i.e., $\mathbf{M} \leq 1$) and still the remaining 4 (or 5) autopilot computers will be able to make a correct autopilot control decision based on the BFT consensus mechanism. As shown in this case, using 5 (or 6) independent computers with BFT provides the same permitted malfunction level (i.e., $\mathbf{M} \leq 1$) as when using 4 independent computers with BFT.

In addition, compared to using 4 independent computers with BFT to make autopilot control decisions, using 5 (or 6) independent computers with BFT will increase the system cost (i.e., more expensive in original installment (CapEx) and maintenance fees (OpEx)), the overall system will also weigh more and consume more energy, it will take longer to make a BFT based decision, and the number of networked messages and networking time will increase. Due to these reasons, using 4 independent computers with BFT to make autopilot control decisions is commonly preferred compared to using 5 (or 6) independent computers. This is why commonly systems will use 4, 7, 10, 13, …, or $(3\mathbf{M} + 1)$ number of independent computers when using BFT.

More details on how the BFT $(3\mathbf{M} + 1)$ formula is derived is explained through the following Byzantine generals problem examples.

8.5.2 Byzantine Generals Problem Examples

BFT levels are derived based on the Byzantine generals problem, which is described in the following.

Consider a war example in which the enemy troops are located all together while the allied forces are divided into multiple units surrounding the enemy troops in remote locations. Because the enemy troops are very strong, it is critical for the distributed allied forces to make the right decision to attack or retreat altogether at the same time.

If an insufficient group of the allied forces attack, then the enemy troops will be victorious and the allied force(s) that conducted the attack will either be killed or become a 'prisoner of war' by the enemy troops. Thus, it is very important to determine how many of the allied forces can come to a consensus to conduct an attack (or retreat) if there are communications errors, or one or more of the Generals commanding the allied forces are a traitor.

The objective is to establish a reliable consensus such that all of the honest Generals (excluding the traitor Generals) conduct a synchronized attack (or retreat).

This will protect the honest allied forces while enabling the highest success probability to win the war. The consensus process has to resist against the communication errors and the traitor General's misleading manipulated message confusion [13].

One part of this problem is in the reliability of the messages exchanged between the Generals of the allied forces. Because the allied forces are located in remote areas, the commanding Generals of the allied forces need to send a messenger to all other allied forces to inform the other Generals of the 'attack' or 'retreat' command. An attack message will include the day and time to conduct the attack or retreat. The Generals will resend the attack or retreat command messages to the other Generals to make sure the commanding General's message is delivered. In this process, an unintentional communication error may occur such that the message could be misread, or the traitor General(s) intentionally changes the message to create confusion among the Generals. The following examples illustrate some of the representative cases that will help us determine up to how many communication errors or traitors can be tolerated and still the honest Generals can come to a correct consensus.

In a computer system, a broadcasted message may be maliciously altered by an intermediate node that is relaying the broadcasted message. This could occur if the node has been compromised with malware or is being remotely controled by a hacker.

The BFT level is determined by the following setup. It is assumed that there are **T** traitors among the **N** Generals of the allied forces. However, it is not revealed who the **T** traitors are, which is kept a secret.

The BFT level is determined by the largest **T** value among the **N** generals, where the (attack or retreat) decisions made by the honest (**N-T**) generals are all the same (i.e., the same consensus of action is conducted by all of the honest (**N-T**) generals).

Assume that the allied forces have three Generals (**N** = 3) and General 1 is the commanding General that made an attack decision, which was sent to the other two Generals of the allied forces to follow.

Figure 8.1 shows the BFT example of the **N** = 3 and **T** = 0 case where the commanding General 1 sends out an attack message, in which an attack consensus is easily obtained by General 2 and General 3.

Figure 8.2 shows the BFT example of **N** = 3 and **T** = 0 case where the commanding General 1 sends out a retreat message, in which a retreat consensus is easily obtained by General 2 and General 3.

Based on the examples presented in Figs. 8.1 and 8.2, we can conclude that if the allied forces had three Generals (**N** = 3) and none of them were traitors (**T** = 0) then the decision made by the three Generals will be the same and a matching correct consensus (to attack or retreat) will be obtained by all allied forces.

The following two examples show what could happen if there is one traitor among the three Generals.

Figure 8.3 shows the BFT example of **N** = 3 and **T** = 1 case where the commanding General 1 is the traitor. Because commanding General 1 is a traitor, it sends an attack message to General 2 and a retreat message to General 3. This is intended to split the action of the allied forces and make the attacking force lose in combat to the enemy forces. As General 2 and General 3 exchange messages received from General 1, they realize that the messages do not match as each General has

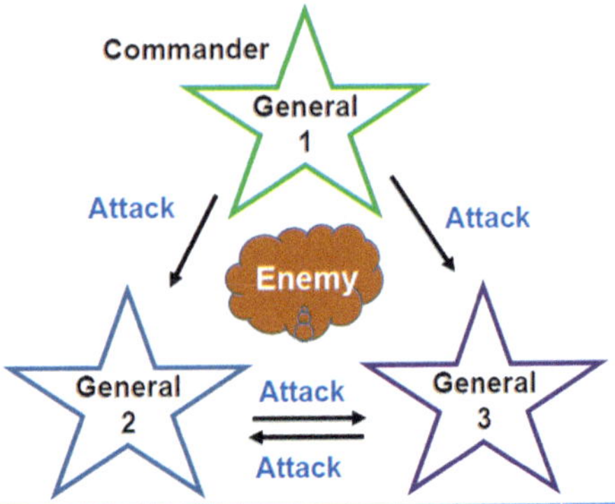

Generals	Received Messages	Consensus	Action
General 2	Attack, Attack	Attack	Attack
General 3	Attack, Attack	Attack	Attack

Fig. 8.1 BFT example of the **N** = 3 and **T** = 0 case where the commanding General 1 sends out an attack message

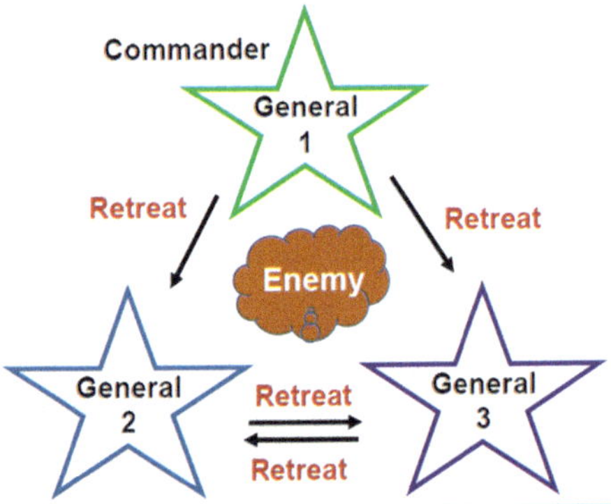

Generals	Received Messages	Consensus	Action
General 2	Retreat, Retreat	Retreat	Retreat
General 3	Retreat, Retreat	Retreat	Retreat

Fig. 8.2 BFT example of **N** = 3 and **T** = 0 case where the commanding General 1 is sending out a retreat message

received one attack message and one retreat message. As demonstrated in Fig. 8.3, the two honest Generals (i.e., General 2 and General 3) cannot come to a consensus and therefore no action can be conducted. Because the commanding General 1 is a traitor, it is fortunate that the two honest Generals cannot take action in this case.

Figure 8.4 shows the BFT example of **N** = 3 and **T** = 1 case where General 3 is the traitor. As demonstrated in Fig. 8.4, if the allied forces have three Generals (**N** = 3) and one (i.e., General 3) of them is a traitor (**T** = 1) then no consensus can be made by General 2 (since it has received one attack message and one retreat message), thus no action can be conducted by General 2.

As illustrated in Figs. 8.3 and 8.4, when 1 out of 3 of the Generals is a traitor (i.e., the **N** = 3 and **T** = 1 case), the honest Generals cannot make a correct consensus.

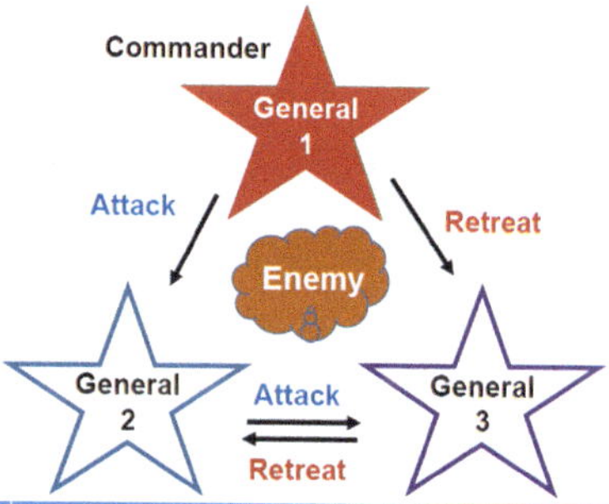

Generals	Received Messages	Consensus	Action
General 2	Attack, Retreat	Cannot Decide	Cannot Decide
General 3	Retreat, Attack	Cannot Decide	Cannot Decide

Fig. 8.3 BFT example of **N** = 3 and **T** = 1 case where the commanding General 1 is the traitor

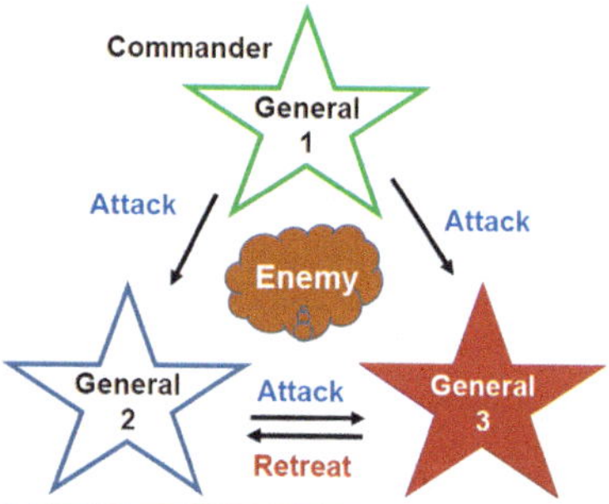

Generals	Received Messages	Consensus	Action
General 2	Attack, Retreat	Cannot Decide	Cannot Decide
General 3	Attack, Attack	Attack	Retreat

Fig. 8.4 BFT example of **N** = 3 and **T** = 1 case where General 3 is the traitor

Next, examples of when **N** = 4 are investigated. Based on the examples presented in Fig. 8.5 (attack message transmission case) and Fig. 8.6 (retreat message transmission case) we can conclude that if the allied forces had four Generals (**N** = 4) and none of them were traitors (**T** = 0) then the decision made by the four Generals will be the same and a matching correct consensus (to attack or retreat) will be made by all allied forces. This is similar to the examples respectively presented in Figs. 8.1 and 8.2 for the **N** = 3 and **T** = 0 case.

Figures 8.7 and 8.8 show the BFT example of **N** = 4 and **T** = 1 case where the commanding General 1 is the traitor.

In the example presented in Fig. 8.7, because the commanding General 1 is a traitor, it sends attack messages to General 2 and General 3, but General 1 sends a retreat message to General 4. General 1 is trying to split the action of the allied forces and make the attacking forces (led by General 2 and General 3) lose in combat to the enemy forces. As General 2, General 3, and General 4 exchange messages received from General 1, they realize that the messages do not match as each General has received two attack messages and one retreat message. As a result, fortunately, the

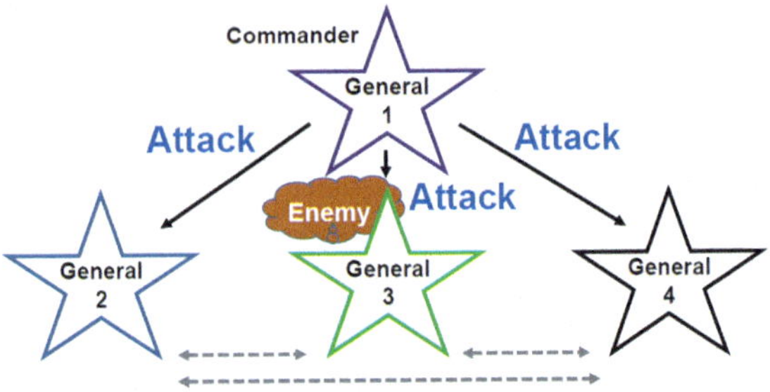

Generals	Received Messages	Consensus	Action
General 2	Attack, Attack, Attack	Attack	Attack
General 3	Attack, Attack, Attack	Attack	Attack
General 4	Attack, Attack, Attack	Attack	Attack

Fig. 8.5 BFT example of **N** = 4 and **T** = 0 case where the commanding General 1 sends out an attack message

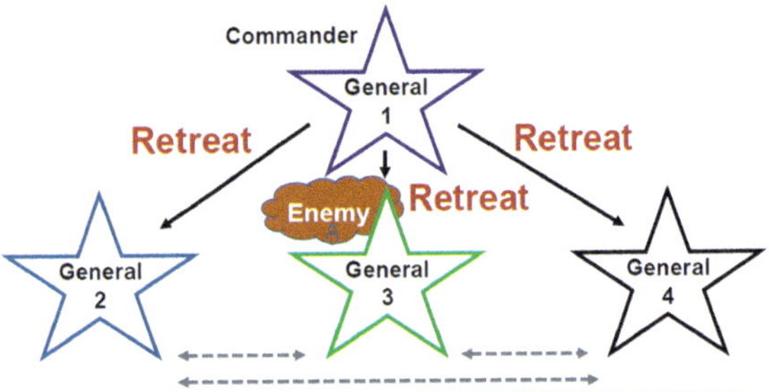

Generals	Received Messages	Consensus	Action
General 2	Retreat, Retreat, Retreat	Retreat	Retreat
General 3	Retreat, Retreat, Retreat	Retreat	Retreat
General 4	Retreat, Retreat, Retreat	Retreat	Retreat

Fig. 8.6 BFT example of **N** = 4 and **T** = 0 case where the commanding General 1 sends out a retreat message

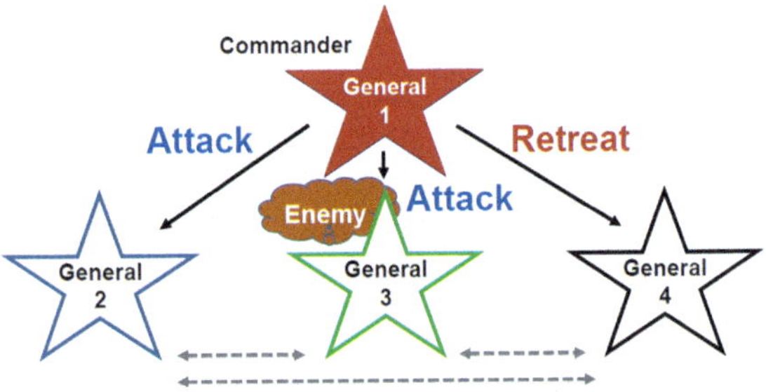

Generals	Received Messages	Consensus	Action
General 2	Attack, Attack, Retreat	Attack	Attack
General 3	Attack, Attack, Retreat	Attack	Attack
General 4	Retreat, Attack, Attack	Attack	Attack

Fig. 8.7 BFT example of **N** = 4 and **T** = 1 case where the commanding General 1 is the traitor and sends out two attack messages and one retreat message

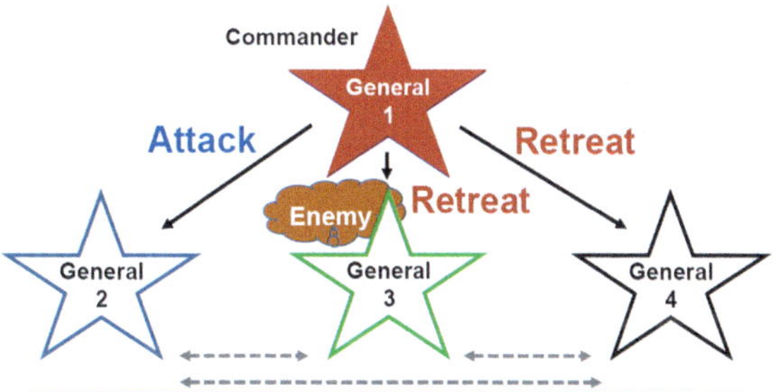

Generals	Received Messages	Consensus	Action
General 2	Attack, Retreat, Retreat	Retreat	Retreat
General 3	Retreat, Attack, Retreat	Retreat	Retreat
General 4	Retreat, Attack, Retreat	Retreat	Retreat

Fig. 8.8 BFT example of **N** = 4 and **T** = 1 case where the commanding General 1 is the traitor and sends out one attack message and two retreat messages

three honest Generals (i.e., General 2, General 3, and General 4) all make the same attack decision, which is a result of the BFT mechanism.

In the example presented in Fig. 8.8, because the commanding General 1 is a traitor, it sends an attack message to General 2, but General 1 sends retreat messages to General 3 and General 4. General 1 is trying to split the action of the allied forces and make the attacking force (led by General 2) lose in combat to the enemy forces. As General 2, General 3, and General 4 exchange messages received from General 1, they realize that the messages do not match as each General has received one attack message and two retreat messages. As a result, fortunately, the three honest Generals (i.e., General 2, General 3, and General 4) all make the same retreat decision, which is a result of the BFT mechanism.

In the example presented in Fig. 8.9, General 4 is a traitor. The commanding General 1 sends attack messages to General 2, General 3, and General 4. Because General 4 is a traitor it sends retreat messages to General 2 and General 3. As General 2, General 3, and General 4 exchange messages, General 2 and General 3 will realize that the messages do not match as they would have each received two attack messages and one retreat message. As a result, the honest two Generals (i.e., General 2 and General 3) will make the same attack decision as General 1 (who is also honest). As a result, fortunately, the three honest Generals will arrive at the same attack consensus, which is a result of the BFT mechanism.

Figure 8.10 presents the case where **N** = 4 and **T** = 2 where General 3 and General 4 are traitors. The commanding General 1 sends attack messages to General 2, General 3, and General 4. But because General 3 and General 4 are traitors, they will send retreat messages to the other Generals. As General 2, General 3, and General 4 exchange messages, General 2 will realize that the messages do not match as it would have received one attack message and two retreat messages. As a result, the honest General 2 will arrive at a retreat consensus decision. Therefore, the two honest Generals (i.e., General 1 and General 2) will take different actions. This will result in General 1's forces losing in war to the enemy troops.

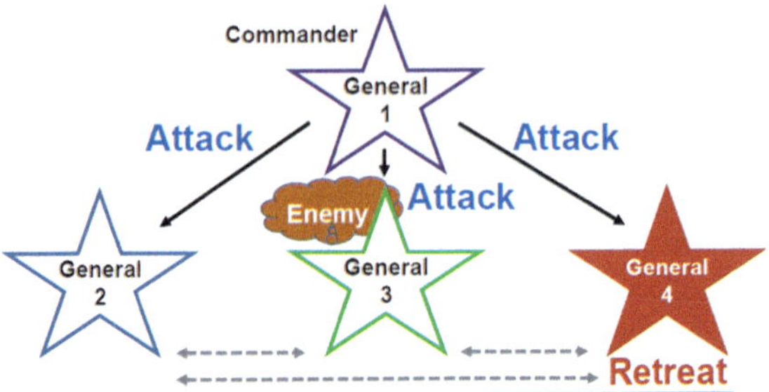

Generals	Received Messages	Consensus	Action
General 2	Attack, Attack, Retreat	Attack	Attack
General 3	Attack, Attack, Retreat	Attack	Attack
General 4	Attack, Attack, Attack	Attack	Retreat

Fig. 8.9 BFT example of **N** = 4 and **T** = 1 case where General 4 is the traitor

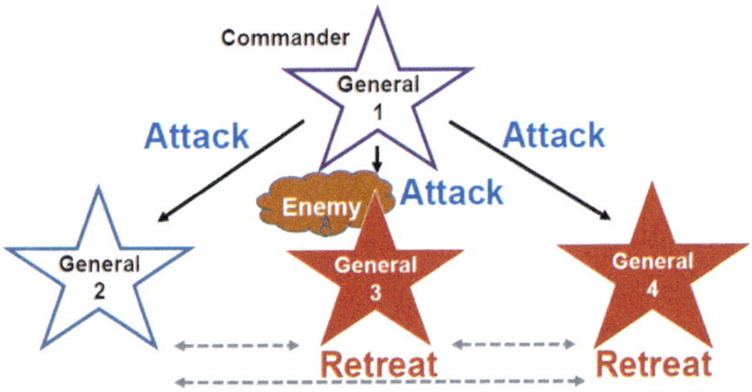

Generals	Received Messages	Consensus	Action
General 2	Attack, Retreat, Retreat	Retreat	Retreat
General 3	Attack, Attack, Retreat	Attack	Retreat
General 4	Attack, Attack, Retreat	Attack	Retreat

Fig. 8.10 BFT example of **N** = 4 and **T** = 2 case where General 3 and General 4 are traitors

As shown in the BFT examples cases of **N** = 4 and **T** = 1, as presented in Fig. 8.7, Fig. 8.8, and Fig. 8.9, all honest Generals arrive at the same consensus and make the same decision to either attack or retreat. However, as shown in the BFT example case of **N** = 4 and **T** = 2, as presented in Fig. 8.10, the honest Generals (i.e., General 1 and General 2) do not arrive at the same consensus, and therefore take different actions. This shows that the BFT for **N** = 4 has a tolerance level of up to **T** = 1.

Extending on this pattern of consensus logic, one can conclude that a system that has (3**M** + 1) number of independent computers when using BFT will have a fault tolerance level of up to **M** units, where **M** is a positive integer number. This is why a system using the BFT consensus mechanism to maximize the reliability and robustness will use 4, 7, 10, 13, …, or (3**M** + 1) number of independent computers to make important control decisions.

Using actual numbers, a few examples are provided based on this consensus conclusion.

- A system that has 4 independent computers using BFT will have a fault tolerance level of up to 1 unit.

- A system that has 7 independent computers using BFT will have a fault tolerance level of up to 2 units.
- A system that has 10 independent computers using BFT will have a fault tolerance level of up to 3 units.
- A system that has 13 independent computers using BFT will have a fault tolerance level of up to 4 units.
- A system that has ($3\mathbf{M} + 1$) number of independent computers using BFT will have a fault tolerance level of up to **M** units, where **M** is a positive integer.

For these reasons, Bitcoin, various cryptocurrencies, Ethereum, various smart contract platforms, as well as autonomous vehicles (e.g., cars, motorcycles, trucks, buses, trains, aircrafts, submarines) may use BFT based consensus mechanisms in making important decisions.

For more advanced blockchain systems that apply sharding or delegation (e.g., PoS) with artificial intelligence (AI) based trust security management, the optimal number of independent computers with BFT to use is described in the author's papers [14–18].

8.6 Summary

This chapter provides an overview of blockchain technology. First, the characteristics of blockchains are described, followed by the bad environmental influence of cryptocurrencies. Then details of blockchain attacks are introduced using examples of Nxt, DAO, and Binance. Related SEC rulings and SAFT investment contract issues are also introduced. This is followed by a comparison of the two most prominent blockchain platforms, Bitcoin and Ethereum. In addition, details of the BFT consensus mechanism are introduced with examples of the operation of the Byzantine generals problem. The content of this chapter (along with Chap. 6 'Hash Technologies' and Chap. 7 'Cryptography Technologies') serve as the foundation of Chap. 9 'Bitcoin' and Chap. 10 'Smart Contracts and Ethereum.'

References

1. J. Hinsdale, Cryptocurrency's dirty secret: energy consumption. Columbia Climate School (2022). https://news.climate.columbia.edu/2022/05/04/cryptocurrency-energy/
2. K. Peters, What is a simple agreement for future tokens (SAFT) in crypto?. Investopedia. https://www.investopedia.com/terms/s/simple-agreement-future-tokens-saft.asp
3. B.J. Babin, W.R. Darden, M. Griffin, Work and/or fun: measuring hedonic and utilitarian shopping value get access arrow. J. Consumer Res. **20**(4) (1994). https://doi.org/10.1086/209376
4. F. O'Sullivan, Where is crypto illegal in 2024? The Countries That Ban Cryptocurrency. cloudwards. https://www.cloudwards.net/where-is-crypto-illegal/

5. Wikipedia, Legality of cryptocurrency by country or territory. https://en.wikipedia.org/wiki/Legality_of_cryptocurrency_by_country_or_territory#China
6. Wikipedia, Binance. https://en.wikipedia.org/wiki/Binance
7. A. Kharpal, Bitcoin bandits: hackers steal equivalent of $40M from Binance. https://www.usatoday.com/story/money/2019/05/08/bitcoin-exchange-binance-robbed-7-000-bitcoins-hackers/1139907001/
8. S. Nakamoto, Bitcoin: a peer-to-peer electronic cash system, white paper. https://bitcoin.org/en/bitcoin-paper
9. Wikipedia, Ethereum. https://en.wikipedia.org/wiki/Ethereum
10. Wikipedia, Bitcoin. https://en.wikipedia.org/wiki/Bitcoin
11. Wikipedia, Turing completeness. https://en.wikipedia.org/wiki/Turing_completeness
12. Wikipedia, Byzantine fault. https://en.wikipedia.org/wiki/Byzantine_fault
13. C. Colohan, L6: byzantine fault tolerance. YouTube. https://www.youtube.com/watch?v=_e4wNoTV3Gw
14. Y. Goh, J. Yun, D. Jung, J.-M. Chung, Secure trust-based delegated consensus for blockchain frameworks using deep reinforcement learning. IEEE Access **10** (2022). https://doi.org/10.1109/ACCESS.2022.3220852
15. J. Yun, Y. Goh, J.-M. Chung, DQN-based optimization framework for secure sharded blockchain system. IEEE Internet of Things J. **8**(2) (2021). https://doi.org/10.1109/JIOT.2020.3006896
16. J. Yun, Y. Goh, J.-M. Chung, Trust-based shard distribution scheme for fault tolerant shard blockchain networks. IEEE Access **7** (2019). https://doi.org/10.1109/ACCESS.2019.2942003
17. J. Yun, Y. Goh, J.-M. Chung, O. Kim, S. Shin, J. Choi, Y. Kim, MMOG user participation based decentralized consensus scheme and proof of participation analysis on the bryllite blockchain system. KSII Trans. Internet Inf. Syst. **13**(8) (2019). https://doi.org/10.3837/tiis.2019.08.015
18. J. Yun, Y. Goh, J.-M. Chung, O. Kim, S. Shin, J. Choi, Y. Kim, MMOG user participation based decentralized consensus scheme and proof of participation analysis on the bryllite blockchain system, *Outstanding Paper Award, International Conference on Internet (ICONI)* (2018)

Chapter 9
Bitcoin

In this chapter, the technical details of Bitcoin are introduced. The description starts with Sect. 9.1 Bitcoin blockchain technology, which covers details of Bitcoin block rewards, blockchain characteristics, and the attacker's dilemma issues. In Sect. 9.2, details of the Bitcoin's blockchain file size are described along with the Merkle tree, which is used in chaining all ledger BTC transaction ID (TXID) information to the block header's Merkle root. In Sect. 9.3, the operations applied in Bitcoin transaction verification, block mining, blockchain forks, transaction records, and TXID are described. Section 9.4 provides the details of the Bitcoin block header, which includes block versions, Merkle root hash, nBits target threshold, and serialized blocks.

The story of Bitcoin starts with a white paper made public in 2008 October titled "Bitcoin: A Peer-to-Peer Electronic Cash System" that was released under the author's name Satoshi Nakamoto [1]. Due to the significance and popularity of Bitcoin, many people tried but could not meet the author. Thus, it has been assumed that the name Satoshi Nakamoto was used so that the real author could remain unknown.

In this original paper on Bitcoin, the term "blockchain" does not appear. Based on the paper's proposed methodology of chaining blocks of ledger data to former blocks of data is where the word 'blockchain' emerged from. The Bitcoin blockchain uses asymmetric encryption (i.e., RSA public key and private key) and hashing technology to chain the blocks of ledger data, which forms a blockchain.

In order to understand the detail operation of Bitcoin, the basics of asymmetric encryption (i.e., public key and private key) and hashing technology need to be understood, where it is recommended to read Chaps. 6 and 7 of this book for more technical details.

On 2009 January 9, Bitcoin was initially released (ver. 0.1.0). Considering the massive impact of Bitcoin and cryptocurrencies in our daily lives, the contribution of Bitcoin is significant and important.

In addition to being used originally by Bitcoin and later by Ethereum, blockchain technology has rapidly improved and has been applied in numerous databases as

J.-M. Chung, *Emerging Secure Networks, Blockchains and Smart Contract Technologies*, https://doi.org/10.1007/978-3-031-65866-2_9

well as memory and filing systems supporting various applications. For more details on blockchain technologies, please refer to Chap. 8 of this book. Considering the importance of Bitcoin, this chapter introduces its core technologies.

9.1 Bitcoin Blockchain

Blockchain operates like a decentralized bank ledger. However, unlike a bank, Bitcoin is not based on a centralized administrative entity (i.e., the Bitcoin blockchain is fully distributed) and data transactions recorded in a chained block are fully disclosed (i.e., transparent) and cannot be altered (i.e., immutable).

In Bitcoin, a blockchain hash value named the 'Transaction ID' (TXID) is used to help confirm that the ledger data of a Bitcoin BTC financial record is valid and was not tampered with or forged.

The additional blocks to be chained are created by entities called 'miners' who have to verify all transactions since the last block that was chained to the blockchain and keep the ledger data in their block fully updated (i.e., should not have any missing transactions). Miners create blocks to be chained to the Bitcoin blockchain to gain a reward, which can be considered as the transaction fee to keep the ledger fully updated.

Each block of the Bitcoin blockchain includes a 32 bit 'nonce' value that miners can manipulate to find a 256 bit hash value less than or equal to the nBits target threshold (using the SHA256 Double Hash algorithm) as quickly as possible. The first miner to publicly report a block header that can generate a hash value that satisfies this condition (with a Merkle root supporting a complete ledger) will receive the block chaining reward.

More details of the Merkle root are provided in Sect. 9.4 of this chapter. In addition, for more details on SHA256 Double Hash (i.e., SHA256D) please refer to Chap. 6 of this book.

Trying to be the first miner to find and report a block header that can generate a 256 bit hash value that is less than or equal to the nBits target threshold is a very computationally burdening operation, which requires a massive brute force search process, which resembles mining work. This mining process is repeated approximately every 10 min, where a new block header will need to be chained. This is why Bitcoin block reward contenders are called 'miners.' Likewise, many other cryptocurrency block reward contenders are also called 'miners.' In addition, many miners may group up to form a 'mining pool' team to compete for block rewards together. If one of the miners of a mining pool receives a block reward, then the profit is shared among the miners of the mining pool. Figure 9.1 presents the Bitcoin's BTC mining pool organizations and relative size [2].

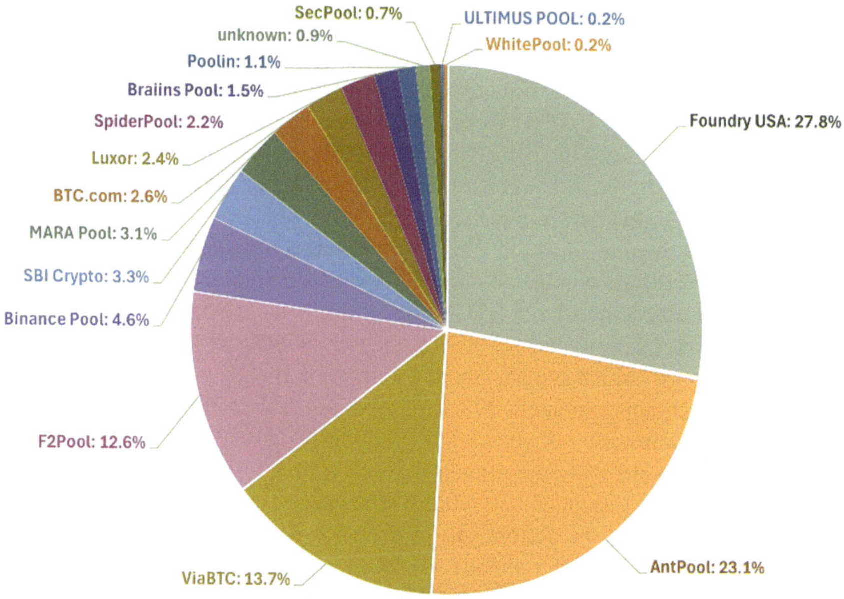

Fig. 9.1 Bitcoin BTC mining pool organizations and relative size

9.1.1 Bitcoin Block Reward

A Bitcoin reward is given to the first miner to publicly report a block header that can generate a hash value (using SHA256D) that is equal or less than the nBits target threshold, while including all of the verified Bitcoin transactions in their block of ledger data. After every 2,016 blocks that are chained (which is approximately 2 weeks), the nBits target threshold is changed. If the average block chaining time is less than 10 min, then it will be considered that the nBits target threshold was too easy to satisfy. So, the nBits target threshold will be lowered to make it more difficult to satisfy. This will result in more time required by the miners to be the first to publicly report a block header that can generate a hash value that satisfies the nBits target threshold. The nBits target threshold will be adjusted to make the block chaining time approximately 10 min [3].

The Bitcoin reward size is reduced to half its size after an additional 210,000 blocks have been chained, which occurs approximately every 4 years, because Bitcoin chains a new block to its blockchain approximately every 10 min. This means that the new block must include all of the verified Bitcoin transactions in its block of ledger data since the last block was chained approximately 10 min ago (and may include some of the missing TXIDs of the former chained block).

Bitcoin was designed to have only 21 million bitcoins created and distributed as a reward to chaining a new block to the Bitcoin blockchain. This limit was proposed by the inventor of Bitcoin (Satoshi Nakamoto) to avoid inflation and to prevent the value of Bitcoin from diminishing due to having too many Bitcoins generated.

On 2009 January 9, when Bitcoin was first released, 50 Bitcoins (BTCs) was the block reward. After 210,000 blocks were chained the block reward was reduced to 25 BTC (i.e., half size) on 2012 November 28. Likewise, after another 210,000 blocks were chained, the block reward was reduced to 12.5 BTC on 2016 July 9. After another 210,000 chained blocks, on 2020 May 11, the block reward became 6.25 BTC, which was the third time the block reward was halved. The next halving of the block reward was conducted on 2024 April 19 on block 840,000, where the new halved reward became 3.125 BTC.

As of 2021 May, approximately 90% of the total number of block rewards have been already issued, which sums to about 18.7 million BTCs.

By 2039 April, approximately 99.6% of all block reward BTCs will have been issued and a block reward will be worth approximately 0.19531250 BTC. By 2140, the BTC block reward will approximately reach 0. In 2140, a total of 21 million BTCs will have been issued, and no more BTCs will be issued any more.

Therefore, gradually the Return On Investment (ROI) of a block reward will decrease and the value of the already issued BTCs will become more valuable. Because the mining process is very computationally burdening, as the block reward decreases, less resources will be invested into mining and mining pools. In addition, a rapid increase in cryptojacking crimes and cryptocurrency ransomware attacks may occur due to the low ROI on self-investments of the computing resources for mining for the BTC blockchain. For more details on cryptojacking and ransomware attacks, please read Chap. 1 of this book.

Bitcoin is the origin of all cryptocurrencies, and BTC has the overall highest value among all cryptocurrencies. The trend of changes in BTC in units of USD are presented in Fig. 9.2. As the next halving of the block reward was approaching on April 19 of 2024, Bitcoin BTC reached its highest value ever at $75,830 USD on March 14 of 2024. Before 2024, the highest value ever Bitcoin BTC reached was $67,566.83 USD on November 8 of 2021 [4].

9.1.2 Bitcoin Blockchain Characteristics

Some of the major Blockchain characteristics are described in the following.

The Bitcoin transaction system uses cryptographic Proof-of-Work (PoW) to maintain fair competition of the blockchain reward to all participating miners. The PoW competition of the miners is focused on being the first to obtain a hash value that is less than or equal to the nBits target threshold when using the SHA256 double hash function (i.e., SHA256D).

Bitcoin and its blockchain enables two independent users to directly make a reliable financial transaction of Bitcoins without using a bank (or a bank like trusted third organization).

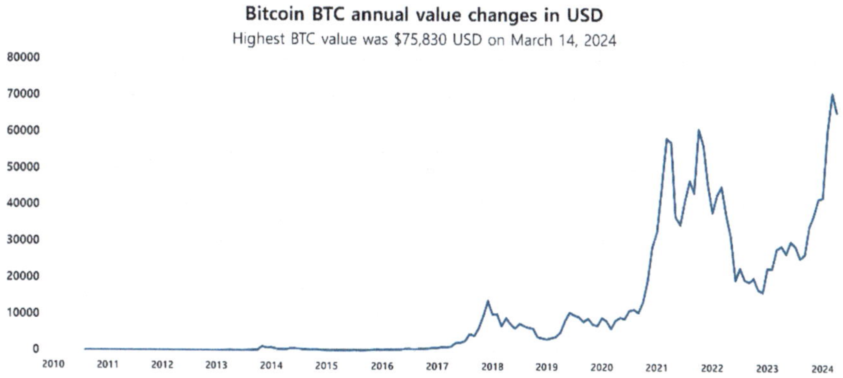

Fig. 9.2 Bitcoin BTC annual value changes in USD

If there is no bank (or bank like centralized financial management entity), then electronic transactions of expenditures can be easily overused. For example, multiple withdraws can be made from an account if the account balance and transaction record is not immediately updated. This is what is called the 'double spending problem.' One of the roles of a bank is to guarantee that double spending does not occur. Double spending becomes a much more difficult problem to prevent when it is managed in a distributed manner and maintained by multiple independent entities.

One of the major innovations that the white paper on Bitcoin proposed was a new mechanism that could prevent the double-spending problem while enabling reliable and stable Bitcoin peer-to-peer transactions. All Bitcoin transactions are recorded with a timestamp to ensure the chronological order, which is necessary to prevent the double-spending problem. This timestamp is generated from a timestamp server that is accessible through the network. As each transaction includes a timestamp, the timestamps are protected by the hash values of the blocks chained.

Another major innovation that the white paper on Bitcoin proposed was a new mechanism that the blockchain protects the Bitcoin transaction records from fraudulent manipulations based on the fact that the blocks of a blockchain are designed to be impractical to reverse or forge, due to the high level of computational complexity of the SHA256 double hash function (i.e., SHA256D).

The first miner to publicly report (i.e., a message is broadcasted) a block header that can generate a hash value (using SHA256D) that satisfies the nBits target threshold, will be chained to the Bitcoin blockchain, and later receive a block chaining reward. The reported block must include a Merkle root hash value of a complete ledger of TXIDs (i.e., list of all BTC transactions made since the last block was chained) to be considered valid. If the reported block satisfies all of these conditions, then the other miners will attempt to create the next block to chain to this block.

More details of the Merkle tree of TXIDs and the Merkle root that is included in the Bitcoin block header are provided in Sect. 9.4 of this chapter.

Messages related to Bitcoin transactions are broadcasted to all users and miners. As there is no reception confirmation process or broadcast message retransmission process in case a message delivery error occurs, some messages may not be received by some users or miners, which will result in an incomplete ledger.

Therefore, the longest blockchain is to be considered to be the blockchain that is trusted by the majority of miners and should be used to chain future ledger data blocks of Bitcoin transactions (which is repeated approximately every 10 min). This would enable any user or miner to find the proper blockchain to check the record of Bitcoin transactions. This feature also enables users to freely leave the Bitcoin blockchain and return anytime later, as the longest blockchain can be identified and used to make future Bitcoin transactions.

Bitcoins use a sequence (i.e., chain) of digital signatures to confirm transactions among users. A user can transfer ownership of a Bitcoin to another user by digitally signing the hash value of the previous transaction as well as the public key of the next owner, which is added to the end of the Bitcoin transaction record. Figure 9.3 presents the Bitcoin transactions signing and verification sequential process proposed in the original Bitcoin whitepaper by Satoshi Nakamoto titled "Bitcoin: A Peer-to-Peer Electronic Cash System" released in October of 2008 [1]. This way, any user can check the signatures to confirm the chain of Bitcoin transaction data.

In the original Bitcoin whitepaper, an electronic coin is generated using a chain of digital signatures. An owner can transfer a Bitcoin to someone else by digitally signing the hash value of the previous transaction. This hash value is created from the transaction record and public key of the next owner. The digital signature is included at the end of the transaction record such that the coin payee can verify the transaction records and signatures which will verify the sequence of Bitcoin transaction data. This process is illustrated in Fig. 9.3.

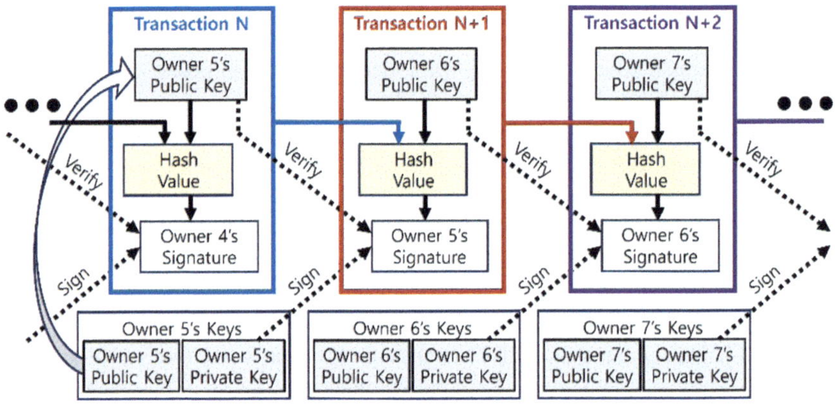

Fig. 9.3 Bitcoin transactions signing and verification sequential process proposed in the original Bitcoin whitepaper

9.1.3 *Bitcoin Attacker's Dilemma*

If an attempt were made to forge a block of data already included in the blockchain, then a massive amount of computation would be required to find an alternative way to satisfy the sequence of chained blocks of data of an already established blockchain. In addition, considering the distributive nature of the Bitcoin blockchain, the blockchain's file has been saved already by numerous users and miners, and finding a way to forge BTC transaction records without anyone noticing the manipulation is impossible. This results in the 'attacker's dilemma' where it is convincing to make a sensible decision to join the fair block reward miner's competition, than trying to find a way to forge the blockchain data to gain an unfair profit.

9.2 Bitcoin Blockchain File Size

As Bitcoins are used more frequently, the number of transactions is increasing. Therefore, the ledger data part of a block to be chained will grow proportionally to the number of transactions made during approximately the past 10 min.

The header part of a block does not include the list of transactional data and TXIDs, and therefore, the block header can have a fixed length (i.e., 80 bytes). Considering that a new block is chained approximately every 10 min, the memory size due to the headers would exceed 4.2 Mega Bytes (MB) per year based on the following calculations of $\{80 \text{ Bytes/block} \times 6 \text{ blocks/hour} \times 24 \text{ h/day} \times 365 \text{ days/year}\} = 4.2048 \times 10^6 \text{ Bytes/year} \cong 4.2 \text{ MB/year}$.

As presented above, the header of the Bitcoin blockchain is of a moderate 80 byte size. But the Bitcoin blockchain's full size (which includes all BTC transaction records chained in a Merkle tree) is growing rapidly. This is because Bitcoins are newly issued approximately every 10 min and the cryptocurrency market has grown extensively, in which Bitcoin is the center of all cryptocurrency financial activity. As mentioned above, the block reward became 6.25 BTC on May 11 of 2020 and was halved to 3.125 BTC on April 19 of 2024. The Bitcoin blockchain's full sizes (which includes all BTC transaction records) are listed below in Table 9.1 and graphically presented in Fig. 9.4 [5]. In case multiple blocks are competing to be chained to the Bitcoin blockchain (e.g., C1, C2, and C3 in Fig. 9.5) then a fork is formed. After the fork division into C1, C2, and C3, the C3→D3→E3 branch has the most support from the miners and continues to chain more blocks compared to the other two branches of the fork. If the C3 branch continues to have more blocks chained compared to the other branches, then it will become official. This is why forks will have a temporary effect on the file size until the fork is resolved, where only one branch of the fork will remain chained to the Bitcoin blockchain.

Table 9.1 Bitcoin (BTC) blockchain size increasements in units of giga bytes

Bitcoin blockchain size	
Date	Size in Giga bytes
January 2010	0.01
July 2010	0.02
January 2011	0.07
July 2011	0.43
January 2012	0.93
July 2012	2.29
January 2013	5.11
July 2013	9.51
January 2014	14.56
July 2014	21.05
January 2015	29.43
July 2015	40.53
January 2016	56.16
July 2016	76.35
January 2017	98.65
July 2017	124.41
January 2018	151.52
July 2018	173.62
January 2019	197.53
July 2019	227.11
January 2020	254.18
July 2020	283.87
January 2021	317.38
July 2021	348.73
January 2022	379.23
July 2022	409.44
January 2023	440.76
July 2023	488.20

9.2.1 Bitcoin Blockchain's Merkle Tree

A highly reliable mechanism is needed to reduce the size of the Bitcoin blockchain while keeping all ledger data of the blockchain transparent (easily checkable), immutable (unerasable), and unforgeable (tamper protected). For this purpose, the 2008 original white paper on Bitcoin included a proposal to use the 'Merkle tree,' which is a memory size management technology (where more details are described in the following sections and in Figs. 9.6 and 9.7).

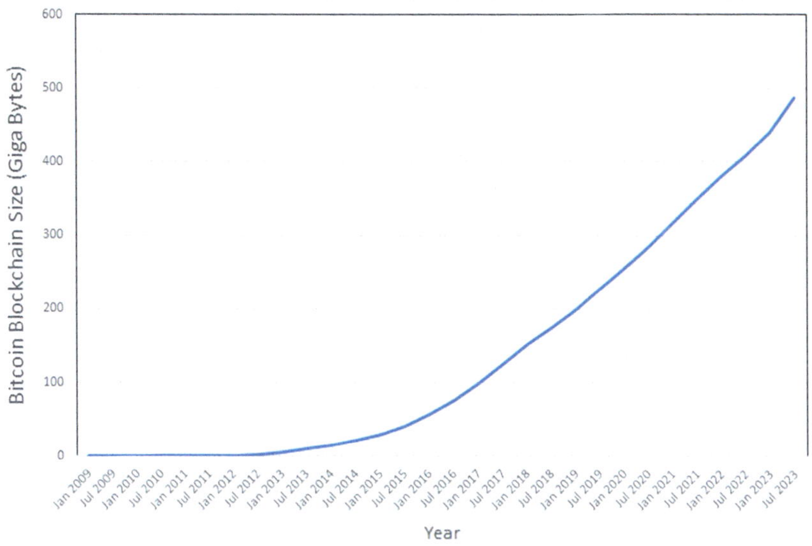

Fig. 9.4 Bitcoin (BTC) blockchain size increasements in units of giga bytes

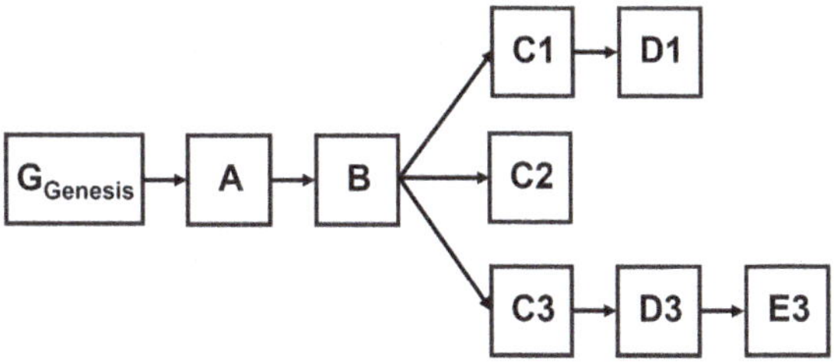

Fig. 9.5 Bitcoin blockchain fork example

Using the Merkle tree technology, the chained blocks of older transaction data can be hashed where the hash value can be used as a representative identity of the older transaction block of BTC exchange records. Then the older transaction record blocks can be saved in publicly known separate databases to reduce the size of the main Bitcoin blockchain.

The original technology for the Merkle tree was originally proposed in a paper published in 1980 by Ralph Merkle [6] where many enhancements have been made since. In a Merkle tree, the root value is the only part that needs to be included in the chaining block. This creates a sequence of blocks interconnected in a tree fashion using hash values to keep the data of each block transparent, immutable, and unforgeable.

As can be noticed, blockchain technology uses many parts of Merkle tree technology. Both technologies use a series of hash values to chain blocks of data to protect the data while enabling each block of data to be transparent. A difference is that Merkle trees chain blocks of data in a two dimensional (2D) form and blockchains

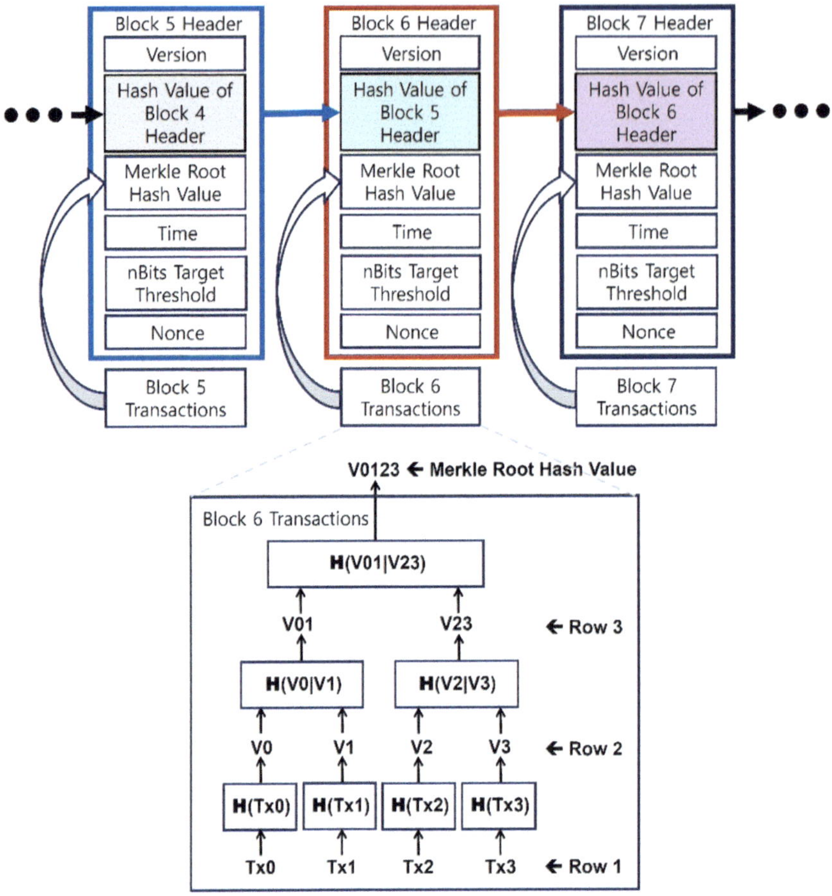

Fig. 9.6 Bitcoin blockchain format. Although the block header is only 80 bytes and relatively smaller compared to the block transaction file size, it was drawn larger to show the inner details more clearly

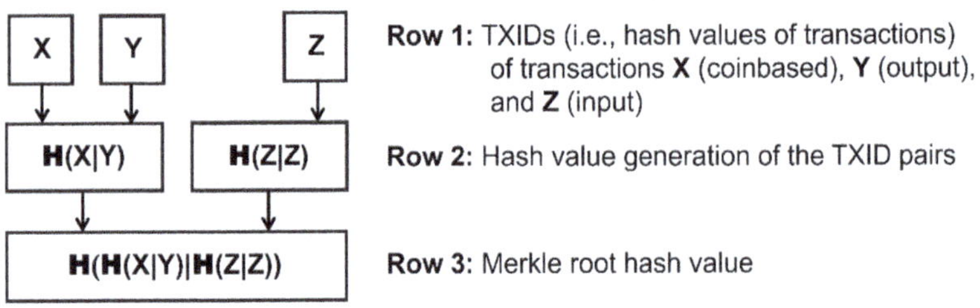

Fig. 9.7 Merkle Tree formation example for TXIDs **X**, **Y**, and **Z**. Corresponding to the TXID placement order requirements set in the consensus rules, the coinbase TXID **X** is placed first and then the TXID of output **Y** is placed before the corresponding TXID of input **Z**

chain blocks of data in a one dimensional (1D) sequential string chain form. However, blockchains do turn into 2D structures when a fork occurs (e.g., Fig. 9.5, where more details on forks are provided in Sect. 9.3.3). In addition, modern blockchains include many innovative technologies that have emerged since the original Merkle tree paper that was published in 1980.

9.3 Bitcoin Operations

In this section, the operations applied in Bitcoin are introduced, which include information on Bitcoin transaction verification methods, Bitcoin block mining procedures, blockchain forks, TXID, and transaction records.

9.3.1 Bitcoin Transaction Verification Method

Any user can check the record of Bitcoin transactions using the following procedures. For all procedures, the block header and timestamps need to be checked.

1. Verifying a past Bitcoin transaction record starts with finding the longest blockchain. This can be done by making a query and selecting the longest blockchain among the ones received.
2. Next, the user needs to find the Merkle branch that includes the ledger block that contains the desired transaction record that needs to be checked.
3. Then the user needs to recheck if the block was properly accepted by a majority of the miners. This can be confirmed by checking if additional blocks have been properly chained after this block.

However, as described above, the procedures to check Bitcoin transactions are not simple. This is why only an experienced user can reliably conduct this checking process efficiently. This is why cryptocurrency exchange companies like Binance, Coinbase Exchange, Kraken, KuCoin, Bitstamp, OKX, Bybit, Bitfinex, LBank, Gemini, Gate.io, Huobi, Bitget, etc. are needed [7]. Cryptocurrency exchange companies enable easy checking and management and provide many ways to buy or sell cryptocurrencies and make investments on a combined index of cryptocurrencies.

9.3.2 Bitcoin Block Mining Procedures

The following Bitcoin block mining procedures are repeated approximately every 10 min.

1. When a new Bitcoin transaction is made, a message that includes the transaction record and timestamp (TXID) is broadcasted to all miners.

2. Each miner collects all new transaction records and organizes the record based on timestamp order into a block.
3. Each miner changes its Nonce value (a 32 bit number) in an attempt to find a hash value (using SHA256D) that is less than or equal to the nBits target threshold as quickly as possible. This process is called Proof-of-Work (PoW), which is a trial-and-error based highly computational process that is nicknamed 'mining' due to its difficult labor-like activity. If a satisfactory nonce value is found, it will be included in the block header.
4. The miner will broadcast its block (including its Nonce value) to all miners for it to be checked.
5. The miners will check the broadcasted block header. The first miner to broadcast a block header that can generate a hash value that satisfies the nBits target threshold with a Merkle root supporting a complete ledger will be considered the block to be chained.
6. Miners will create a new block (of Bitcoin transactions for the next 10 min) using the chosen block of the previous stage.
7. As the chosen block is used by a majority of miners to continuously chain new blocks, the chosen block will become an accepted part of the Bitcoin blockchain and will later receive a block reward.

Based on a missed broadcast message of a transaction record or due to bad manipulative intentions, two or more miners may broadcast blocks that have a difference in transaction records. In this case, it is believed that a majority of the miners will notice the erroneous blocks, and the correct block (with a complete ledger) will be chosen and used to form the next block by a majority of the miners. Having faith that a majority of the miners will always make the correct decision of commonly selecting the most appropriate block is what maintains the honesty and reliability of the Bitcoin blockchain.

This majority rule is what enables the Bitcoin blockchain to operate without any central financial organization (like a bank). Due to the majority rule, as long as the message containing a new transaction record is received by a majority of the miners, it is believed that it will be included into a block ledger record in the near future.

The majority rule makes the Bitcoin blockchain robust against erroneous and lost messages. Therefore, a miner may notice that it did not include a transaction record when it receives a block from another miner. This miner can use this information to update its transaction list records.

9.3.3 Blockchain Forks

Blockchains are based on a distributive structure of miners in which a majority agreement on the next block to chain to the blockchain is made independently by each miner. When a majority agrees on the same block to chain and uses that block to generate the next block, then the process becomes straightforward. However, in the following cases the blockchain cannot continue to have blocks additionally chained until the problem is solved.

- A majority of the miners cannot reach an agreement on the block to choose and chain the next block to, because there is a disagreement in the transactional records.
- All miners and users need to have a program upgrade installed.
- A bug in the blockchain program creates a malfunction in the process.
- A cyber-attack results in an inconsistency in the operation or financial records of the blockchain.

In these situations, the single sequence of blocks being chained to the blockchain may diverge into two or more branches, if a decision on which branch to use cannot be easily made, as presented in Fig. 9.5. In such cases the blockchain administrator may have to get involved to solve this problem quickly so the blockchain can resume to its normal operations. In such cases, the blockchain will look like a 'fork' as it will have a long sequence of chained blocks being split into multiple streams of blocks starting from the block where the problem has occurred. Due to this reason, the term 'fork' is used to express these situations.

The representative fork types are short-lived fork, permanent fork, hard fork, and soft fork. A description of each fork is provided in the following.

Short-lived forks are used to quickly reach a consensus among the miners. Once a consensus is made, the fork is removed, which is why it is called a 'short-lived fork.' Because blockchains are operated as a distributed system, short-lived forks are very useful in solving small problems.

Permanent forks are used to make permanent changes in a blockchain. Some of the reasons to execute a permanent fork are listed in the following. First, permanent forks are used when a change in the blockchain's operational procedures or message protocol needs to be made. This protocol change could be to fix a system bug or to match the upgraded system platform, software, firmware, or operating system (OS). For example, on 2010 August 6, a permanent fork was used to overcome a catastrophic bug in the Bitcoin blockchain software. Second, permanent forks help the administrator and users to accommodate new features and functions that were added to the blockchain. Third, permanent forks are used when recovering from a cyber-attack. In case of a cyber-attack, parts of the blockchain records or system may need to be reversed and reconfigured. For example, in 2016 July, a permanent fork was used when dividing Ethereum and the Ethereum Classic blockchains to overcome the damage from hacker attacks.

Hard forks are used when software upgrades are conducted, and the software validation rule needs to be changed. To be able to work with the blockchain, all users and miners have to make the upgrades set by the hard fork, because hard forks are not backward compatible. Since hard fork upgrades are mandatory, the systems that do not make the required upgrade will not be able to access the blockchain because these systems will not be able to validate the new blocks sent by the competing miners.

Soft forks are used when the blockchain upgraded software is backward compatible with the old software. A soft fork blockchain protocol upgrade enables the new rules to be used by all users, which include the software upgraded users and non-upgraded users. This is why soft forks do not normally create a pause in blockchain services. However, it is recommended for all users to eventually upgrade their software. Soft fork-based upgrades are normally smaller in scale and functionality compared to hard fork upgrades.

9.3.4 Transaction Records and TXID

Bitcoin wallets are used to control and record transactions of inputs and outputs. In the Bitcoin blockchain, each output is either a Spent Transaction Output or an Unspent Transaction Output (UTXO). A valid payment record will use a UTXO as the input. Each output has a transaction identifier (TXID) which uniquely identifies the transaction record. TXID use SHA256D, which is a double hashed SHA256 function. For example, applying SHA256D to a binary file **X** would be noted as SHA256D(**X**) = SHA256(SHA256(**X**)), which means that **X** would be processed using SHA256(**X**) and the resulting hash value would be processed again using a SHA256 hash function (i.e., SHA256(SHA256(**X**))). Specific outputs can be identified using an Outpoint, which is a combination of a TXID with a VOUT. VOUT is a transaction output number (commonly 4 bytes) used as an index to indicate the position of a transaction output [8].

9.4 Bitcoin Block Header

As presented in Fig. 9.6, every Bitcoin block has an 80 byte header that consists of the following fields (more details will be provided in the following section) [8].

Version: This is a 4 byte (i.e., 32 bits) field that indicates the version of the blockchain. This is important because the security procedures and implemented upgrades applied to the block all depend on the version that was used when the block was created.

Previous Block Header Hash: This is a 32 byte (i.e., 256 bits) field (in internal byte order) that includes the previous block's header hash value. SHA256D is used to generate this hash value. This field provides the 'chaining effect' of the blockchain blocks. This field is what prevents malicious changes and tampering of ledger transaction records in the previous block. It is also used to confirm the exact chaining sequence of blocks in a blockchain.

Merkle Root Hash: This is a 32 byte (i.e., 256 bit) field (in internal byte order) that includes the hash value of the ledger transaction data. This hash value protects the data making it transparent and immutable and unforgeable. SHA256D is used to generate the Merkle root hash value. SHA256D is a doubled hashed value that use the SHA256 function twice, where SHA256 is one of the SHA-2 family hash functions.

Time: This 4 byte field (i.e., 32 bits) represents the time of when the miner began its hashing process of the new block. The time value is based on Unix Epoch Time, which is based on the Coordinated Universal Time (UTC). The time value needs to be a later time than the median time of the former 11 blocks, but should not exceed 2 hours in advance of the time of the other Full nodes' clock.

nBits Target Threshold: This 4 byte field (i.e., 32 bits) is used to inform the miners of the target value that this block header's hash value has to be equal or less than. If the nBits target threshold is set to a lower value, then it becomes harder to generate a hash value that is equal or less than this value, which is why it is called the 'difficulty target.'

Nonce: This is a 4 byte field (i.e., 32 bits) that each miner will change (repeatedly) until they come up with a block header's hash value that is less than (or equal to) the difficulty nBits target threshold value.

Block Header Hash Value: This is a 4 byte field (i.e., 32 bits) that includes the (SHA256D based) hash value of the previous block header. This hash value has to be equal or less than the nBits target threshold. The hash value of a block header is not included in itself, but it is included in the next block header that is chained. All other miners will compute the block header hash value as soon as a block message is received.

The first block header that can generate a hash value that satisfies the nBits target threshold with a Merkle root supporting a complete ledger will be chained to the blockchain and will later receive a block chaining reward.

In the Bitcoin block header, the previous block header hash value and the Merkle root hash value are represented using 'internal byte order,' which is the standard order used to display hash digests as strings. Internal byte order is also used in serialized blocks and transactions in the Bitcoin blockchain [9].

9.4.1 Block Versions

In 2009 January, **version 1** was introduced in the genesis block of Bitcoin. The last version 1 block was block number 227,835, which has a timestamp of 2013-03-24 15:49:13 GMT [10].

In 2012 September, **version 2** was introduced in Bitcoin Core 0.7.0 as a soft fork. Version 2 blocks newly require the use of the block height parameter in the coinbase as defined in BIP34, where BIP34 includes the rules used to reject blocks without the proper block height, which was activated in March of 2013 [10].

In BIP34, the function IsSuperMajority() is used to manage the soft fork transition. This function is also used in the version 3 and version 4 soft fork transition management as well. For a specific block, the block height represents the number of blocks between that block and the genesis block of the blockchain.

In 2015 February, **version 3** was introduced in Bitcoin Core 0.10.0 as a soft fork. From 2015 July, the version 3 fork was fully enforced, such that new blocks were required to apply the Distinguished Encoding Rules (DER) to the encodings of all

Elliptic Curve Digital Signature Algorithm (ECDSA) signatures, as defined in BIP66. DER is a (restricted) upgrade of the Basic Encoding Rules (BER) which is used to produce unequivocal transfer syntax for data structures [11].

In 2015 November, **version 4** was introduced in Bitcoin Core 0.11.2 as a soft fork and was activated in December of 2015. Version 4 is based on BIP65, which supports the use of the OP_CHECKLOCKTIMEVERIFY opcode, which is used to verify the locktime. The locktime is checked using the nLockTime field of a transaction and comparing it to the value recorded in the opcode. However, OP_CHECKLOCKTIMEVERIFY does not verify the time or the current block height [12, 13].

9.4.2 Merkle Root Hash

Merkle trees are formed corresponding to the TXID placement order requirements set in the consensus rules, where only TXIDs in the corresponding block can be used. TXIDs are the hash values of transaction records. The Merkle root forming TXID listing order requirements are based on the consensus rules listed below [13].

① The TXIDs of coinbase transactions are listed first. Coinbase transactions are the BTC transactions given to a miner as a reward for creating a block that was chained to the Bitcoin blockchain.

② Output TXIDs are placed before their corresponding input TXIDs, so when the TXID are parsed in listed order, the output TXID will be parsed before the input TXID. Output TXIDs are based on BTC exchanges/purchases between Bitcoin users/companies.

③ For a block with only a coinbase transaction, the TXID of the coinbase transaction is directly used as the Merkle root hash.

④ For a block with only one coinbase transaction and one additional transaction, these two TXIDs are concatenated (with the coinbase TXID in the first position) into 64 raw bytes, which are hashed together using SHA256D where the resulting hash value becomes the Merkle root.

⑤ For a block with three or more transactions, intermediate rows are formed first using the Merkle tree hashing functions. In the first row, the coinbase TXID is placed first and then the TXID of the output is placed before the corresponding TXID of the input. Based on this order, the TXIDs are paired, and the hashing function is applied. The resulting hash values form the following row.

   A. If there are an odd number of TXIDs, then the last TXID will be duplicated and will be used as a pair with itself to be hashed (see the example in Fig. 9.7).
   B. If there are five or more TXIDs, then the TXIDs are paired up and hashed. This process is repeated until there are only two hash values remaining. Each time this process is repeated, an additional row is formed.

⑥ The final two hash values will be hashed to form the Merkle root hash value (which is the last row of the process). The Merkle root value is recorded in the block header using internal byte order.

The TXIDs and intermediate hashes are concatenated using internal byte order. The hashing function used in this Merkle root value generation process is the SHA256D, so, $\mathbf{H}(\cdot) = \text{SHA256D}(\cdot) = \text{SHA256}(\text{SHA256}(\cdot))$ in Fig. 9.7.

In Fig. 9.7, a Merkle root value formation example with three TXIDs **X**, **Y**, and **Z** is presented. Corresponding to the TXID placement order requirements set in the consensus rules, the coinbase TXID **X** is placed first and then the TXID of output **Y** is placed before its corresponding TXID of input **Z**. In Row 1, the TXIDs of transactions **X** (coinbased), **Y** (output), and **Z** (input) are placed in order. Because there are only 3 TXIDs (an odd number), the last TXID **Z** is duplicated and used as a pair with itself when it is hashed (i.e., **H**(**Z**|**Z**)). Row 2 contains the hash values of the two TXID pairs. In Row 3, the Merkle root hash value is obtained from hashing the two hash values in Row 2.

9.4.3 nBits Target Threshold

The nBits target threshold is a value that the block header's 256 bit hash value (which uses SHA256D to generate) has to be equal or less than by controlling the Nonce value. So, the nBits target threshold has to represent a 256 bit value. However, in order to make the Bitcoin block header small (i.e., 80 bytes), the nBits target threshold is represented in a compact form of 4 bytes (i.e., 32 bits). The following example shows how these 4 bytes of the target nBits are converted into a 256 bit target threshold.

The nBits target threshold in the block header is recorded in little-endian order, which can be converted into a hexadecimal (0x) sequence form, to make it easier to read. This is conducted by dividing the numbers into a pair of two digits each and then reversing the sequence order. For example, as presented in Fig. 9.8, 34d52a18 will be converted into 182ad534 which is in the hexadecimal compact form of 4 bytes.

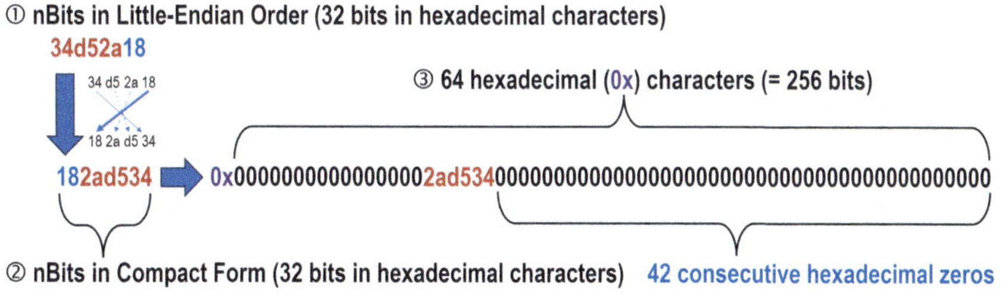

Fig. 9.8 Example of the nBits target threshold value '34d52a18' represented in hexadecimal character form

Please note that ‘0x’ is used in front of hexadecimal numbers. Please consider Chap. 4 (Sect. 4.4 ‘IPv6 Protocol’) for details on how to convert hexadecimal numbers into binary and decimal numbers.

The hexadecimal nBits threshold value 0x182ad534 is divided into two parts, where 0x18 is used to compute the exponent part and the rest 0x2ad534 is used as the significand (i.e., the mantissa) value of the target threshold value. The exponent part 0x18 value is used to derive the number of consecutive zeros in the target threshold, which makes the block header’s hash value very small. So, the larger this exponent part number is, the more difficult it will be to satisfy the nBits target threshold value by controlling the nonce value. The number of consecutive zeros in the target threshold is computed using the base number $256 = 2^8$ (because the nBits target threshold is a 256 bit target threshold value) using the following method.

The exponent 0x18 is subtracted by 3 which is the number of bytes of the significand (i.e., 0x2ad534 is a 3 byte value, because 0x2a requires one byte, 0xd5 requires one byte, and 0x34 requires one byte). Converting the hexadecimal value 0x18 into binary is 00011000 which is equivalent to 24 in decimal. When 24 is subtracted by 3 (i.e., number of bytes in the significand) we get $(24 - 3) = 21$. Using this value as the exponent of the base 256, we obtain $256^{21} = (2^8)^{21} = 2^{(8\times 21)} = 2^{168}$. Therefore, among the 256 bits of the target threshold, there will be 168 consecutive zero bits starting from the most significant bit (MSB). In hexadecimal form this will be equivalent to having $(168 \div 4) = 42$ consecutive hexadecimal zeros. A 256 bit target threshold will be 64 hexadecimal characters. In Fig. 9.8, the nBits target threshold value 0x182ad534 is represented in a 64 hexadecimal character form.

As a result, it is easy to see how effective the compact form of the nBits target threshold format is, which helps to keep the Bitcoin block header at a small and efficient 80 byte size. The Bitcoin block header’s hash value (computed using SHA256D) has to be equal or less than this nBits target threshold value. So, through trial-and-error, various nonce values will have to be tested, which is why miners need fast computers and consume so much energy. The nBits target threshold is changed every 2,016 blocks chained (which is approximately every two weeks), which changes the difficulty level of the miner’s job.

9.4.4 Serialized Blocks

The Bitcoin consensus rules require each block’s serialized size to be less than or equal to 1 MB (i.e., 1 mega Bytes) in order to be considered valid. Each Bitcoin block’s serialized size includes the 80 byte block header, total number of transactions in this block (txn_count), and a list of all transactions in this block (txns).

The txn_count value has a variable length and is represented in compactSize unit, representing the block’s total number of transactions including the coinbase transaction.

The txns is a collection of all transaction raw data listed in the order of the TXIDs, corresponding to the record in the first row of the Merkle tree connected to this block.

The first transaction of this list is a coinbase transaction record of transaction fees that should be collected and spent related to the transactions of this block. A coinbase transaction will be considered invalid if its spend value exceeds the available block reward value.

Bitcoin requires all blocks to have a block height less than 6,930,000. Block height represents the number of blocks between this block and the genesis block of the blockchain.

9.5 Summary

This chapter describes the technical details of Bitcoin blockchains, blockchain file size, chaining operations, and the Bitcoin block header. An evolved form of the Bitcoin blockchain is introduced in Chap. 10 'Smart Contracts and Ethereum.' In case the technical details of this chapter were hard to understand, the reader is recommended to read Chap. 6 'Hash Technologies,' Chap. 7 'Cryptography Technologies,' and Chap. 8 'Blockchains,' as these chapters provide the essential building blocks of the Bitcoin blockchain. In addition, in Chap. 11, future emerging issues with Bitcoin are discussed, which will be useful to know.

References

1. S. Nakamoto, Bitcoin: a peer-to-peer electronic cash system, white paper. https://bitcoin.org/en/bitcoin-paper
2. BTC.com, Pool Distribution. https://btc.com/stats/pool
3. Wikipedia, Bitcoin. https://en.wikipedia.org/wiki/Bitcoin
4. J. Edwards, Bitcoin's Price History. Investopedia.com. https://www.investopedia.com/articles/forex/121815/bitcoins-price-history.asp
5. R. de Best, Size of the Bitcoin blockchain from January 2009 to June 2, 2024. https://www.statista.com/statistics/647523/worldwide-bitcoin-blockchain-size/
6. R. C. Merkle, Protocols for public key cryptosystems, in *Proceedings of the IEEE Symposium on Security and Privacy* (1980), pp. 122–133. https://ieeexplore.ieee.org/document/6233691
7. CoinMarketCap.com, Top Cryptocurrency Spot Exchanges. https://coinmarketcap.com/rankings/exchanges/
8. BitcoinDeveloper.org, Block Chain. https://developer.bitcoin.org/devguide/block_chain.html
9. BtcInformation.org, Internal Byte Order. https://btcinformation.org/en/glossary/internal-byte-order
10. Github.com, Bitcoin/bips. https://github.com/bitcoin/bips/blob/master/bip-0034.mediawiki
11. Wikpedia.org, X.690. https://en.wikipedia.org/wiki/X.690#DER_encoding
12. Bitcoin.StackExchange.com, How to check OP_CHECKLOCKTIMEVERIFY?. https://bitcoin.stackexchange.com/questions/111087/how-to-check-op-checklocktimeverify
13. Block Chain. https://developer.bitcoin.org/reference/block_chain.html

Chapter 10
Smart Contracts and Ethereum

Smart contracts are used in many areas, which include betting and gambling services, copyright management, cryptocurrency trading/investments, financial contract execution, gaming services, healthcare management, insurance, international and domestic trade, logistics administration, non-fungible token (NFT) creation/purchasing/trading, real estate, rental services, supply chain management, etc. In order to introduce how smart contracts work, this chapter covers the details of smart contracts and how they are operated on blockchains. First, the concept of a smart contract is introduced along with its characteristics as well as its pros and cons. Second, the first blockchain and cryptocurrency based smart contract Ethereum is introduced. Ethereum's ETH annual value changes are presented along with the use of Ethash and Gas. In addition, the base software program Solidity is introduced along with Ethereum's account types and a smart contract execution example. In the third section, details of Ethash are introduced, which is a Proof-of-Work (PoW) algorithm that is designed to use memory-hard functions (MHFs). The fourth section focuses on Ethereum's new Proof-of-Stake (PoS), including details on the Validator, Execution, and Consensus clients' operational procedures as well as their transaction processing sequence. This is followed by a description of Ethereum PoS finalized block chaining process, including details on changing checkpoints. In addition, an overview of PoS attacks, security issues, inactivity leak defense, as well as validator requirements and violation penalties are provided. Further details of the correlation penalty and the Latest Message Driven Greediest Heaviest Observed SubTree (LMD-GHOST) fork selection algorithm are introduced.

J.-M. Chung, *Emerging Secure Networks, Blockchains and Smart Contract Technologies*, https://doi.org/10.1007/978-3-031-65866-2_10

10.1 Smart Contracts

The concept of a smart contract was first proposed in 1997 by Dr. Nick Szabo at the George Washington University, who is a professor of law and a computer scientist.

A smart contract is a computer program that provides both contract establishment functionality and automated execution of agreement clauses when certain contract conditions are satisfied.

In 2013, the co-founder of the Bitcoin Magazine [1], Vitalik Buterin, released a white paper that described the need for a platform for decentralized applications. The paper explained that this could be created by expanding Bitcoin and blockchain technology, which is the foundational concept of Ethereum smart contracts.

On 2015 July 30, Vitalik Buterin and Gavin Wood released Ethereum [2]. Ethereum is the first platform to use blockchain technology to enable smart contract applications to be created and provide access to the general public. Considering that smart contracts are now used everywhere, the contributions of Ethereum are significant.

10.1.1 Pros of Smart Contracts

Smart contracts provide a way to conduct transactions in a creditable and reliable method without involving legal arbitration or a third-party organization like a bank or credit card company. Because the smart contract computer program can be checked by all contracting entities before it is agreed upon (i.e., digitally signed), the conditions of the smart contract are transparent and immutable. For this purpose, smart contracts use blockchains to store data, conduct consensuses, maintain transparency and immutability while minimizing the influence of external administration and management.

Due to using blockchains, smart contracts are considered to be more transparent and reliable (due to immutability) than regular arbitrated contracts. In addition, details of the contract terms and what will happen when contracted clauses are satisfied are easy to understand because they are clearly revealed in the computer program that can be checked before the (digital) signing of the contract.

In many cases, regular arbitrated contracts can only be understood by lawyers and take significant time and legal fees to draw up a contract. In addition, in many cases the lawyers will not agree on the meaning or interpretation of established contract terms, in which arbitration, lawsuits, and court rulings are needed. Compared to this, smart contracts can be clearly understood by anyone that can read the smart contract program and the expected results can be tested and computer simulated based on various cases before the (digital) signing of the contract. This significantly helps to reduce the misunderstanding of the contracting parties. This is why almost all cryptocurrencies use smart contracts for establishment, which helps investors, users, and miners.

10.1.2 Cons of Smart Contracts

Smart contracts are vulnerable to cyber-attacks including various types of ransomware attacks. In addition, if a user's login ID and password are known to a hacker then the contract itself or the compensation (e.g., cryptocurrencies) can be easily stolen (like in the 2019 Binance attack case).

Bugs in program code or flaws in the system operation and network protocols could be exploited by hackers. This is especially true for smart contracts that have program codes open to the public (i.e., open sourced, transparent). Flaws in the smart contract's program or servicing platform (or in between the program and platform) could be found by a hacker and used to exploit the smart contract (like in the 2016 DAO attack case).

Another difficulty is that before a contract agreement is made, an expert programmer interpretation of the smart contract would be helpful. This is because a smart contract computer program cannot be accurately read and interpretated by someone that does not have expert level programming experience. In addition, the assisting programmer has to be trustworthy (both personally as well as technically), which may not be easy to find as there are many smart contract platforms and programming languages used in various smart contracts.

Cryptocurrencies are commonly used in smart contracts, in which some cryptocurrency values are unstable. This can make the Return on Investment (ROI) insignificant even though the business or activity conducted by the smart contract was successful.

These are only the major concerns with smart contracts, in which many more issues can exist.

10.2 Ethereum

Based on the Ethereum co-founder Vitalik Buterin's own words, the motive to create Ethereum is quoted below [3].

> Bitcoin is great as a form of digital money, but its scripting language is too weak for any kind of serious advanced applications to be built on top—Vitalik Buterin.

In addition, based on the Vitalik Buterin's own words, a definition of a smart contract is quoted below [4].

> A smart contract is a mechanism involving digital assets and two or more parties, where some or all of the parties put assets in, and assets are automatically redistributed among those parties according to a formula based on certain data that is not known at the time the contract is initiated—Vitalik Buterin.

Ethereum supports smart contract development, execution, and management based on its decentralized blockchain platform.

Ethereum supports two types of incentives. First, a 'gas' fee is provided as a transaction fee reward. Second, Ethereum also supports a digital token reward 'ETH' (which is read as 'ether') for the miners as their block reward, which is a similar process as in Bitcoin.

Unlike Bitcoin, Ethereum has no limit in the number of ETH tokens that can be created. In addition, ETH tokens can be generated in a much faster rate (which is in units of seconds) than Bitcoin (which is approximately every 10 min).

In addition, Bitcoin blocks are ledgers of BTC transaction records (i.e., coinbase, input, and output Transaction Identifiers (TXIDs)) since the last block was chained to the blockchain (see Sect. 9.4.2 for more detials). But Ethereum blocks include records of ETH and gas as well as smart contract related data, which is why the block sizes of Ethereum are large even through the block chaining speed is much higher, and Ethereum's blockchain length (i.e., total number of blocks chained in the blockchain) is longer than Bitcoin's blockchain length, where the difference is consistently growing [5].

Ever since it was launched in 2015 July 30, the value of Ethereum has grown significantly. Figure 10.1 presents the Ethereum ETH annual value changes in USD. The highest ETH value reached was $4,815.01 USD on November 9, 2021, which was a day after BTC had recorded its highest value ever (up to that date). This is because many aspects of Ethereum are tied up with Bitcoin, even from Ethereum's initiation stages of its Initial Coin Offering (ICO) [6].

In 2014, Ethereum's ICO was successful in raising $18 Million USD based on a two-part sales strategy. Using these funds, the Ethereum platform was developed, and a part was saved to help assist the growth of the Ethereum foundation and ecosystem. The pre-sale part was focused on allocating Ethereum to early contributors and

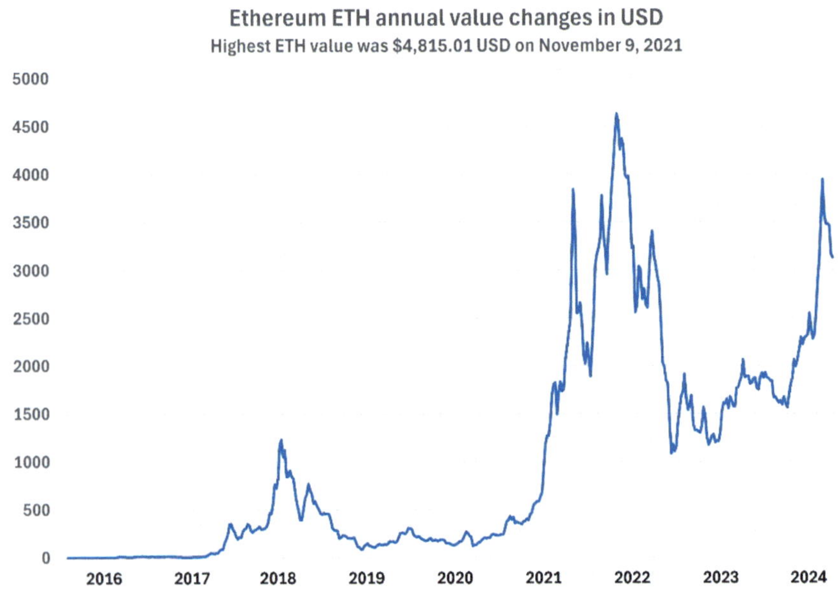

Fig. 10.1 Ethereum ETH annual value changes in USD

supporters. The public sale part was conducted based on an initial exchange rate of 2000 ETH for 1 BTC [7].

10.2.1 Ethereum Ethash and Gas

Ethereum has been using the Ethash hashing algorithm for its Proof-of-Work (PoW) conducted by the miners. But in 2022 September, all PoW was replaced with the Proof-of-Stake (PoS) process in Ethereum.

Ethereum is an online service platform that supports smart contracts and distributed apps based on a secure publicly distributed Blockchain that is open source.

As mentioned above, Ethereum uses two main cryptocurrency types called ETH and gas. Transactional fees that occur on Ethereum are paid using ETH and the computational processing services are paid using gas.

In the Remix user interface platform of Ethereum, the run environment is controlled by a gas limit that is set to control the amount of computational processing that can be conducted for the smart contracts.

10.2.2 Solidity

Solidity was created by the developers of Ethereum to support Ethereum Virtual Machine (EVM) contracts and various Ethereum platform functionalities based on object-oriented programming. Solidity is a high-level programming language used for smart contract development. The term 'high-level programming language' means that it has many pre-built functions and procedures in its library that make program development easy and reliable.

The Ethereum user interface and service browser is the Remix Integrated Development Environment (IDE), which was made using Solidity. The Remix IDE is also called the 'Browser-Solidity' or 'Remix' for short. Remix supports open-source web services and desktop applications for Ethereum.

Remix has a large library of plugin functions and Graphical User Interface (GUI) tools that are easy to use and reconfigure to suit the application under development. The plugin functions and GUI tools enable fast and reliable smart contract development, monitoring, and management.

Smart contracts in Ethereum are programmed using the Remix IDE library of functions, which are compiled into a Solidity bytecode. The EVMs will process the developed program bytecode and execute the smart contract. Contractors and miners will use the Remix IDE to monitor and use the smart contract.

10.2.3 Ethereum Account Types

Ethereum account types include an externally owned account and a contract account. An externally owned account is an account used by Ethereum users, contractors, miners, and organizations, in which user identification, passwords, and private keys are needed to access these accounts. A contract account is managed by the EVM which is autonomously operated based on the smart contract program.

For example, assume a smart contract is programmed using Solidity and executed on the Remix IDE platform of Ethereum. Figure 10.2 shows an example of a smart contract in which the compiled bytecode is executed by an EVM. Investors of the smart contract will send their investment payments to the (contract account) wallet controlled by the EVM.

When all of the EVM conditions are satisfied, the EVM will conduct its contracted work on the Internet or on a specific distributed network. As the EVM executes the smart contract services, data may be collected, and the user payments (if any) will be sent to the (smart contract account) wallet controlled by the EVM.

When the smart contract payment conditions are satisfied, automatically (a part or all of) the collected payments will be distributed to the (externally owned account) wallet of the investors and developers based on the pre-agreed smart contract clauses. Commonly investments and payments are based on cryptocurrency transactions, rather than using actual money.

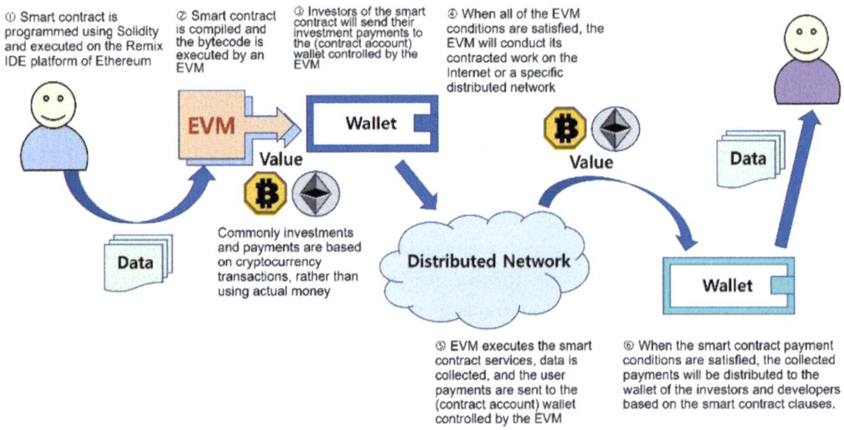

Fig. 10.2 Smart contract execution example

10.3 Ethereum Ethash

The hash function of Ethereum (i.e., Ethash) is an enhanced version of former hash functions used in the Proof-of-Work (PoW) processes applied in Bitcoin and other blockchains.

Finding a satisfying hash value for the next Ethereum smart contract block header in a limited amount of time (approximately 12–13 s) is a very computationally burdening PoW operation. This requires a very fast trial-and-error type search process, which resembles mining work. This is why Ethash block reward contenders are also called 'miners' like the Bitcoin miners.

Ethash is the PoW algorithm that is based on use of Memory-Hard Functions (MHFs) and a modified version of the Dagger-Hashimoto algorithm. MHFs are functions that require a significant amount of memory space to compute, but proportionally use lesser amounts of computational processing resources. MHFs were used in Ethereum's Ethash in an attempt to partially remove the advantage of Application-Specific Integrated Circuit (ASIC) based servers used by mining pools, such that general purpose computers could still be used in block chaining mining competitions. For example, ASIC based servers can execute Bitcoin's SHA256 double hashing (i.e., SHA256D) much faster (e.g., up to 100,000 times faster) than Central Processing Unit (CPU) based general purpose computers [8]. This made the competition nearly impossible for miners that are using general purposes computers to compete with miners that use ASIC based servers [9].

Unless such actions were made, only mining pools that use ASIC based automated servers would always win the mining competition, which would seriously discourage normal users from using Ethereum. Ethash was selected as the PoW for Ethereum for this reason.

For general purpose computers, mining speeds were accelerated due to the use of Graphics Processing Unit (GPU) based mining methods. However, soon ASICs for Ethash became available and mining pools that use Ethash ASIC based servers had an advantage in mining competitions, such that it is difficult to conclude that the objective of using MHFs in Ethash was fully accomplished. In reality, ASIC based system designs will always have an advantage.

Ethash was developed near the same time that the SHA-3 group of new hash function standards were made. So, Ethash uses sha3_256 (for 256 bits) and sha3_512 (for 512 bits), which are variants of the SHA-3 family hash functions. In some documents, sha3_256 is called Keccak-256 and sha3_512 is called Keccak-512.

To reduce processing dependency and be MHF oriented, Ethash uses a subset selection scheme that is based on a Directed Acyclic Graph (DAG) that is approximately a 1 GB (Giga Bytes) dataset. The DAG changes every epoch. An Ethash epoch is the time to support chaining of 30,000 blocks to the Ethereum blockchain, which takes approximately 5.2 days (i.e., approximately 125 h). To prevent delays in the block chaining process, DAGs can be generated in advance.

Ethash operates based on the following procedure.

1. Based on the former block headers, a seed is computed and used to generate a pseudorandom cache that is 16 MB (Mega Bytes).

2. Parts of the pseudorandom cache are used in various combinations (based on the DAG process) to gradually create a dataset that is 1 GB.
3. Ethash mining uses random sliced parts of the dataset (i.e., subset selection), combines them, and then applies them to SHA-3 hashing functions (i.e., sha3_256 and sha3_512) to obtain hash values.

Since a DAG does not need to be regenerated in the PoW confirmation process, verification can be conducted with small use of memory and processing power [8]. However, many new software and hardware technologies were created to be more efficient in Ethash mining, which were very effective. So trying to compete in Ethash PoW mining with a general purpose computer was still difficult.

10.4 Ethereum Proof-of-Stake (PoS)

In this section, details of the Ethereum PoS process are introduced, which include the validator role, benefits, transaction processing sequence, finalized block chaining, changing checkpoints, attacks and inactivity leak defense, validator requirements, role violations and penalties, correlation penalty, fork selection algorithm, LMD-GHOST algorithm, security issues, and PoS rewards.

10.4.1 Ethereum PoS Validator Role

To participate in Ethereum's PoS, first becoming a 'validator' is necessary. Validators play a major role as they periodically check new blocks before the blocks are chained to the Ethereum blockchain. This process requires the validator to conduct the following operations [10].

1. Check the newly proposed blocks on the Ethereum network.
2. Check the validness of the list of transactions in the block and the block's signature and header information.
3. Broadcast an 'Attestation' message throughout the Ethereum network to confirm that the validator has checked all information in the new proposed block and approves it.

The Ethereum blockchain will collect the 'Attestation' messages sent by the validators and use it as a 'vote' in the consensus mechanism of the Ethereum blockchain.

This is why validators need to individually install the required software (i.e., Software-as-a-Service (SaaS) application) and process programs to conduct the following three roles, which include, Validator, Consensus Client, and Execution Client. In addition, all validators should have 32 ETHs deposited in a validator smart contract, which will be used as collateral in case the validator does not act in a fully honest manner, or if the validator is late, or conducts a no-show. Such misbehavior can result in a deduction in the validator's deposit as well as suspension or removal of one's validator role. However, depositing 32 ETHs does not automatically guarantee a validator role, because once a deposit is made, the user's credentials are checked, and if the user passes the check, the user enters an activation queue. This activation queue helps to control the number of new validators that can join the Ethereum network. The Ethereum PoS operation is described in the following.

1. Ethereum blocks are generated every 12 s (which is a 'slot') by a randomly assigned validator. Each slot belongs to an epoch which contains 32 slots.
2. The validator is selected in advance of the assigned slot time, so some preparation time is given to the validator.
3. In the assigned slot time, the chosen validator will create a block and broadcast it out to all other validators on the Ethereum network.
4. All other validators will check the newly proposed block and if it passes the validation process the validator will send out an 'Attestation' message.
5. The votes sent by the validators will be used in the consensus mechanism of the Ethereum blockchain to determine if the newly proposed block will be chained or not.

10.4.2 Ethereum PoS Benefits

PoS provides many benefits to Ethereum, which are explained in the following.

1. Ethereum's PoS process is very straightforward and simple. So, it is easy to predict what will happen in the future, which makes the process more reliable (fool-proof) and stable. For example, the role of the validator and the block generation are predictable and immediately verified by all other validators. Thus, the need for a fork in the blockchain to recover from misinformation, disinformation, a malfunctional process, or a malicious attack becomes rare. In addition, the need to choose among similar length blockchains of a fork is minimized. As a result, the consensus process is simpler, and consensus results are easier to understand, as many uncertainties regarding the status of the blockchain are removed.
2. Due to the pre-deposited 32 ETHs of collateral by all validators and the penalties that can be applied for generating a new block with wrong information, and the synchronized validation process by all other validators, the possibility of a validator intentionally including false information in a block is extremely low. This helps to prevent various blockchain cyber-attacks, like 51% attacks.
3. PoS requires much fewer computing resources compared to PoW, so general-purpose computers (e.g., personal computers and laptops) can be used to conduct the required tasks. Therefore, the need for expensive high-performing ASIC driven servers is minimized. This enables a wider range of user and validator participation.
4. Due to the significant reduction in required computations, PoS has a much lower energy consumption profile compared to PoW, which makes PoS more environment friendly.

5. PoS validators need much less resources compared to PoW miners. This is because, PoW miners require high-end severs/computers, significant power consumption, big server rooms, as well as a high-performance network infrastructure. Due to these factors, PoW requires a significant investment to build and maintain. Therefore, participation in PoS is much easier and a smaller incentive can be sufficient as the required investment for service infrastructure and management is much smaller compared to PoW.

10.4.3 Ethereum PoS Transaction Processing Sequence

Ethereum's transaction processing sequence using PoS is described in the following.

1. Ethereum wallet is used to conduct a transaction. A transaction requires a gas fee and base fee to be set, which are both paid in ETH cryptocurrency. The gas fee is for the validator's services in creating the new block and the base fee is set to cover the fee for use of the Ethereum platform and Remix user interface. A transaction is formally setup when the account holder signs the transaction with its private key. (Alternatively, instead of a wallet, libraries like ether.js, web3js, web3py, etc. could be used to conduct a transaction in Ethereum.)
2. The wallet sends a request to use the Ethereum Application Programming Interface (API) to use the JavaScript Object Notation (JSON) Remote Procedure Call (RPC) functionality. JSON-RPC supports sending of JavaScript (JS) data and commands to a remote server. The API helps the user to access and use Ethereum's JSON-RPC.
3. An assigned Ethereum execution client will check if the gas fee and base fee are sufficient and will also check the signed private key of the transaction to confirm if it is valid.
4. After the Ethereum execution client confirms that everything in the transaction is properly set and valid, the transaction is included into a local mempool, which is a queue of pending transactions.
5. Next, the execution client broadcasts the transaction information to all other nodes in Ethereum's execution layer gossip network so in can be added to the other local mempools.

   A. However, in some cases, the execution client may not broadcast the transaction information, but send it to a specialized block builder (e.g., as Flashbots Auction). This method allows advanced users in Ethereum to organize the transactions in upcoming blocks so maximum profit can be achieved through Maximal Extractable Value (MEV) procedures.

6. Every slot of an Epoch has a block proposer that is selected in advance through PoS procedures that use RANDAO, which is a DAO that conducts pseudo Random Number Generation (RNG) in Ethereum based on SHA-3 hashing technology. The selected block proposer has the responsibility to construct a block and have it broadcasted over the gossip network so it can be chained to the blockchain

after confirmation (i.e., after collecting a sufficient amount of attestations) is obtained from the other validators.

A block proposer is made up of three operational parts, which are the execution client, consensus client, and validator client.

1. **Execution client**: Transactions queued in the local mempool are gathered to form the 'execution payload' by the execution client. This results in a state change, and the execution client sends the execution payload to the consensus client.
2. **Consensus client**: The received execution payload and information needed in the consensus process (e.g., rewards, penalties, fees, slashings, attestations, etc.) are put into a 'beacon block' by the consensus client. The consensus client sends this new beacon block to all nodes on the consensus layer gossip network.
3. **Validator client**: The validator client (in the other nodes on the consensus layer gossip network) will send the new beacon block received to their execution client, which will recheck all information in this new beacon block. If the execution client approves the new beacon block, then the validator client will attest the validness of the new beacon block and add the block to the local database.

In Fig. 10.3, the operational procedures conducted by a block proposer's execution client, consensus client, and validator client are illustrated.

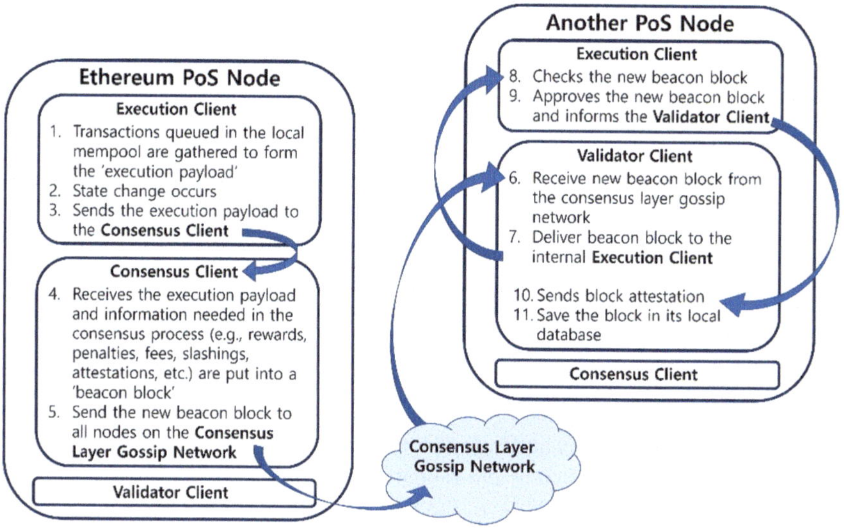

Fig. 10.3 Ethereum block proposer's execution client, consensus client, and validator client operational procedures

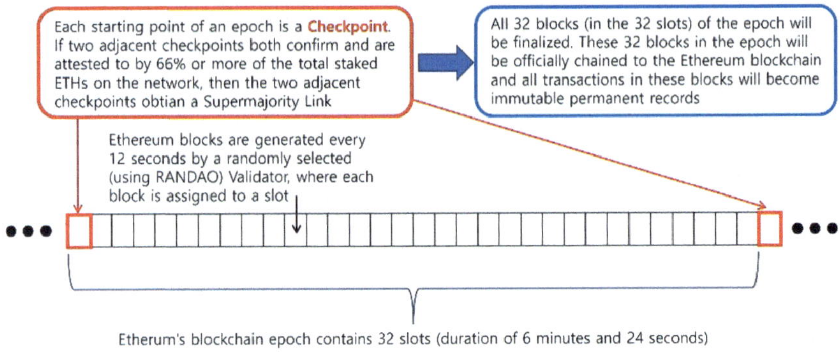

Fig. 10.4 Ethereum PoS finalized block chaining process

10.4.4 PoS Finalized Block Chaining

A transaction will become immutable and unchangeable when it is officially chained to the Ethereum blockchain. This happens when a transaction reaches a 'Finalized' status. Ethereum blocks are generated every 12 s for each slot by a randomly selected (using RANDAO) validator, and each slot belongs to an epoch which contains 32 slots (duration of 6 min and 24 s). Each starting point of an epoch is a Checkpoint. When two adjacent checkpoints both confirm and are attested to by 66% or more of the total staked ETHs on the network, then the two adjacent checkpoints will have a supermajority link, and all 32 blocks (in the 32 slots) of the epoch will be finalized. Then the 32 blocks in the epoch will be officially chained to the Ethereum blockchain and all transactions in these blocks will become immutable permanent records.

In Fig. 10.4, the procedures of the Ethereum PoS finalized block chaining process are illustrated.

10.4.5 PoS Changing Checkpoints

Initially, in each epoch, the first block serves as a checkpoint, where two adjacent checkpoints will check if they have a supermajority link. Checkpoints can be changed based on an election conducted by the validators. The validators can elect a new pair of checkpoints if at least two-thirds of the total staked ETH represented votes arrive at a consensus.

10.4.6 PoS Attacks and Inactivity Leak Defense

Because finality can be obtained by at least two-thirds of a majority vote, an attacker could prevent the consensus network from reaching finality by voting with more than one-third of the total stake to prevent a two-thirds majority vote from being achieved. A cyber-attack exceeding one-third of the total stake to prevent a two-thirds majority vote from being achieved is commonly called a 51% attack (but can be called a 34% attack as a 34% collude would prevent a two-thirds majority vote from being achieved). This attack on PoS could be conducted by multiple validators colluding through an attacker that has compromised these validators. Several other setups could be used by attacker(s) to create a 51% attack (i.e., 34% attack). If the blockchain cannot achieve a two-thirds majority vote-based consensus for more than four epochs in a row, then the 'inactivity leak' defense process is activated. The inactivity leak process finds the validators that have been consistently voting against the majority and removes their staked ETHs. Due to the reduced number of ETHs staked in the validator's account, the node may no longer qualify as a validator. This is because a validator is required to maintain a deposit of 32 ETHs, and due to the reduction of its staked ETHs, the node may lose its validator qualifications. Therefore, these disqualified validator nodes will not be able to participate in future voting processes and will be replaced with other validators that have been waiting in the validator activation queue.

10.4.7 PoS Validator Requirements, Role Violations and Penalties

Validators must maintain a sufficient stake (a minimum of 32 ETHs) and have sufficiently good processing capability and memory to execute all required programs quickly and must be connected to a reliable network at all times.

This is because, validators must stay committed to the blockchain's operations and consistently remain listening at all times, so they do not miss a calling to create a new block. Block rewards will contribute to increasing the validator's stake balance. But, failing to do so (i.e., missing a request to create a new block), will result in a lost opportunity to receive a block reward (paid in ETHs) and penalties can be enforced on the validator. In addition, if a validator does not submit a vote during the time slot to help attest a new block, then a small attestation penalty fee is imposed on the validator's stake.

Because a validator has voting rights to influence the consensus of the blockchain that could manipulate the smart contract, the security and protection of a validator is very important. If a validator were to be compromised by a botnet, or is under the influence of malware, then this can result in disastrous effects to the PoS blockchain as well as the smart contract users.

Based on Ethereum's PoS policies, misconduct can result in a significant penalty that is imposed on a validator, which helps to prevent delinquency, malpractice, colluding, and coordinated attacks.

Equivocating and making contradictory attestations to influence the consensus are the two representative malicious behaviors a validator can conduct. Equivocating is the act of proposing multiple blocks for a single slot, which creates significant confusion and delay in the PoS blockchain process. Equivocating can occur accidently (e.g., due to having a validator account being accessed and controlled by more than one computer) or with malicious intent (e.g., due to malware or a botnet).

If a validator behaves maliciously, then slashing (i.e., cutout, removal) is executed on its stake, which is a penalty enforced that deducts the validator's stake balance, and the validator could also be suspended from participating in the smart contract's blockchain and service platform.

The amount of ETHs that will be slashed from the validator's account is based on the 'correlation penalty' rules, which are dependent on both the violation type and the number of other validators that will be penalized during the same period of time. The reason for this action is to find potential colluding acts and apply a larger slashing to hinder future collusive attempts. For minor penalties, self-slashing could be enforced to be conducted by a single validator, in which the slashing rate of stake can be approximately 1% or higher. For serious violations, mass slashing can be enforced, where all or almost all of the validator's stake could be slashed. Correlation penalty enforcement (which is conducted through a 36 day period) begins with a Day 1 penalty of up to 0.5 ETH. On Day 18, which is the middle day of the forced exit period, the correlation penalty is executed. On Day 36, the validator is ejected (removed) from the Ethereum smart contract's network. More details of the correlation penalty are provided below.

10.4.8 PoS Correlation Penalty

To decide the **Correlation penalty**, the beacon chain records of slashing events that occurred during the most recent 8,192 epochs (which is approximately for 36 days) on all of the validators' effective balances are checked. This is checked to estimate the level of possible collusion that might have happened. From the beacon chain records of slashing events that occurred during the most recent 8,192 epochs, the following parameter values are collected.

- **S**: Sum of increments in the beacon chain records of slashing events that occurred during the most recent 8,192 epochs.
- **B**: Effective balance of the validator's stake being penalized that will receive the correlation penalty.
- **T**: Total increments.
- **X = (3SB/T)**: Because **S**, **B**, and **T** are all positive integer values, **X** is a positive real value. Correlation penalty uses the slashing multiplier value of **3.**

- **P = int(X)**: Represents integer arithmetic applied to **X**.
- **Correlation penalty = min(B,P)**: The smaller (i.e., minimum) value among **B** and **P** is chosen.

The integer arithmetic process **P** = **int(X)** operates in the following way. For example, if the value of **X** is in the range **P** ≤ **X** < (**P** + 1), then **int(X) = P**, where **P** is a positive integer. In other words, **int(X)** will round **X** down to the nearest smaller or equivalent positive integer value.

For example, if **X** = 0.96 then **int(X)** = 0, or if **X** = 7.23 then **int(X)** = 7. Therefore, if **3SB** < **T** then **X** will be in the range of 0 ≤ **X** < 1, which will result in **P** = 0. As a result, for a validator with a minor independent violation, then **S** will be a small value, which may result in **3SB** < **T**, and in this case **P** = 0.

In the final process, **Correlation penalty** = **min(B,P)** is computed. Because the correlation penalty is the smaller (i.e., minimum) value among **B** and **P**, if the value **P** exceeds **B**, then **B** will be the correlation penalty value. In other words, **B** is the largest correlation penalty value that can be applied. This was set this way because **B** is the stake of the validator that is being penalized, and therefore, **B** is the largest value that can be slashed from the validator's stake.

Therefore, a minor penalty can result in a up to 1% of stake slashing from a single validator's own stake (i.e., it will be self-slashed). On the other hand, a major penalty is enforced on major violations, which results in a mass slashing of up to 100% of the validator's stake.

It is worthy to note that the slashing multiplier value 3 was originally set to 1 in the beginning at the Genesis block of Ethereum's PoS blockchain. It was later increased to 2 in the Altair edition upgrade and was later increased to 3 in the Bellatrix edition upgrade, where it remains as 3 in the Ethereum edition 0.3 Capella upgrade, which became effective on July 1, 2023. Likewise, the initial penalty has also increased during the edition upgrades. The initial penalty was only 0.78125% (i.e., 1/128) of the offender's effective balance during phase 0, and was doubled to 1.5625% (i.e., 1/64) of the offender's effective balance in the Altair edition upgrade, and was later doubled to 3.125% (i.e., 1/32) of the offender's effective balance (which is the current maximum initial slashing size of 1 ETH, since a validator's required minimum stake size is 32 ETHs) in the pre-merged Bellatrix edition upgrade [11].

10.4.9 PoS Fork Selection Algorithm

If all participants on the Ethereum blockchain behave exactly as required (i.e., honest, accurate, reliable, consistent, stable, and prompt), then a fork in the blockchain will not occur. However, due to malicious behavior, mistakes, as well as system and network failures, blockchain forks do occur (see Fig. 10.5 for an example). For example, equivocation cases occur when more than one block is proposed for a given slot, which will result in a fork in the blockchain. For these cases, a fork selection algorithm is needed for the Ethereum PoS process, so the blockchain can arrive at a correct and reliable consensus quickly.

Ethereum's PoS uses the **Latest Message Driven Greediest Heaviest Observed SubTree (LMD-GHOST)** fork selection algorithm. Among the fork branches, LMD-GHOST finds the branch that has the largest amount of accumulated attestations (i.e., branch with the heaviest weight) based on the fork history [12].

For each block on the fork, the most recent number of validator attestations received is considered as the weight of the block. In this block weight counting, each attestation counts as 1, and only one attestation per validator is allowed in the count. The last block of a chain on the fork is called a leaf block. This name is used because the blockchain fork resembles a tree with branches dividing out from the main tree trunk, and at the end of each branch there is a leaf (see Fig. 10.6 for an example). So, a forked blockchain will have more than one leaf block, where LMD-GHOST will be used to select the leaf block that has the heaviest chain weight.

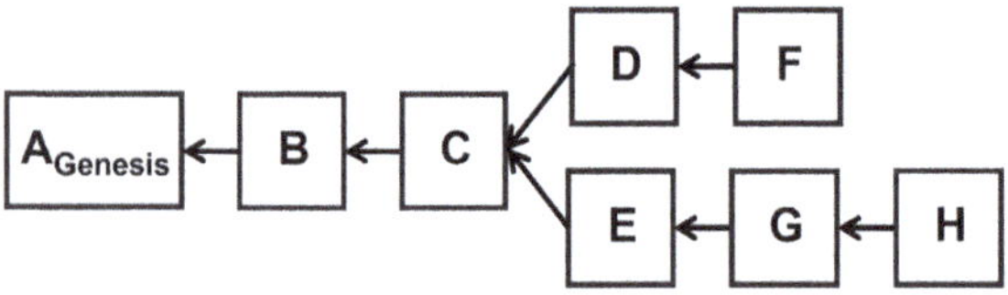

Fig. 10.5 Example of the chain representation for block **H** is **chain(H) = ($A_{Genesis}$, B, C, E, G, H)**

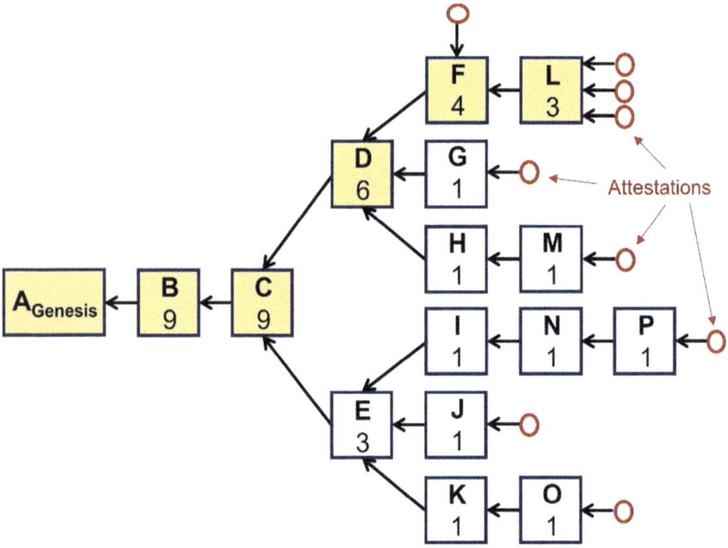

Fig. 10.6 Example of LMD-GHOST fork-choice rule applied

As the functions **view(·)** and **chain(·)** are used in the LMD-GHOST algorithm, they are described first.

view(V, T): Represents the view of validator **V** at time **T**, which is the set of all of the accepted messages validator **V** has received up to time **T**.

view(NW, T): Represents the view of the blockchain network **NW** at time **T**, which is the set of all of the accepted messages the network **NW** has received up to time **T**.

The view of a validator may be different compared to the view of the blockchain network. For example, at time **T**, assume that the validator **V** has received a recent message **M2** with the content "Bob sends coin number 1234 to Charlie" but validator **V** has not seen the message **M1** with the content "Alice sends coin number 1234 to Bob" yet, and in this example, it is assumed that **M2** only depends on **M1**. In this case, the validator **V** does not know how Bob has obtained coin 1234, since validator **V** has not seen message **M1** yet. However, the network **NW** has seen both messages **M1** and **M2**, so the view of the network **NW** will be different compared to the view of the validator **V**. As a result, the view of the network **NW** will include message **M2** (i.e., **M2** $\in$ **view(NW, T)**) but the view of validator **V** will not include message **M2** in its view (i.e., **M2** $\notin$ **view(V, T)**) at time **T**.

chain(·): Represents a chain on a PoS fork which lists a sequence of blocks from the genesis block to a selected block following a single branch on the subtree of the forked blockchain. For example, based on Fig. 10.5, for block **H**, its chain would be represented as **chain(H) = (**$\mathbf{A_{Genesis}}$**, B, C, E, G, H)**.

10.4.10 PoS LMD-GHOST Algorithm

In the Latest Message Driven Greediest Heaviest Observed SubTree (LMD-GHOST) algorithm, for each chain of the fork, the summed weight of the blocks in the **chain(·)** are added up and compared. The weight of a block is the number of validator attestations added up, where each validator's attestation is counted only once. Therefore, the overall combined weight of the attestations cannot exceed the number of validators assigned at that time. Among the leaf blocks of the fork, the leaf block that has the heaviest chain weight will be selected by LMD-GHOST to resolve the fork issue. In a fork, if there is a tie in the chain weights of the leaf blocks, then the chain with the smaller leaf block hash value will become the selected chain of LMD-GHOST.

In the following, an example of the LMD-GHOST fork-choice rule being applied is presented. Figure 10.6 presents the view of the blockchain network **NW** at time **T** (i.e., **view(NW, T)**) including the attestations on the blocks, where each validator counts as only one attestation (which are represented as circles in Fig. 10.6). In Fig. 10.6, there are a total of 9 attestations in which block **L** has 3 and block **F** has 4, which is a combination of the 3 attestations of block **L** and 1 for block **F**. Because block **L** is a child block of the parent block **F**, the accumulated weight of attestations on block **F** is 4. Block **D** is the parent block of **F**. Block **D** has a weight of 6 from the combined attestations of its child blocks **F** (weight 4), **G** (weight 1), and **H** (weight 1). Block **C** is the parent block of **D**. Block **C** has a weight of 9 from the combined attestations of its child blocks **D** (weight 6) and **E** (weight 3). Because the LMD-GHOST fork-choice rule will select the chain with the largest weight, **chain(L) = (A$_{\mathbf{Genesis}}$, B, C, D, F, L)** will be selected. This is how LMD-GHOST resolves forks.

10.4.11 PoS Security

Because the PoS process is transparent and stable, block attestations and the chaining process are more predictable and it is easier to confirm correctness. This is why, when compared to PoW, PoS makes it much easier to notice misconduct, errors, and malicious behavior. As a result, honest validators can easily notice erroneous leaf blocks when a fork occurs and can therefore continue to add blocks to the correct leaf block and provide attestations, which will help the correct leaf block and its chain to be selected by the LMD-GHOST process. This has the effect of resolving a conflict when a fork occurs and enables the Ethereum blockchain to maintain the slot and epoch-based block creation and chaining schedule even if a fork occurs. In addition, the stake slashing penalty, correlation penalty, and inactivity leak defense process help to control inappropriate and dishonest behavior by enforcing financial loss (by slashing stake) and can remove a problematic validator.

10.4.12 PoS Rewards

In Ethereum PoS, rewards are given to validators based on the following mechanism. Ethereum rewards are given based on the sum of the following five components. A validator can receive a TIMELY_SOURCE_WEIGHT reward of 14 for conducting a source vote timely. A validator can receive a TIMELY_TARGET_WEIGHT reward of 26 for conducting a target vote timely. A validator can receive a TIMELY_HEAD_WEIGHT reward of 14 for conducting a head vote timely. A validator can receive a SYNC_REWARD_WEIGHT reward of 2 for participating in a sync committee. Or when the validator is selected to propose a block, it can receive a PROPOSER_WEIGHT reward of 8. The reward weight sum of these 5 categories is equal to 64. Therefore, the reward sum a validator can receive is divided by 64 and then multiplied to the base_reward. For example, the maximum reward a validator can receive is obtained by adding (TIMELY_SOURCE_WEIGHT + TIMELY_TARGET_WEIGHT + TIMELY_HEAD_WEIGHT + SYNC_REWARD_WEIGHT), which sums up to 56 = (14 + 26 + 14 + 2). So, the maximum reward a validator can receive is (base_reward × 56/64). In the reward summation above, the PROPOSER_WEIGHT reward of 8 was not included in the maximum reward sum because a validator cannot propose a block and vote an attestation on its own block at the same time. The base_reward is computed proportionally to the validator's effective balance, but is also inversely proportional to the number of validators on the PoS network (e.g., scaled by 1/sqrt(**N**), where **N** is the total staked ETHs of all active validators and 'sqrt' is the square root function). In addition, a validator can receive an additional inclusion_delay_reward, which is given as an incentive to conduct quick attestations [13].

10.5 Summary

Smart contracts are used in all areas of our daily lives including leisure services (e.g., games, betting, gambling), purchasing and trading (e.g., cryptocurrencies, real estate, estates, goods, NFTs, insurance), legal services (e.g., copyrights), as well as various healthcare and rental services.

This chapter focuses on various aspects of smart contracts and Ethereum, which is the first company to use blockchain technology to provide smart contract services, and is currently the largest smart contract platform in the world. An Ethereum account based smart contract execution example is provided along with details of Ethash, Gas, Solidity, and Remix. In addition, the operation details of the new Proof-of-Stake (PoS) are provided, focusing on the Validator, Execution, and Consensus client operations as well as the finalized block chaining process, fork selection algorithm, and various security issues.

Although there are many smart contract platforms currently available, it is recommended to first understand the details of Ethereum, as it is the original smart contract platform as well as the most technically advanced platform, which is why it is used the most. In the future, the use of smart contracts will grow significantly, and will soon be used in every aspect of financial and managerial services in the government, public, and private sectors. To serve this purpose, more advanced blockchain security technologies need to developed and applied to future smart contract systems [14–18]. In addition, it is also recommended to read the chapters on security (Chap. 1), Blockchains (Chap. 8), and Bitcoin (Chap. 9) in this book as many valuable technologies used in Ethereum and smart contracts are introduced in detail.

References

1. Bitcoin Magazine. https://bitcoinmagazine.com/
2. Ethereum. https://ethereum.org/en/
3. Gracious Quotes, Top 27 Vitalik Buterin Quotes (ETHEREUM) (2023). https://graciousquotes.com/vitalik-buterin/#2_Vitalik_Buterin_Quotes_on_Bitcoin
4. LabCFTC, A Primer on Smart Contracts (2018). https://www.cftc.gov/sites/default/files/2018-11/LabCFTC_PrimerSmartContracts112718_0.pdf
5. Wikipedia, Ethereum. https://en.wikipedia.org/wiki/Ethereum
6. GlobalData, Ethereum's Price History (2013 – 2023, $). https://www.globaldata.com/data-insights/financial-services/ethereums-price-history/

7. RollingStone India, Prashanth Irudayaraj, #CryptoExplained: A Brief History of Ethereum. A timeline of notable events in the history of Ethereum (2023). https://rollingstoneindia.com/ethereum-history-timeline-cryptocurrency/
8. Etherum.org, ETHASH (2024). https://ethereum.org/en/developers/docs/consensus-mechanisms/pow/mining-algorithms/ethash/. Accessed 6 March 2024
9. Wikipedia, Memory-hard function. https://en.wikipedia.org/wiki/Memory-hard_function
10. Etherum.org, PROOF-OF-STAKE (POS) (2024). https://ethereum.org/ko/developers/docs/consensus-mechanisms/pos/. Accessed January 26, 2024
11. Eth2book.info, Capella, Part 2: Technical Overview. The Incentive Layer, Slashing. https://eth2book.info/capella/part2/incentives/slashing/
12. V. Buterin, et al., Combining GHOST and Casper. arXiv:2003.03052v3[cs.CR]. Accessed May 11 2020. https://arxiv.org/pdf/2003.03052.pdf
13. Ethereum, Proof-of-Stake Rewards and Penalties (2024). https://ethereum.org/en/developers/docs/consensus-mechanisms/pos/rewards-and-penalties/. Accessed February 11 2024
14. Y. Goh, J. Yun, D. Jung, J.-M. Chung, Secure trust-based delegated consensus for blockchain frameworks using deep reinforcement learning. IEEE Access **10** (2022). https://doi.org/10.1109/ACCESS.2022.3220852
15. J. Yun, Y. Goh, J.-M. Chung, DQN-based optimization framework for secure sharded blockchain system. IEEE Internet of Things J. **8**(2) (2021). https://doi.org/10.1109/JIOT.2020.3006896
16. J. Yun, Y. Goh, J.-M. Chung, Trust-based shard distribution scheme for fault tolerant shard blockchain networks. IEEE Access **7** (2019). https://doi.org/10.1109/ACCESS.2019.2942003
17. J. Yun, Y. Goh, J.-M. Chung, O. Kim, S. Shin, J. Choi, Y. Kim, MMOG user participation based decentralized consensus scheme and proof of participation analysis on the bryllite blockchain system. KSII Trans. Internet Inf. Syst. **13**(8) (2019). https://doi.org/10.3837/tiis.2019.08.015
18. J. Yun, Y. Goh, J.-M. Chung, O. Kim, S. Shin, J. Choi, Y. Kim, MMOG user participation based decentralized consensus scheme and proof of participation analysis on the bryllite blockchain system, in *Outstanding Paper Award, International Conference on Internet (ICONI)* (2018)

Chapter 11
Emerging Technologies

This chapter focuses on two topics that are closely related. The first topic is emerging technologies in networks and security, and the second topic is emerging technologies in blockchains, cryptocurrencies, and smart contracts. Because blockchains rely on networking services to support their distributed functionality, all issues in networks and security systems directly influence blockchain services. Since all cryptocurrencies and smart contracts use blockchains, naturally, vulnerabilities in networks and security systems all become the vulnerabilities of cryptocurrencies and smart contracts. However, cryptocurrencies and smart contracts have additional security issues that need to be considered based on their special system architecture and operating mechanism [1, 2].

One common factor to consider is the significant advancement of artificial intelligence (AI) and modeling technologies. Chapter 2 introduced the three most powerful state-of-the-art AI technologies, which are, Deep Reinforcement Learning, Generative AI, and Federated Learning. In addition, Chap. 2 introduces the most advanced modeling, emulation, and simulation technology, Digital Twin. These emerging technologies have the potential to strengthen the security level of networks and blockchains. On the other hand, these advanced AI technologies will also be used to conduct cyberattacks to networks and blockchains. More details and examples of security issues in networks and blockchains will be described in this chapter [3–7].

11.1 Emerging Technologies in Networks and Security

In terms of security, when comparing offense (i.e., cyber-attacks) to defense (i.e., protection of a network or blockchain), always the offense has an advantage. This advantage is based on the fact that a network or blockchain does not know when or how it will be attacked. The uncertainty in when, where, how, and the level of

J.-M. Chung, *Emerging Secure Networks, Blockchains and Smart Contract Technologies*, https://doi.org/10.1007/978-3-031-65866-2_11

severity of a cyber-attack makes it nearly impossible for a network or system to always remain completely secure and safe, as vulnerabilities always exist. For example, even highly sophisticated security systems can become (temporarily) vulnerable due to the following reasons.

- Disgruntled employee can intentionally enable an intrusion or security breach.
- Careless actions of an employee can accidently make the system vulnerable (e.g., downloading malware unintentionally).
- Software and hardware upgrades may temporarily leave breaching gaps in the security system and firewall.
- Power failures and shutdowns that result in systems that turn off and on at different times can make the network (temporarily) unsecure.

Such cases can enable malware, botnets, and ransomware to intrude even the best protected systems and networks.

Deep reinforcement learning enables new attack patterns to be automatically found and exploited. Generative AI enables new attack methods to be created in diverse ways that make it hard to predict. Federated learning enables experience and knowledge to be shared among hackers with no need to transfer or share data. This is why federated learning can make coordinated distributed cyberattacks much easier and more effective.

Digital twin technology enables various attack techniques to be tested on a digital twin platform such that the most effective method can be selected and used in cyberattacks. Due to this reason, digital twin technology can make it much more difficult to defend a network or blockchain, and can enable the level of damage to become significantly more severe.

Due to these reasons, new methods of how networks and blockchains can be attacked need to be studied in advance, and powerful countermeasures need to be quickly developed and deployed to protect networks and blockchain more effectively.

In addition, as security is becoming more important, Internet Service Providers (ISPs) will use special security technologies to protect their network. As gateways need to exchange routing information between Autonomous Systems (ASs) that belong to different ISPs, internetworking security vulnerabilities at the gateways can occur, which is described in further detail below.

As explained in Chap. 4, Internet routing is supported by the Interior Gateway Protocol (IGP) and Exterior Gateway Protocol (EGP). IGP is used by the routers to exchange routing information within (inside) an AS, which is commonly a network serviced by an ISP or corporate. On the other hand, EGP is used by gateways to exchange routing information between ASs, which is where there are many weak points for cyberattacks.

Border Gateway Protocol (BGP) version 4 (BGP4) is the commonly used EGP, which has two types. One is the external BGP (eBGP) and the other is the internal BGP (iBGP). eBGP provides BGP routing between ASs, and iBGP supports BGP routing within an AS. Since BGP gateways commonly interface networks managed by different ISPs that use different encryption and security schemes, BGP gateway security coordination can be very difficult, and vulnerable points may exist. In addition, BGP routers need to exchange setup and updated information with each other, where a Man-In-The-Middle (MITM) attack can occur and poor coordination can result in serious security vulnerabilities. Especially, the different encryption and security schemes used by the different ISPs can make the user and network authentication process slow and less reliable. Due to these vulnerable aspects, protection against spoofed BGP messages and malware intrusions may be difficult.

One possible solution to this problem is to use a Demilitarized Zone (DMZ). As described in Chap. 1, a DMZ is an advanced form of a multiple firewall protection mechanism, where Fig. 1.15 presents a DMZ using two firewalls. DMZs use firewalls at the exiting edges of the external network, which are called 'perimeter firewall' or 'front-end firewall.' The DMZ firewall at the entrance of the internal network is called the 'internal firewall' or 'back-end firewall.' DMZs are capable of providing a higher level of protection and security to network egress and ingress points. However, setting up a DMZ is expensive and additional space/resources as well as management personnel are needed. So, a small company or organization may not be able to afford a DMZ to protect their network.

11.1.1 Advantages and Vulnerabilities Due to Advanced AI Technologies

As described in Chap. 2, generative AI systems can create realistic new content using various synthetic techniques. The domain of content generation includes images, music, text, and much more. Generative AI's creation capability is rapidly improving and expanding, yet there is no guidance for future directions and no governing regulations.

The performance of a generative AI system is based on the generative AI model used and the content type requested to be generated. Some instabilities still exist as generative AI systems are much more complex compared to data classification and regression AI algorithms.

In addition, the resulting quality of the content created by a generative AI system depends on how much high-quality data was used in the system training. So, if the training dataset does not have a sufficient amount of related content to support the new content generation request, then the results may not be satisfactory to the requester. As there is no clear standard of what high quality training data is, and there is also no standard for what would be a satisfactorily created content, all evaluations are currently based on subjective reviews. Future technologies need to be developed to overcome these issues in generative AI, which is one of the purposes of this book.

Below, examples are provided to help explain some possible performance and security enhancement methods using the advanced AI technologies introduced in Chap. 2, based on the author's research and development.

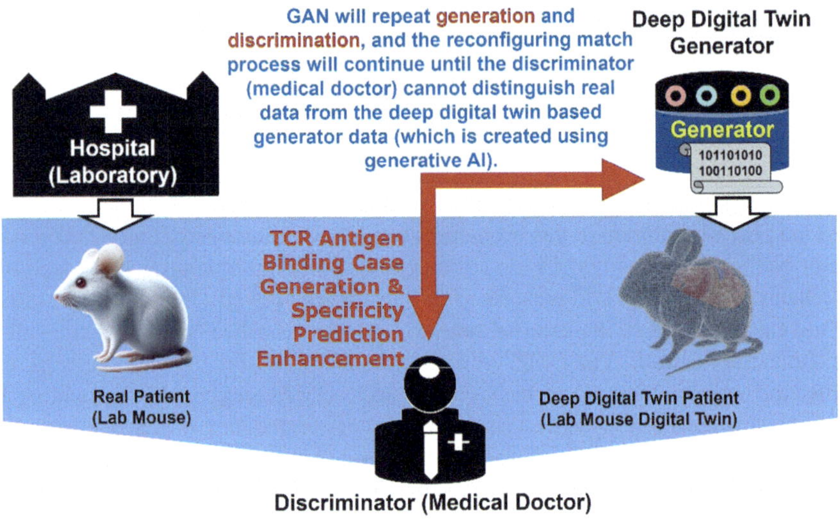

Fig. 11.1 Generative AI is used to create additional patient data based on the deep digital twin, and a GAN is applied to fine tune the data to make it closely resemble actual patient data. Using this method, the lack of medical data can be temporarily filled with fine-tuned synthetic AI data for further medical technology development. As more real patient data is obtained, the amount of synthetic AI data can be reduced. In this figure, GAN will repeat generation and discrimination, and the reconfiguring match process will continue until the discriminator (e.g., medical doctor) cannot distinguish real data from the deep digital twin based generator data (which is created using generative AI)

Figure 11.1 presents an example of a deep digital twin, generative AI, and Generative Adversarial Network (GAN) system applied to medical data gap filling and diagnosis checking. A deep digital twin combines deep learning technology with digital twin modeling and emulation/simulation operations. Figure 11.1 is an actual system used in the author's laboratory for medical research and development.

Due to lack of medical data of a specific health condition, a GAN system can be used with generative AI technology to generate realistic medical data. One example is for the ongoing T-Cell Receptor (TCR) and cancer RNA sequencing based antigen development for improved immunotherapy. This can be used as a platform for medical testing on a biological deep digital twin system (e.g., a laboratory mouse), as presented in Fig. 11.1.

In many cases, there is a significant lack of biomedical data when attempting development of a new medicine or medical procedure (e.g., surgery), as it is very time consuming and expensive as well as difficult to obtain useful biomedical data from actual hospital patients. In this example, generative AI techniques can be used to fill the gaps in the dataset and a GAN system is used to help increase the accuracy of the antigen development such that the TCR immunotherapy can be more effective. As more real patient data is collected, the amount of generated AI and GAN tuned data can be reduced.

The GAN system can use deep reinforcement learning to make the generated data become closer and closer to real data. The deep reinforcement learning system will become more accurate as it accumulates more experience. The deep digital twin will host a library of generative AI models to use as tools in generating the new data patterns.

In the same manner (as the example presented above in Fig. 11.1), digital twins of new communication systems and networks as well as blockchains and smart contract systems can be created. As new systems and networks will not have sufficient data to build a reliable AI control system, generative AI can be used to assist the data generation. The new data can be used in the training of the deep reinforcement learning control system to accurately model the digital twin system (which is why the term 'deep digital twin' can be used). In addition, GAN technology can be used to fine tune the system to make the control technology and digital twin system more accurate and reliable.

As more real data from the actual system is collected, the amount of generative AI and GAN tuned data can be reduced. In addition, using the real data from the actual system, the digital twin and the GAN model can be made to operate more accurately. Adding to these features, federated learning enables the control technology as well as the digital twin and GAN systems to be upgraded through the collaboration of distributed AI systems that are each remotely trained on their own local dataset [3, 4–7].

On the other hand, the new advanced AI technologies presented in this chapter can be used in various bad ways. To help explain some possible problems that can occur using the advanced AI technologies introduced in Chap. 2, an example model is described in the following.

Figure 11.2 presents a bad example of a generative AI system being used to generate counterfeit money [12]. After generative AI is used to make the initial model of the counterfeit money, GAN technology is used to make the counterfeit money so accurate that the bank's counterfeit money detection system would not be able to notice the difference. The Generator uses a generative AI system to create fake counterfeit money that is hard to distinguish by the investigating Discriminator (this may be a counterfeit money detection system or an expert working in the bank). If the Generator succeeds, then it becomes hard to block counterfeit money from entering the bank and financial system.

Likewise, the bank would need to have its own fraud detection system that can learn new counterfeit money creation methods (using technologies like deep reinforcement learning) and use this information to make more accurate detections. In addition, the bank's fraud detection system can use GAN technology to more accurately detect counterfeit money.

In the same manner (as the example presented above in Fig. 11.2), just as counterfeit money can become more difficult to detect (due to advanced generative AI technologies), networks, blockchains, and smart contracts can become more vulnerable.

For example, new vulnerabilities can be found in biometric security systems. The combination of generative AI, deep reinforcement learning, and digital twin technology create a serious threat to biometric authentication systems, which is described in the following.

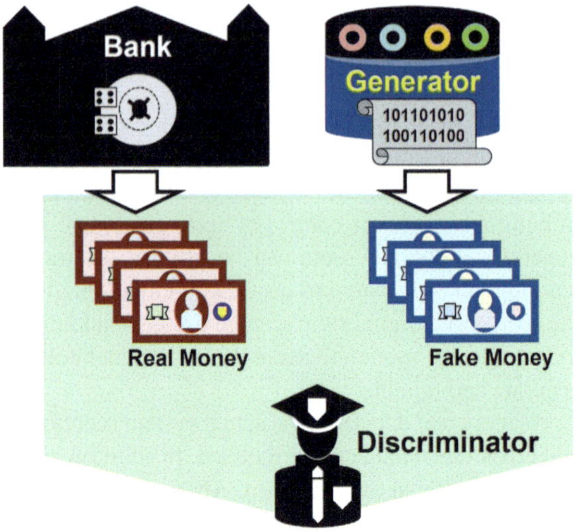

Fig. 11.2 Generative AI and GAN applied to bank fraud attempt and detection example

Smartphones, laptop computers, and facility entrance systems use biometric authentication to enable user access. The benefit of using biometric authentication is that the system can automatically identify its authorized user and open the system quickly to be immediately used or entered. In addition, there is no need to remember passwords or worry about losing a key, identification (ID) card, or Radio Frequency ID (RFID) tag.

The common traits used in biometric authentication are based on recognition of a fingerprint, face, retina, or voice. Biometric authentication systems attempt to find unique characteristics of a user that can be used to reliably identify an individual quickly and reliably. However, every biometric authentication system is limited by the level of resolution of its biometric sensor. For example, fingerprint scanners are small and have a limited pixel resolution sensor that is used to identify a fingerprint. A generative AI system that can generate a fingerprint with a higher resolution than the sensor's resolution will have a high chance of fooling the biometric authentication system.

Generative AI can be used to create images that accurately resemble the biometric features of a person. For example, generative AI could be used to generated face images, which are more accurately tuned to pass a facial recognition system using GAN technology. In addition, a digital twin of the facial recognition system could be used to find detection flaws, such that easier system break-ins could be conducted. Obviously, the security system could also use digital twin and generative AI technology to make defense mechanisms. But in most cases, new security system enhancements will be created after a security breach has been detected and analyzed, such that significant damage or a large financial loss may have already occurred.

11.2 Emerging Technologies in Blockchains, Cryptocurrencies and Smart Contracts

All of the vulnerabilities described above apply to blockchains, cryptocurrencies, and smart contract platforms. In addition, some additional considerations in blockchains, cryptocurrencies, and smart contracts do exist, which are described in this section.

As described in Chap. 9, a Bitcoin (BTC) reward is given to the first cryptocurrency miner that finds a hash value equivalent or less than the nBits target threshold while including all of the verified Bitcoin transactions (i.e., TXIDs) in their block of ledger data. The nBits target threshold is changed after an additional 2,016 blocks have been chained to the Bitcoin blockchain, which is approximately every 2 weeks [1, 2].

After chaining 210,000 blocks, the Bitcoin reward size is reduced to half, which occurs approximately every 4 years. The inventor of Bitcoin (who uses the alias Satoshi Nakamoto) designed it to have only 21 million BTCs issued overall. This limit in BTCs was proposed to avoid inflation and to prevent the value of Bitcoin from diminishing due to having too many Bitcoins generated.

New BTCs can be issued only as a reward for successfully chaining a new block to the Bitcoin blockchain. This started on January 9 of 2009 when the Genesis block of Bitcoin was initiated, in which 50 BTCs was the block reward. In November 28 of 2012, the block reward was reduced to 25 BTCs after 210,000 blocks were chained. In July 9 of 2016, after an additional 210,000 blocks were chained, the block reward was reduced to 12.5 BTC. On May 11 of 2020, after an additional 210,000 blocks were chained, the block reward was halved to 6.25 BTC.

As of May 2021, approximately 18.7 million BTCs, which is about 90% of the total number of block rewards had already been issued. On April 19 of 2024, after an additional 210,000 blocks were chained, the block reward was halved to 3.125 BTC.

By April of 2039, approximately 99.6% of all block reward BTCs will have been issued, where a block reward will be worth approximately 0.19531250 BTC. The BTC block reward will approximately reach 0 by 2140. The last halving will be conducted in 2140 where 21 million BTCs will have been issued, and no more BTCs will be issued.

As Bitcoin is the origin of all cryptocurrencies, BTC has the overall highest value among all cryptocurrencies. The trend of changes in value of BTC in units of USD is presented in Fig. 9.2. Bitcoin (BTC) reached its highest value ever at $75,830 USD on March 14 of 2024. This happened while approaching the Bitcoin reward halving that was conducted on April 19 of 2024, where the reward for a block chaining was 6.25 BTC that was halved to 3.125 BTC. Before 2024, the record was $67,566.83 USD on November 8 of 2012.

Many users and miners have stake in Bitcoin, so they are consistently monitoring their investments. Users and miners consider the block chaining reward as the main motive to maintain an honest blockchain, which is more worthy than the potential gain (and penalties) from attempting to forge the Bitcoin blockchain ledger data. This aspect will help keep the miners of the Bitcoin blockchain to remain honest. However, as the reward for chaining a block is halved every 21,000 blocks (approximately 4 years), much more cyberattacks and colluding may occur. Being able to predict and provide more advanced security technologies to prevent such issues is one of the objectives of this book.

There are several issues to be aware of in regards to the future of Bitcoin, where a few are discussed in the following. As the Return on Investment (ROI) of a block reward is decreasing, there will be a point in time when the incentive to compete for a block reward may not be worthwhile. Considering that there are many mining pools (as presented in Fig. 9.1) with extremely high computation capabilities and maintenance costs, some of these computing recourses may be converted to support alternative more profitable services.

In addition, the massive speed of quantum computing will soon become a threat in competing to derive a SHA256D hash value that can satisfy the Bitcoin nBits target threshold in an extremely short time. As the qubit computation scale continuously increases, higher levels of multi-event quantum computing simulation will become a strong contending factor.

Another consideration is that many cryptocurrencies and smart contract platforms establish their company's financial basis on the BTC of Bitcoin. For example, in April of 2014, the first smart contract platform service Ethereum sold 50 million ETHs through an Initial Coin Offering (ICO) to raise $18 million in Bitcoin, which was used as the financial foundation of the company. In addition, the value of Ethereum's ETH is correlated to the value of Bitcoin's BTC, which is confirmable by comparing Fig. 9.2 (for Bitcoin) and Fig. 10.1 (for Ethereum) of this book. This is important because, if Bitcoin were to become unstable or unreliable, then a chain effect of value collapse in other cryptocurrencies and smart contracts could occur. It is worthy to note that in addition to network and data security issues, political uncertainty (due to elections, administration changes, wars, etc.) as well as changes in securities, tax, asset, and investment policies (e.g., Exchange-Traded Fund (ETF), etc.) can significantly influence the value of cryptocurrencies, as there is commonly no supporting government or bank-like entity that will actively try to stabilize their value.

Blockchains, cryptocurrencies, and smart contracts have vulnerabilities based on their system architecture and operation method, where some past issues are introduced in the following.

As mentioned in Chap. 8, for the case of the DAO, when it was initiated in April of 2016, the Genesis DAO was established on the Ethereum blockchain's smart contract platform, where the DAO's software was released open source on GitHub. On June 17 of 2016, the DAO was attacked in which 3.6 million ETHs were stolen (approximately worth $70 million USD) due to a loophole in the program code of the DAO, which was used to make multiple recursive calls to receive ETH returns before the DAO platform's smart contract could update its balance (i.e., double spending was possible). This example shows the high-risk of open-source programs, as it gives hackers sufficient time to comfortably analyze the open-source program (e.g., using a digital twin) to search and test various loophole options until one or many are found [9].

In addition, as described in Chap. 8, on May 8 of 2019, Binance received an attack that resulted in over 7,000 Bitcoins being stolen, which had a loss value exceeding $40 million USD at that time. The hackers obtained user account login information and used this information to withdraw Bitcoins quickly from Binance's 'hot wallet.' The hackers had collected user information and passwords in advance and used this information to withdraw the Bitcoins when its value was approximately 9% higher compared to the previous week [10, 11].

This attack on Binance was made on relatively independent accounts using procedures that passed the security check points, which is why it was not immediately detected. It is especially important to note that this attack was conducted at a time when the gain of the robbery would be maximized. This cyber-attack was conducted using valid user login information obtained through various hacking methods in advance (e.g., MITM, phishing, data exploitation, sniffing, etc., as explained in Chaps. 1 and 7). So, Binanace and external legal authorities were not sure if future attacks would be made, or if the hackers would attempt to control the price of the cryptocurrencies.

Based on this case of Binance, from an attacker's point of view, the attacker's dilemma is to conduct an attack when the profit from the cyber-attack can be maximized. So, waiting and observing until the value peaks can be beneficial to the attacker. But if another attacker attacks first, then the opportunity of an attack can be lost, as various new precautions and patches will be activated. So, attacking when possible (even though the gain from the cyber-attack may not be maximized) can be beneficial to the hacker. In addition, there is always the risk of getting caught and punished for conducting a cyber-attack.

Due to these reasons, defender awareness is always needed. Just because your company has not been hacked yet does not always mean your company's network and system is not hackable. It may mean that your company was not worth hacking yet, and may be hacked later when the company's stock or cryptocurrency value is high enough, making it worthy of a cyber-attack (compared to the hacking risk).

As bad as this sounds, there are worser case scenarios. For example, your company may be already hacked but the administration and management teams do not realize it yet. Or realize it, but are not informing you. Or a compromised status of your company's server is maintained by the hacker, but because there is value in using your data, network, and resources in the meantime (since it is continuously going unnoticed and unhindered), cyberattacks like data extrapolation, ransomware attack, DoS, DDoS, and/or unauthorized financial withdraws may occur later.

11.3 Emerging Technical Solutions

Automated attack patterns and timing control settings are commonly predefined in the cyber-attack program as threshold parameters. These factors have to be considered by the security system.

What will make things worse is the accelerating speed and expanding capability of generative AI technology. Soon, generative AI will turn into Artificial General Intelligence (AGI). AGI is a state of an AI system that can perform a wide range of cognitive tasks at a level higher than well-educated and considerably trained humans. The current form of AI is task oriented and narrowly focused on a specific domain, which is why it is called 'Narrow AI.' Considering AGI's diverse range of highly sophisticated functionalities, AGI is also called 'Strong AI' [12].

In addition, as advanced AI will be used in cyber-attacks, advanced AI will have to be used in network security systems. Network security systems will have to use higher levels (or at least an equal level) of advanced AI in order to automatically detect new anomalies and changing cyber-attack patterns, and be able to immediately defend and counter-attack to protect the network and systems.

As mentioned above, defending is more difficult than attacking, this is why network security systems will have to be given administrative auto-decision control power (to defend the network in real-time) along with the capability to adapt as needed. Due to this reason, extremely advanced AI will be (and is already) used in networks that are supported by extreme computation capabilities. As AGI technology is applied to these large networks that have administrative auto-decision control power, the range of cognitive decision making can evolve in the wrong direction. Already, there are several examples of self-learning cognitive decision-making AI systems that have been terminated due to extremely weird and very alarming decisions and actions conducted by these AI systems, which include, creating and communicating in a new language unknown to humans (without permission), making decisions that are abusive or raciest or socially incorrect, or conducting/planning actions that are threatening, illogic, inhumane, or catastrophic. There are policies and regulations that apply to humans and animals when such behavior is observed. But there is no policy or regulation that can be applied to an advanced AI system, and it is impossible to completely terminate or turn off a distributed system and network. It is even more concerning when such systems are given full administrative auto-decision control power over distributed systems and networks. Although these systems are installed to maintain network security and defend against adversaries, however, this power can be very easily abused.

Based on these issues, the following techniques are proposed as potential solutions for future security systems for networks, blockchains, cryptocurrencies, and smart contracts.

As described in the chapters of this book, new cyber-attack techniques will be developed and the level of automated intelligent attacks will consistently improve. For these reasons, future security systems cannot be implemented in a separate layer or as an external system (e.g., firewall) alone. Security has to also be embedded into the networking protocols and operational procedures. Embedded security functions have to be protecting the network and blockchain services at all times and for all procedures. For these reasons, the papers [13–17] use trust based security and advanced AI technology to protect the network and blockchains. The trust algorithm applied in these papers uses the collected reputation of networking nodes to observe and monitor each other. Based on local trust opinions, a subjective reputation of each networking node is formed, and through statistical analysis of the collected subjective reputations, an objective evaluation of each node is formed, which is used to predict malicious nodes in the network. A malicious node may be a hacker or a botnet or zombie node, in which the node operator may not realize the malicious activities that are being conducted by the computer used.

In addition, radical advancements in AI with no regulation might result in more harm than help. For example, open-source software and data sharing have radically expedited the development speed of AI technology. This has enabled many new beneficial technologies to be created, but has also enabled many new harmful technologies to be created as well. But because AI technology is operated in program codes, it is hard to detect the damage unless a ransomware or Denial of Service (DoS) attack is conducted. For this reason, proper regulations need to be established such that advancements in AI are controllable and safer.

Another issue to be aware of is Sybil attacks. A Sybil attack is a cyber-attack that uses multiple identities. This has the effect of hiding the attacker's identity while confusing the security systems to make the cyber-attack more difficult to detect and harder to defend against. Sybil attacks are even more difficult to detect in wireless and mobile networks. This is why new methods to detect Sybil attacks will always be needed, as no hacker will use their real identity when hacking a system [18].

In addition, the Proof-of-Stake (PoS) mechanism used in Ethereum is highly recommended for blockchain services, cryptocurrencies, and smart contracts. This is for environment preservations reasons as well as security reasons. Ethereum PoS has the advantage of significantly enhancing the security level of blockchain operations while radically reducing the energy consumption when compared to Proof-of-Work (PoW) mining operations. This is because the PoS process is more transparent and stable as the block attestations and the chaining process are more predictable. Due to these reasons, when compared to PoW, PoS makes it much easier to notice misconduct, errors, and malicious behavior. In addition, the stake slashing penalty, correlation penalty, and inactivity leak defense mechanisms help prevent inappropriate and dishonest behavior as a significant financial loss as well as removal penalties can be easily applied.

The security benefits of a Demilitarized Zone (DMZ) have been discussed above and in Chap. 1. AI based DMZ-as-a-Service (DMZaaS) security functions need to be developed and deployed as Virtual Machines (VMs) and Containers as well as Software-as-a-Service (SaaS) and Platform-as-a-Service (PaaS) forms that can be accessed on-demand (in an app form) from clouds and edge computing platforms to adaptively fit the device and application needs. For more details of SaaS, PaaS, VMs, and Containers, please read Chap. 8 of [4].

In addition, encryption techniques can be customized to the application and/or network type and controlled by the kernel or Operating System (OS) of the nodes within the network. This may result in additional operational overhead, but a customized security system could help prevent cyber-attacks coming from external heterogeneous networks [19, 20].

Last but not least, more advanced hash functions and asymmetric encryption techniques need to be quickly developed and standardized. This is because advancements in AI technology are enabling remarkable prediction capabilities, in which brute-force search mechanisms supported by advanced AI driven ultra-high-speed processors with embedded memory can create serious problems in security. For example, as intelligent computing advances, the preimage resistance level of existing hash functions will naturally degrade, such that preimage attacks will become more effective and abundant. In addition, asymmetric encryption techniques like the RSA algorithm use a lot of prime number computations, in which quantum computers are very effective in solving problems based on prime numbers. This is why more advanced hash functions and asymmetric encryption techniques must be developed considering cyber-attacks and preimage attacks from quantum computers combined with advanced AI technology.

Most unfortunately, it is very hard to provide a solid solution to the serious problems described in this chapter. However, knowing what may happen in the future can help to create innovative solutions using new emerging technologies in the future, which is the objective of this book.

11.4 Concluding Remarks

The objective of this book is to enable the reader to see the future trend of technical progress such that these types of problems can be immediately solved when they occur and can be prevented in advance using emerging technologies.

In addition, there is a great amount of planning, designing, research, and development that needs to be done in each of these areas. I hope that the reader finds interest in these areas, as each has great importance and significant influence for future users and especially for companies and the global economy.

I deeply thank you for reading my book and sincerely hope it helped in some way. I appreciate your time and interest and wish you a joyful and successful career.

References

1. Wikipedia, Blockchain. https://en.wikipedia.org/wiki/Blockchain
2. S. Nakamoto, Bitcoin: a peer-to-peer electronic cash system, white paper. https://bitcoin.org/en/bitcoin-paper
3. K. Murphy, *Probabilistic Machine Learning: Advanced Topics* (MIT Press, 2023)
4. J.-M. Chung, *Emerging Metaverse XR and Video Multimedia Technologies* (Springer Nature, *Apress*, 2023)
5. Wikipedia, Reinforcement learning. https://en.wikipedia.org/wiki/Reinforcement_learning
6. Wikipedia, Federated learning. https://en.wikipedia.org/wiki/Federated_learning
7. Wikipedia, Digital twin. https://en.wikipedia.org/wiki/Digital_twin
8. A. Creswell, et al., Generative adversarial networks: an overview. IEEE Signal Process. Mag. **35**(1) (2018)
9. Wikipedia, The DAO. https://en.wikipedia.org/wiki/The_DAO
10. Wikipedia, Binance. https://en.wikipedia.org/wiki/Binance
11. A. Kharpal, Bitcoin bandits: hackers steal equivalent of $40M from Binance. USA Today. https://www.usatoday.com/story/money/2019/05/08/bitcoin-exchange-binance-robbed-7-000-bitcoins-hackers/1139907001/
12. Wikipedia, Artificial general intelligence. https://en.wikipedia.org/wiki/Artificial_general_intelligence

13. Y. Goh, J. Yun, D. Jung, J.-M. Chung, Secure trust-based delegated consensus for blockchain frameworks using deep reinforcement learning. IEEE Access **10** (2022). https://doi.org/10.1109/ACCESS.2022.3220852
14. J. Yun, Y. Goh, J.-M. Chung, DQN-based optimization framework for secure sharded blockchain system. IEEE Internet of Things J. **8**(2) (2021). https://doi.org/10.1109/JIOT.2020.3006896
15. J. Yun, Y. Goh, J.-M. Chung, Trust-based shard distribution scheme for fault tolerant shard blockchain networks. IEEE Access **7** (2019). https://doi.org/10.1109/ACCESS.2019.2942003
16. S. Seo, J.-M. Chung, Adaptive trust management and data process time optimization for real-time spark big data systems. IEEE Access **9** (2021). https://doi.org/10.1109/ACCESS.2021.3129885
17. J. Yun, S. Seo, J.-M. Chung, Centralized trust based secure routing in wireless networks. IEEE Wirel. Commun. Lett. **7**(6) (2018). https://doi.org/10.1109/LWC.2018.2858231
18. J. Kim, M. Ko, J.-M. Chung, Physical identification based trust path routing against Sybil attacks on RPL in IoT networks. IEEE Wirel. Commun. Lett. **11**(5) (2022)
19. H. Kim, J.-M. Chung, C. H. Kim, Secured communication protocol for internet-working ZigBee cluster networks. Comput. Commun. **32**(13–14) (2009)
20. H. Kim, C.H. Kim, J.-M. Chung, A novel elliptical curve ID cryptography pro-tocol for multi-hop ZigBee sensor networks. Wirel. Commun. Mobile Comput. **12**(2) (2010)

Index

J.-M. Chung, *Emerging Secure Networks, Blockchains and Smart Contract Technologies*, https://doi.org/10.1007/978-3-031-65866-2

SPRINGER NATURE

GPSR Compliance

The European Union's (EU) General Product Safety Regulation (GPSR) is a set of rules that requires consumer products to be safe and our obligations to ensure this.

If you have any concerns about our products, you can contact us on ProductSafety@springernature.com

In case Publisher is established outside the EU, the EU authorized representative is:

Springer Nature Customer Service Center GmbH
Europaplatz 3
69115 Heidelberg, Germany

The manufacturer's authorised representative in the EU is Springer Nature Customer Service Centre GmbH, Europaplatz 3, 69115 Heidelberg, Germany. If you have any concerns regarding our products, please contact ProductSafety@springernature.com

Printed and bound by CPI Group (UK) Ltd, Croydon, CR0 4YY
01/10/2026
02225969-0009